甘蔗制糖工业分析

Industrial Analysis of Cane Sugar Production

主　编　黄　凯　尹　丽　张立颖
副主编　蒙丽丹　吴乃锦　雷光鸿　雷艳萍
　　　　梁　勇　莫　蓓　宁方尧
主　审　黄　洁

北京理工大学出版社
BEIJING INSTITUTE OF TECHNOLOGY PRESS

内容提要

本书共分四个模块，主要内容涵盖制糖工业分析的基本知识，制糖过程中原料、材料、中间制品的主要成分，产品的主要理化指标检测及生产报表的计算，主要生产指标的分析方法等。在编写过程中，编者深入制糖生产企业调研、收集资料，力求做到理论联系实际，及时反映制糖生产检测的新动态，按照食糖质量标准贯彻实施，符合制糖企业有效控制产品质量和生产成本的需求。

本书可作为国内外制糖专业的高职教材及制糖企业的培训教材，也可供有关科研、管理部门的科技人员参考。

版权专有　侵权必究

图书在版编目（CIP）数据

甘蔗制糖工业分析 / 黄凯，尹丽，张立颖主编 . -- 北京：北京理工大学出版社，2024.5
ISBN 978-7-5763-3192-9

Ⅰ. ①甘…　Ⅱ. ①黄…②尹…③张…　Ⅲ. ①甘蔗制糖－工业分析　Ⅳ. ① TS244

中国国家版本馆 CIP 数据核字（2023）第 236817 号

责任编辑：阎少华	**文案编辑**：阎少华
责任校对：周瑞红	**责任印制**：王美丽

出版发行 /	北京理工大学出版社有限责任公司
社　　址 /	北京市丰台区四合庄路 6 号
邮　　编 /	100070
电　　话 /	（010）68914026（教材售后服务热线）
	（010）63726648（课件资源服务热线）
网　　址 /	http://www.bitpress.com.cn
版 印 次 /	2024 年 5 月第 1 版第 1 次印刷
印　　刷 /	河北鑫彩博图印刷有限公司
开　　本 /	787 mm×1092 mm　1/16
印　　张 /	20
字　　数 /	450 千字
定　　价 /	79.00 元

图书出现印装质量问题，请拨打售后服务热线，负责调换

甘蔗制糖工业分析

　　《甘蔗制糖工业分析》是一本特色鲜明、实践性强的专业课程教材。本书依据食糖制造工（分析）国家职业标准和广西工业职业技术学院制定的课程教学大纲和课程标准，结合《食糖制造工（分析）》和"食品检验管理"1+X证书的部分内容，注重课程思政元素的融入，并秉承二十大精神，努力做到质量优先，效益兼顾的管理理念。

　　随着我国经济、工业和科技实力的不断提升，越来越多的技术被输出到国外，广西的制糖装备和技术就是其中之一，为了响应"一带一路"倡议，帮扶第三世界国家的制糖技术，广西工业职业技术学院作为《食糖制造工》国家职业标准的主要制定者之一，开发了双语教材。本书作为制糖工业分析教材，最大的特点在于将制糖工业分析常用部分翻译成英文，方便国内外制糖技术人员交流使用。本书可供高等职业院校学生、应用型本科学生以及国外制糖技术人员学习使用。

　　在编写过程中"坚持以人民为中心的发展思想"，引入最新的制糖分析检验技术——近红外快速检测，在实现中间制品和最终产品快速检测的同时，解放劳动人民双手，提高企业工作效率。

　　全书共分四个模块，主要内容涵盖制糖工业分析的基本知识，制糖生产过程中原料、材料、中间制品的主要成分，产品的主要理化指标检测及生产报表的计算，主要生产指标的分析方法等方面。在编写过程中，我们深入制糖生产企业调研、收集资料，力求做到理论联系实际，及时反映制糖生产检测的新动态，按照食糖质量标准贯彻实施，符合制糖企业有效控制产品质量和生产成本的需求。本书的编写分工如下：模块1、模块2由黄凯、尹丽、蒙丽丹、雷光鸿、雷艳萍、梁勇编写，模块3由蒙丽丹、雷光鸿、吴乃锦编写，模块4由张立颖、雷艳萍、莫蓓、宁方尧编写；本书由黄洁主审。

　　本书可作为国内外制糖专业的高职教材及制糖企业的培训用书，也可供有关科研、管理部门的科技人员参考。同时，由于编者水平有限，编写时间仓促，书中难免有错误和不足之处，恳请读者批评指正。

在此特别感谢学校领导的关心和指导，以及制糖企业的科技人员、学校食品智能加工技术专业全体同仁的热情帮助和大力支持，为本书的编写提供了良好的条件和支持。

最后诚挚地希望本书能够对读者有所帮助，促进国内外制糖工业的发展。

编 者

目录

甘蔗制糖工业分析

模块 1　化验室基本知识 ··· 1
　任务 1.1　仪器识别及滴定操作 ··· 1
　任务 1.2　分析天平的使用练习 ··· 3

模块 2　原料及在制品日常分析 ·· 19
　任务 2.1　密度法、折光法测在制品锤度 ······································ 19
　任务 2.2　在制品的糖度及简纯度测定 ··· 48
　任务 2.3　蔗糖分及重力纯度的测定 ·· 61
　任务 2.4　兰－艾农法测定蔗汁还原糖 ··· 70
　任务 2.5　中和汁硫熏强度的测定 ·· 82
　任务 2.6　糖汁 pH 值的测定 ··· 85
　任务 2.7　蔗汁磷酸值的测定 ··· 87

模块 3　成品糖理化指标分析 ·· 91
　任务 3.1　白砂糖蔗糖分测定 ··· 91
　任务 3.2　白砂糖还原糖分测定 ··· 95
　任务 3.3　白砂糖干燥失重的测定 ·· 100
　任务 3.4　白砂糖电导灰分的测定 ·· 102
　任务 3.5　白砂糖色值及混浊度的测定 ······································· 108
　任务 3.6　白砂糖二氧化硫含量的测定 ······································· 114
　拓展学习——近红外测定成品糖 ·· 117

模块 4　化验室标准溶液的配制 ··· 119

Module 1　Basic knowledge of laboratory ································· 120

Task 1.1　Instrument identification and titration operation ················ 120

Task 1.2　Analytical balance use exercise ································· 122

Module 2　Daily analysis of raw materials and Semi-product ············· 144

Task 2.1　Measuring Brix of WIP by density method and Refractometry ········ 144

Task 2.2　Determination of pol and apparent purity of Semi-product ········· 184

Task 2.3　Determination of sucrose content and gravity purity ············· 201

Task 2.4　Determination of reducing sugar in sugarcane juice by Lane-Eynon method ······ 214

Task 2.5　Determination of sulphur strength of neutralized juice ············· 230

Task 2.6　Determination of sugar juice pH ································· 234

Task 2.7　Determination of phosphate value of sugarcane juice ············· 236

Module 3　Analysis of physical and chemical index of finished sugar ······ 242

Task 3.1　Determination of sucrose content in white granulated sugar ········ 242

Task 3.2　Determination of reducing sugar in white granulated sugar ········ 247

Task 3.3　Determination of loss on drying of white granulated sugar ········ 254

Task 3.4　Determination of conductivity ash of white granulated sugar ······ 257

Task 3.5　Determination of colour and turbidity of white granulated sugar ···· 265

Task 3.6　Determination of sulfur dioxide content in white granulated sugar ···· 272

Extended learning—Determination of finished sugar by near infrared spectroscopy ······ 276

Module 4　Preparation of standard solution in laboratory ················ 278

附录一　甘蔗制糖工业分析常用数据表 ·· 279

附录二　糖生产许可证审查细则（2006 年版） ······································ 313

参考文献 ·· 314

模块 1
化验室基本知识

任务 1.1　仪器识别及滴定操作

● 企业案例

小明是新入职的分析检验人员，一天，化验室主任给了他 HCl 溶液和 NaOH 溶液各一瓶，让他选用合适的滴管和指示剂对以上溶液进行中和滴定。假如你是小明，你会怎么做呢？

● 任务目标

通过本任务的学习，学生达到以下目标：
（1）熟悉实验室常用的玻璃仪器并熟知化验室基础知识；
（2）掌握滴定管的滴定操作技术；
（3）学会观察与判断滴定终点。

● 素质目标

养成细致谨慎的习惯。

● 任务描述

熟读并知晓本模块所必需的化验室基本知识，使用 50 mL 酸式和碱式滴定管对 HCl 溶液和 NaOH 溶液进行滴定操作。

● 实施条件

（1）仪器：50 mL 酸式和碱式滴定管各 1 支，250 mL 锥形瓶 3 个，250 mL 和 400 mL 烧杯各 1 个，10 mL 和 100 mL 量筒各 1 个。

（2）试剂：0.1 mol/L HCl 溶液、0.1 mol/L NaOH 溶液、酚酞指示剂、甲基橙指示剂。

● 程序与方法

（1）认真阅读本模块的化验室基础知识，并在课前完成相关练习题。

（2）滴定练习。

将准备好的酸式滴定管洗净，旋塞涂好凡士林，检漏，以 0.1 mol/L HCl 溶液润洗 3 次（每次 5～10 mL），再装入 HCl 溶液至"0"刻度以上，排除滴定管下端的气泡，调节液面至"0.00" mL 处。

将准备好的碱式滴定管洗净，检漏，以 0.1 mol/L NaOH 溶液润洗 3 次（每次 5～10 mL），再装入 NaOH 溶液至"0"刻度以上，排除滴定管下端的气泡，调节液面至"0.00" mL 处。

1）从滴定管放出溶液：从滴定管准确放出 20 mL HCl 溶液于 250 mL 锥形瓶中。

2）滴定：在上述盛 HCl 溶液的锥形瓶中加入 2 滴酚酞指示剂，从碱式滴定管中用 NaOH 溶液进行滴定。注意观察滴落点周围颜色的变化。

3）滴定终点的判断：开始滴定时，滴落点周围无明显的颜色变化，滴定速度可稍加快。当滴落点周围出现暂时性的颜色变化（浅粉红）时，应一滴一滴地加入 NaOH 溶液。随着颜色消失减慢，应缓慢滴入溶液。接近终点时，颜色扩散到整个溶液，摇动 1～2 次才消失，此时应加一滴，摇几下。最后加入半滴溶液，并用蒸馏水冲洗锥形瓶壁。一直滴到溶液由无色突然变为浅粉红色，并在 30 s 内不消失即终点，记下读数。

为了练习正确判断滴定终点，在锥形瓶中继续准确加入少量 HCl 溶液，使颜色褪去，按上述方法再用 NaOH 溶液滴定至终点。如此反复多次，直至能比较熟练地判断滴定终点，且终点读数 NaOH 溶液的用量相差不超过 0.02 mL。

按上述方法，在 250 mL 锥形瓶中准确加入 20 mL NaOH 溶液，加入 2 滴甲基橙指示剂，用 0.1 mol/L HCl 溶液滴定至溶液由黄色变为橙色为止，反复练习。

● 原始数据记录

1. 酚酞作指示剂

序次	1	2	3
耗用 0.1 mol/L NaOH 溶液 /mL			

2. 甲基橙作指示剂

序次	1	2	3
耗用 0.1 mol/L HCl 溶液 /mL			

● 思考

如何能又快又好地完成中和滴定？

任务 1.2　分析天平的使用练习

● **企业案例**

小明完成了酸碱中和滴定任务后，化验室主任给了他一台分析天平和电子天平，要求他对一些样品进行称量。

● **任务目标**

通过本任务的学习，学生达到以下目标：
（1）熟悉分析天平的构造；
（2）掌握电子天平的称量操作，并能正确称取试样。

● **素质目标**

养成耐心的好习惯。

● **任务描述**

使用分析天平进行称量，并记录数据。

● **实施条件**

（1）仪器：分析天平、托盘天平、药匙、小烧杯、锥形瓶。
（2）试样：粉状或粒状。

● **程序与方法**

（1）熟悉分析天平的构造；
（2）完成教师指定的试样称量任务；
（3）准确记录数据。

● **思考**

分析天平逐渐被电子天平淘汰，印证了人类发展的什么规律？在现阶段，学习分析天平的使用又可以达到什么目的？

1.2.1 任务相关知识——化验室安全基本知识

1.2.1.1 化学实验室的安全知识

（1）化验员工作态度要严肃认真，必须熟悉业务，操作严守规程。

（2）化验室内要配备各种防火器材，如砂袋、砂箱、灭火器等，化验员必须熟知各种防火器材的使用及灭火对象。

（3）易燃、易爆、有毒物品应专人保管。

（4）化验室内的所有电气设备均应绝缘良好，仪器应妥善接地。

（5）一切能产生毒性或刺激性气体的实验均应在通风橱内进行，头部不要伸进通风橱内，并应配备防毒面具。

（6）所盛放药品的试剂瓶、试样、溶液都要贴上标签。绝对不要在容器内装入与标签不相符的物品。

（7）谨慎处理易燃和有剧毒的物质。使用此类物质时，应在通风条件良好并远离火源的地方进行。金属汞易挥发，若不小心打破温度计，应将汞珠尽快收集起来，并用硫粉盖在液体汞上，使金属汞转化为不挥发的硫化汞。

（8）加热过程中不能离开工作岗位。试管加热前，应将外壁的水滴擦干；加热时勿将试管口朝向他人或自己，不要俯视正在加热的液体。

（9）打开盐酸、硝酸、氨水及过氧化氢等试剂瓶塞时，小心气体骤然冲出。嗅闻气味时不要将鼻子直接接近瓶口，而应用手扇闻。使用浓酸、浓碱和洗液时，应避免接触皮肤或溅在衣服上，更应注意保护眼睛。

（10）稀释硫酸时，必须在硬质耐热的烧杯中或者锥形瓶中进行。切记：只能将硫酸缓缓地注入水中，边倒边搅拌。温度过高时，应等冷却或降温后再继续进行，严禁将水倒入硫酸中。因硫酸稀释时会生成大量的热，同时浓硫酸的密度大于水，将水倒入硫酸中，水必然会浮在硫酸的上面，与硫酸混合时产生的热可能使溶液沸腾溅出而伤人。

（11）使用各种电气设备时，必须注意电压、电流与功率的匹配，切勿用湿手接触电源插头。熟悉实验室水、电、气的安装情况，总闸位置及灭火器材存放位置，以便应急使用。

（12）禁止使用化验室的器皿盛装食物，也不要用茶杯、食具盛装药品，更不要将烧杯或量杯当茶具使用。

1.2.1.2 化验室意外事故的处理

（1）割伤处理：在伤口上涂抹碘酒后，敷贴创可贴，重者及时送医院。

（2）烫伤处理：在伤口上涂抹烫伤药物或用10% $KMnO_4$ 溶液润湿伤口至皮肤变为棕色，也可用5%的苦味酸溶液涂抹伤口。

（3）酸碱腐蚀：衣物或皮肤溅有酸碱时，均应用干布或吸水纸吸干，并立即用大量水冲洗。酸灼伤时，局部用水冲洗后，再用饱和碳酸氢钠、稀氨溶液或肥皂水处理；碱灼伤时，局部用水冲洗后，则采用2%～5%醋酸或3%硼酸溶液处理。若酸溅入眼睛，则首先用大量水冲洗，然后用 1%～3% 碳酸氢钠溶液冲洗，最后用大量水冲洗。严重时，经上述处理后，立即送医院治疗。若碱溅入眼睛，则应用大量水冲洗，然后用 3% 硼酸溶液冲洗并立即送医院。

（4）吸入溴、氯、氯化氢等有毒气体时，可吸入少量酒精与乙醚混合的蒸气解毒，同时应到室外呼吸新鲜空气。吸入硫化氢、一氧化碳气体，应立即到室外呼吸新鲜空气。

（5）若有毒物入口，可内服一杯稀硫酸铜溶液，再用手指伸入咽喉部，促使呕吐，然后立即送医院。

（6）若遇触电事故，则首先切断电源，尽快用绝缘物，如干燥的木棍或竹竿等，使触电者脱离电源。必要时进行人工呼吸，并立即送医院抢救。

1.2.1.3 化验室的防火与灭火常识

1. 引起化验室火灾的主要原因

（1）易燃物质离火源太近。
（2）电线老化、插头接触不良或电气故障等。

2. 容易产生火灾的几种物质混合或接触

（1）活性炭与硝酸铵混合；
（2）沾染了强氧化剂（如氯酸钾）的衣物；
（3）抹布与浓硫酸；
（4）可燃性物质（木材或纤维等）与浓硝酸；
（5）有机物与液氧；
（6）铝与有机氯化物；
（7）磷化氢、硅烷、烷基金属及白磷等与空气接触。

3. 灭火方法

化验室内一旦着火或发生火灾，切勿惊慌，应冷静、果断地采取扑灭措施并及时报警（表 1-1）。

表 1-1 燃烧物灭火方法说明

燃烧物	灭火方法	说明
纸张、纺织品或木材	砂、水、灭火器	需降温和隔绝空气
油、苯等有机溶剂	CO_2、干粉灭火器、石棉布、干砂等	适用于贵重仪器的灭火
醇、醚等	水	需冲淡、降温和隔绝空气
电表及仪器燃烧	CCl_4、CO_2 等灭火器	灭火材料不能导电，切勿用水和泡沫灭火器灭火
活泼金属（如钾、钠等）及磷化物与水接触	干砂土、干粉灭火器	绝不能使用水或泡沫灭火器、CO_2 灭火器
身上的衣物	就地滚动，压灭火焰或脱掉衣服，用专用防火布覆盖着火处	切勿跑动，否则将加剧燃烧

1.2.2 玻璃器皿的使用

1.2.2.1 玻璃器皿的种类、规格及洗涤方法

1. 玻璃器皿的种类

因为玻璃具有很高的化学稳定性、热稳定性和很好的透明度、一定的机械强度和良好的绝缘性能，所以，在各类化验室中大量使用玻璃仪器。

玻璃的化学成分主要是 SiO_2、CaO、Na_2O、K_2O。引入 B_2O_3、Al_2O、ZnO、BaO 等，各种成分不同的玻璃具有不同的性质和用途。

特硬玻璃和硬质玻璃，其 SiO_2 和 B_2O_3 含量较高，均属于高硼硅酸盐玻璃二类。它们具有较高的热稳定性，在化学稳定性方面耐酸、耐水性能好，耐碱性能稍差。一般仪器玻璃和量器玻璃为软质玻璃，其热稳定性及耐腐蚀性稍差。

玻璃的化学稳定性较好，但并不是绝对不受侵蚀，而是其受侵蚀的量符合一定的标准。因玻璃被侵蚀而有痕量离子进入溶液中和玻璃表面吸附溶液中的待测分析离子，是微量分析要注意的问题。氢氟酸对玻璃有很强的腐蚀作用，故不能用玻璃仪器进行含有氢氟酸的实验。碱液特别是浓的或热的碱液对玻璃有明显的腐蚀作用，储存碱液的玻璃仪器，如果是带磨口的，还会造成磨口粘在一起无法打开。因此，玻璃容器不能长时间存放碱液。

石英玻璃属于特种玻璃仪器，其理化性能与玻璃相似，具有优良的化学稳定性和热稳定性，但价格比较昂贵。

2. 常用的玻璃仪器

化验室所用到的玻璃仪器种类很多，玻璃仪器按用途分，大致可以分为容器类、量器类及其他常用器皿三大类。各种不同专业的化验室还用到一些特殊的玻璃仪器。

这里主要介绍一般通用的玻璃仪器及一些磨口玻璃仪器的知识。常用玻璃仪器的名称、规格、用途见表 1-2。

表 1-2 常用玻璃仪器的名称、规格、用途一览表

仪器	规格	一般用途	使用注意事项
试管	如 25 mm×150 mm 10 mm×75 mm	反应容器。便于操作、观察，用药量少	1）试管可直接用火加热，但不能骤冷。 2）加热时用试管夹夹持，管口不要对人，且要不断移动试管，使其受热均匀，盛放的液体不能超过试管容积的 1/3。 3）小试管一般用水浴加热
离心管	如 25 mL、15 mL、10 mL	少量沉淀的辨认和分离	不能直接用火加热

续表

仪器	规格	一般用途	使用注意事项
烧杯	如 1 000 mL、250 mL、100 mL、50 mL、25 mL	反应容器。反应物较多时使用	1）可以加热至高温。使用时应注意勿使温度变化过于剧烈。 2）加热时底部垫石棉网，使其受热均匀
烧瓶	如 500 mL、250 mL、100 mL、50 mL	反应容器。反应物较多，且需要长时间加热时使用	1）可以加热至高温。使用时应注意勿使温度变化过于剧烈。 2）加热时底部垫石棉网，使其受热均匀
锥形瓶	如 500 mL、250 mL、100 mL	反应容器。摇荡比较方便，适用于滴定操作	1）可以加热至高温。使用时应注意勿使温度变化过于剧烈。 2）加热时底部垫石棉网，使其受热均匀
碘量瓶	如 250 mL、100 mL	用于碘量法	1）勿擦伤塞子及瓶口边缘的磨砂部分，以免产生漏隙。 2）滴定时打开塞子，用蒸馏水将瓶口及塞子上的碘液洗入瓶中
容量瓶	如 1 000 mL、500 mL、250 mL、100 mL、50 mL、25 mL	配制准确浓度的溶液时使用	1）不能受热。 2）不能在其中溶解固体
漏斗	如 6 cm 长颈漏斗、4 cm 短颈漏斗	用于过滤或倾注液体	不能用火直接加热
滴管	材料：尖嘴玻璃管与橡皮乳头构成	1）吸取或滴加少量（数滴或 1~2 cm^3）液体。 2）吸取沉淀的上层清液以分离沉淀	1）滴加时，保持垂直，避免倾斜，尤忌倒立。 2）管尖不可接触其他物体，以免沾污

续表

仪器	规格	一般用途	使用注意事项
量筒、量杯	量筒：如 100 mL、50 mL、25 mL、10 mL 量杯：如 100 mL	用于液体体积计量	不能加热
吸量管、移液管	吸量管：如 10 mL、5 mL、1 mL 移液管：如 50 mL、10 mL、5 mL、1 mL	用于精确量取一定体积的液体	不能加热
滴定管、滴定管架	滴定管分碱式（a）和酸式（b）、无色和棕色。如 50 mL、25 mL	1）滴定管用于滴定操作或精确量取一定体积的溶液。2）滴定管架用于夹持滴定管	1）碱式滴定管盛碱性溶液，酸式滴定管盛酸性溶液，二者不能混用。2）碱式滴定管不能盛氧化剂。3）见光易分解的滴定液宜用棕色滴定管。4）酸式滴定管活塞应用橡皮筋固定。
洗瓶	材料：塑料。如 500 mL	用蒸馏水或去离子水洗涤沉淀和容器	
滴定板	材料：白色瓷板 规格：按凹穴数目分 12 穴、9 穴、6 穴等	用于点滴反应、一般不需分离的沉淀反应，尤其是显色反应	1）不能加热。2）不能用于含氢氟酸和浓碱溶液的反应
干燥器	18 cm、15 cm、10 cm	1）定量分析时，将灼烧过的坩埚置于其中冷却。2）用于存放样品，以免样品吸收水汽	1）灼烧过的物体放入干燥器前温度不能过高。2）使用前要检查干燥器内的干燥剂是否失效

3. 玻璃器皿的洗涤方法

在化验分析工作中，洗净玻璃仪器不仅是一个必须做的实验前的准备工作，还是一个技术性的工作。仪器洗涤是否符合要求，对分析工作的准确度和精密度均有影响。不同分析工作（如工业分析、一般化学分析、微量分析等）有不同的仪器洗净要求，现以一般定量化学分析为主，介绍玻璃仪器的洗涤方法。

（1）洗涤仪器的一般步骤。

1）用水刷洗。准备用于洗涤各种形状仪器的毛刷，如试管刷、烧杯刷、瓶刷、滴定管刷等。用毛刷蘸水刷洗仪器，冲掉可溶性物质及刷掉表面的灰尘。

2）用去污粉、肥皂或合成洗涤剂刷洗。去污粉是由碳酸钠、白土、细砂等混合而成的，其碱性可去油污，以固体物质的摩擦作用可去掉多种污物，缺点是去油效果不佳、会损伤玻璃，故滴定管等仪器及比色皿等光学玻璃的透光面严禁用去污粉擦洗。

近年来多用合成洗涤剂来刷洗玻璃仪器，它有较强的去油污能力。洗净的仪器倒置时，水流出后仪器壁面不挂水珠。可用少量纯水刷洗仪器 3 次，洗去自来水带来的杂质，即可使用。

（2）各种洗涤液的使用。针对仪器沾污物的性质，采用不同洗涤液能有效地洗净仪器，见表 1-3。要注意在使用各种性质不同的洗涤液时，一定要把上一种洗涤液除去后再用另一种，以免相互作用，生成的产物可能更难洗净。

表 1-3　几种常用的洗涤液

铬酸洗液： 将研细的重铬酸钾 20 g 溶于 40 mL 水中，慢慢加入 360 mL 的浓硫酸	用于去除器壁残留油污。用少量洗液刷洗或浸泡一夜，洗液可重复使用
工业盐酸（浓或 1∶1）	1）用于洗去碱性物质及大多数无机物残留。 2）碱－乙醇洗液不要加热
碱性洗液： 氢氧化钠 10% 水溶液或乙醇溶液	水溶液加热（可煮沸）使用，其去油效果特别好，注意煮的时间太长会腐蚀玻璃
碱性高锰酸钾洗液： 将 4 g 高锰酸钾溶于水中，加入 10 g 氧化钠，用水稀释至 100 mL	清洗油污或其他有机物，洗后容器沾污处有褐色二氧化锰析出，再用浓盐酸或草酸洗液、硫酸亚铁、亚硫酸钠等还原剂除去
草酸洗液： 将 5～10 g 草酸溶于 100 mL 水中，加入少量浓盐酸	洗涤高锰酸钾洗液洗后产生的二氧化锰，必要时加热使用
碘－碘化钾溶液： 将 1 g 碘和 2 g 碘化钾溶于水中，用水稀释至 100 mL	洗涤用硝酸银滴定液后留下的黑褐色沾污物，也可用于擦洗沾污过硝酸银的白瓷水槽
有机溶剂： 苯、乙醚、丙酮、二氯乙烷等	1）可洗去油污或可溶于该溶剂的有机物质，用时要注意其毒性及可燃性。 2）用乙醇配制的指示剂溶液的积垢可用盐酸－乙醇（1∶2）洗液洗涤

	续表
乙醇、浓硝酸（不可事先混合）	一般方法很难洗净的有机物可用此法：在容器内加入不少于 2 mL 的乙醇，加入 10 mL 浓硝酸，静置片刻后，立即发生激烈反应，放出大量热及二氧化氮，反应停止后再用水冲洗，操作应在通风橱中进行，不可塞住容器，做好防护

铬酸洗液因毒性较大，近年来用量逐渐减少，多以合成洗涤剂、有机溶剂等去除油污，尽可能不用铬酸洗液，但有时仍要用到，因此也列入表内。

这些洗涤方法都是应用物质的物理（互溶性）及化学性质达到洗净仪器的目的。在化验室中可以把废酸、回收的有机溶剂等分别收集起来加以利用。

（3）砂芯玻璃滤器的洗涤。

1）新的滤器使用前应以热的盐酸或铬酸洗液边抽滤边清洗，再用蒸馏水洗净。可正置或反置用水反复抽洗。

2）针对不同的沉淀物采用适当的洗涤剂，先溶解沉淀，或反复用水抽洗沉淀物，再用蒸馏水冲洗干净，然后 110 ℃烘干，最后保存在无尘的柜中或有盖的容器中。若不然，积存的灰尘和沉淀堵塞滤孔很难洗净。

（4）特殊要求的洗涤方法。在用一般方法洗涤后用蒸汽洗涤是很有效的。有的实验用仪器要求用蒸汽洗涤。方法是在烧瓶上安装一个蒸汽导管，将要洗的容器倒置在上面用蒸汽吹洗。

某些痕量分析对仪器洗涤的要求很高，要求洗去 ppb（ppb 为浓度的一种表示方式，用溶质质量占全部溶液质量的十亿分比来表示的浓度，也称十亿分比浓度）级的杂质离子。洗净的仪器还要浸泡在 1∶1 HCl 或 1∶1 HNO_3 中数小时（多至 24 h），以免吸附无机离子，然后用纯水冲洗干净。有的仪器需要在几百摄氏度的温度下灼烧，以达到痕量分析的要求。

1.2.2.2 玻璃器皿的准备和使用

1. 玻璃仪器的干燥和保管

（1）玻璃仪器的干燥。玻璃仪器应保管在干燥、洁净的地方，做实验经常要用到的仪器应在每次实验完毕后洗净干燥备用。用于不同实验的仪器对干燥有不同的要求，一般定量分析中用的烧杯、锥形瓶等仪器洗净即可使用，而很多用于有机化学实验或有机分析的仪器要求是干燥的，有的要求没水迹，有的则要求无水。应根据不同的要求来干燥仪器。

1）晾干。不急于使用且要求一般干燥的仪器，可在用纯水刷洗后，放在无尘处倒置控去水分，然后自然干燥。

2）烘干。洗净的仪器控去水分，放在电烘箱内烘干，烘箱温度为 105～110 ℃，烘 1 h 左右。也可放在红外灯干燥箱中烘干。此法适用于一般仪器。加热时，应逐渐升温，避免骤冷、骤热。称量用的称量瓶等在烘干后要放在干燥器中冷却和保存。带实心玻璃塞及厚壁仪器烘干时要注意慢慢升温并且温度不可过高，以免烘裂。注意计量仪器不能受热，不能用来储存浓酸或浓碱，不可放在烘箱中烘干。

硬质试管可用酒精灯加热烘干，要从底部烤起，把试管口向下，以免水珠倒流把试管炸裂，烘至无水珠后把试管口向上赶净水汽。

3）热（冷）风吹干。对于急于干燥的仪器或不适于放入烘箱的较大的仪器可使用吹干的办法，通常用少量乙醇、丙酮（或最后再用乙醚）倒入已控去水分的仪器中摇洗（溶剂要回收），然后用电吹风机吹，开始用冷风吹 1～2 min，当大部分溶剂挥发后吹入热风至完全干燥后，再用冷风吹去残余的蒸汽，不使其在容器内冷凝。此法要求通风良好，防止中毒，不可接触明火，以防有机溶剂蒸气爆炸。

4）不可将热溶液或热水倒入厚壁仪器中。

（2）玻璃仪器的保管。在储藏室里玻璃仪器要分门别类地存放，以便取用。经常使用的玻璃仪器放在实验柜中，要放置稳妥，高的、大的放在里面。以下提出一些仪器的保管方法。

1）移液管。洗净后用干净的滤纸包住两端。如果用于要求较高的实验，则全部用滤纸包起来，以防受沾污。

2）滴定管。用纯水刷洗后注满纯水，上盖玻璃短试管或塑料套管，或者倒置夹在滴定管夹上。

3）比色皿。在小瓷盘或塑料盘下垫滤纸，洗净倒置晾干后收于比色皿盒或洁净的器皿中。

4）带磨口塞的仪器。容量瓶或比色管等最好在清洗前就用小线绳或橡皮筋把塞和管口拴好，以免打破塞子或互相弄混。需长期保存的磨口仪器要在塞与口之间垫一张纸片，以免粘住。长期不用的滴定管要除掉凡士林后垫上纸，用皮筋拴好活塞保存。磨口塞间如有砂粒，不要用力转动，以免损伤其精度。同理，不要用去污粉擦洗磨口部位。

5）成套仪器，如索氏萃取器、气体分析器等用完要立即洗净，放在专门的纸盒里保存。

总之，本着对工作负责的精神，对所用的一切玻璃仪器要清洗干净，按要求保管，养成良好的工作习惯，不要在容器里遗留油脂、酸液、腐蚀性物质（包括浓碱液）或有毒药品，以免造成隐患。

（3）打开磨口塞的方法。当磨口活塞打不开时，用力易拧碎，要针对不同情况采取相应的措施来打开。

1）凡士林等油状物质粘住活塞，可以用电吹风或微火慢慢加热使油类黏度降低，熔化后用木棒轻敲塞子即可打开。

2）因长时间不用粘住的活塞，可泡在水中，几小时后可打开。

3）因碱性物质粘住的活塞，可将仪器在水中加热至沸腾，再用木棒轻敲塞子。

4）装有试剂的试剂瓶塞打不开，要靠积累一些经验，采取适当的方法来打开磨口瓶塞。

5）瓶内是腐蚀性试剂，如浓硫酸等，要在瓶外放好塑料圆桶以防瓶破裂，操作者需戴有机玻璃面罩，操作时不要使脸部离瓶口太近，打开有毒蒸气的瓶口（如液溴），要在通风橱内操作，可用木棒轻敲瓶盖，也可洗净瓶口，用洗瓶吹出一点蒸馏水润湿磨口，再轻敲瓶盖。

6）因结晶或碱金属盐沉积及强碱粘住的瓶塞，可把瓶口泡在水中或稀盐酸中，经过一段时间浸泡后再尝试打开。

2. 滴定分析仪器的准备和使用

滴定分析是根据滴定时所消耗的标准溶液的体积及浓度计算分析结果的。因此，除了要准确地确定标准溶液的浓度外，还必须准确地测量溶液体积。溶液体积测量的误差是滴定分析中误差的主要来源。体积测量如果不准确（如误差大于 0.2%），其他操作步骤即使都很准确也是徒劳的。因此，为了使分析结果能符合所要求的准确度，就必须准确地测量溶液的体积，一方面取决于所用容量仪器的容积是否准确；另一方面取决于能否正确使用这些仪器。

在滴定分析中测量溶液准确体积所用的容量仪器有滴定管、移液管、吸量管及容量瓶等。滴定管和移液管为"量出"式量器，量器上标有"A"字样，但我国目前统一用"Ex"表示"量出"，用于测定从量器中放出液体的体积。一般容量瓶为"量入"式量器，量器上标有"E"字样，但我国目前统一用"In"字样表示"量入"，用于测定注入量器中液体的体积。还有"量出"式容量瓶，瓶上标有"A"或"Ex"字样，它表示在标明温度下，液体充满到标线刻度后，按一定方法倒出液体时，其体积与瓶上标明的体积相同。

滴定管是用于准确测量滴定时放出的溶液体积的量器，它是具有刻度的细长玻璃管。按容量及刻度值不同，滴定管分为常量滴定管、半微量滴定管、微量滴定管；按要求不同，可分为"蓝带"滴定管、棕色滴定管（用于装高锰酸钾、硝酸银、碘等标准溶液）；按构造不同，可分为普通滴定管和自动滴定管；按用途不同，可分为酸式滴定管及碱式滴定管。

带有玻璃磨口旋塞以控制液滴流出的是酸式滴定管（简称酸管），用来盛放酸类或氧化性溶液。但不能放碱性溶液，因为磨口旋塞会被碱腐蚀而黏住。带有玻璃珠的乳胶管控制液滴，碱式滴定管（简称碱管）下端连接尖嘴玻璃管，用于盛放碱性溶液和非氧化性溶液，不能放 $KMnO_4$、$AgNO_3$、I_2 等溶液，以防胶管被氧化而变质。

滴定管进行滴定时，应该将滴定管垂直地夹在滴定管架上。酸式滴定管的使用：左手无名指和小拇指向手心弯曲，轻轻地贴着出口管，用其余三指控制活塞的转动，但应注意不要向外拉旋塞，以免推出旋塞造成漏液，也不要过分往里扣，以免造成旋塞转动困难而不能操作自如。碱式滴定管的使用：左手无名指及小拇指夹住出口管，大拇指与食指在玻璃珠所在部位向一旁挤乳胶管，玻璃珠移至手心一侧，使溶液从玻璃珠旁边空隙处流出，不要用力捏玻璃珠，也不能使玻璃珠上、下移动；不要捏到玻璃珠下部的乳胶管，以免空气进入而形成气泡，影响读数。停止滴定时，应先松开大拇指和食指，最后松开无名指与小拇指。

滴定速度控制必须掌握三种滴液方法：逐滴连续滴加，即一般的滴定速度，"见滴成线"的方法；只加一滴，要做到需加一滴就只加一滴的熟练操作；使液滴悬而不落，即只加半滴，甚至不到半滴的方法。滴定至终点时应立即关闭旋塞，并注意不要使滴定管中的溶液有稍许流出，否则影响最终读数。读数要求读到小数点后第二位，即估计到 ±0.01 mL，如读数为 25.33 mL，并立刻记录数据。

滴定结束后，滴定管内剩余的溶液应弃去，不得将其倒回原试剂瓶中，以免沾污整瓶溶液。随即洗净滴定管，倒置在滴定管架上。

1.2.3 分析数据的记录

测定的结果都是包含误差的近似数据，即真实值是不可能测出的。所以在记录、计算时应以测量可能达到的精度为依据来确定数据的位数和取位。如果参加计算的数据的位数取少了，就会降低测定结果的精度；如果位数取多了，易使人误认为测量精度很高，且增加了不必要的计算工作量。

1.2.3.1 有效数字的运算规则

1. 有效数字

有效数字是指在实验工作中实际能测量到的数字。实验中使用的仪器刻度的精确程度总是有限的。例如，50 mL 量筒，最小刻度为 1 mL，在两刻度间可再估计一位，因此，实际测量能读到 0.1 mL，再估计一位，可读至 0.01 mL，由于最后一位是估计的，所以这一位数字称为可疑数字。例如，读滴定管的液面位置数时，甲可能读为 21.32，乙可能读为 21.33，丙可能读为 21.31。由此可见 21.3 是滴定管上显示出来的，即滴定管的最小刻度为 0.1，而最后一位数字是由实验者估计出来的。因实验者不同，而得到不同估计值，但这一位估计数字是客观存在的，因此它是有效数字。也就是说，有效数字是实际测到的准确数字加一位可疑数字。

由上述可知，有效数字与数学的数有着不同的含义。数学上的数只表示大小，有效数字则不仅表示量的大小，而且反映了所用仪器的准确程度（仪器的最小刻度）。例如，"取 6.5 g NaCl"，这不仅说明 NaCl 质量为 6.5 g，而且表明用感量 0.1 g 的台秤称即可；若是"取 6.500 0 g NaCl"，则表明一定要在分析天平上称取。

仪器读数的有效数字位数，由仪器的性能决定。例如，分析天平可称准至 0.000 2 g；滴定管可读准至 0.01 mL 等。在记录测量数值时，必须而且只保留一位估计值。对于数字式仪表，所显示的数字是有效数字，不用估算。

2. 有效数字的位数

有效数字的位数越多，则测定的相对误差越小。因此，记录测量数据时，不能随便写，不然就会夸大或降低准确度。

从数字左边第一个非零数字开始直到最右边所含数字的个数称为这个数的有效数字的位数。

例如：0.023 有两位有效数字，230.40 有五位有效数字。

要注意的是，在左边第一个非零数字之前的所有零都不是有效数字，这些零仅仅是为了标出小数点的位置。但是，位于最后一个非零数字后面的那些零都是有效数字。"0"在数字中起的作用是不同的。有时是有效数字，有时不是，这与"0"在数字中的位置有关。

（1）"0"在数字前，仅起定位作用，"0"本身不是有效数字，如 0.027 5 中，数字 2 前面的两个 0 都不是有效数字，这个数的有效数字只有 3 位。

（2）"0"在数字中，是有效数字。如 2.006 5 中的两个 0 都是有效数字，2.006 5 有 5 位有效数字。

（3）"0"在小数的数字后，也是有效数字。如 6.500 0 中的 3 个 0 都是有效数字。0.003 0

中数字 3 前面的 3 个 0 不是有效数字，3 后面的 0 是有效数字。因此，6.500 0 是 5 位有效数字。0.003 0 是 2 位有效数字。

（4）以"0"结尾的正整数，有效数字的位数不定。如 54 000，可能是 2 位、3 位或 4 位甚至 5 位有效数字。这种数应根据有效数字的情况改写为指数形式。如为 2 位，则写成 5.4×10^4；如为 3 位，则写成 5.40×10^4 等。

总之，要能正确判别并书写有效数字，测定过程中记录的数据都应该是有效数字。下面列出了一组数字，以及有效数字位数：

7.400 0	54 609	5 位有效数字
33.15	0.070 20%	4 位有效数字
0.027 6	2.56×10^{-1}	3 位有效数字
49	0.000 40	2 位有效数字
0.003	4×10^{-5}	1 位有效数字
63 000	200	有效数字位数不定

3. 有效数字的运算

在数的运算中，一般来说，两数相加或相减，应使它们有相同的精确度，两数相乘或相除，应使它们有相同的准确度，即每一个数都保留同样位数的有效数字，计算结果也是如此。

近似运算中应该注意以下几点。

（1）相加或相减时，和或差的有效数字保留的位数应以小数点后位数最少的数字为依据。

例如，0.031 2+23.34+2.503 81 应以 23.34 为依据，将其他数字按四舍五入的原则取到小数点后第二位，然后相加。

$$\begin{array}{r} 0.03 \\ 23.34 \\ +)\ 2.50 \\ \hline 25.87 \end{array}$$

（2）在做乘除运算时，有效数字的位数取决于相对误差最大的数或者有效数字位数最少的数。

例如：

$$\frac{0.023\ 4 \times 4.033 \times 71.07}{127.5} = 0.052\ 604\ 2$$

计算结果应取 0.052 6，即与 0.023 4（有效数字位数最少的数）的位数相同。

但在运算过程中，每步运算的结果可比有效数字位数最少的数多保留一位。

例如，$0.023\ 4 \times 4.303 = 0.100\ 690\ 2$。

可取 0.100 7 继续运算（比 0.023 4 多一位有效数字）：

$$0.100\ 7 \times 71.07 = 7.156\ 749$$

可取 7.157 进行下一步运算：

$$7.157 \div 127.5 = 0.056\ 133 \cdots\cdots$$

结果应取 0.056 1。

（3）在分析计算中，常常会碰到一些分数，如某物质的当量等于其分量的 1/2，或从 250 mL 试液中吸取 25 mL，即吸出 1/10。这里的"2""10"都可视为足够有效，但不能把它们是一位或两位数认作具有一位或两位有效数字，从而作为判断计算结果的有效数字位数的根据。这一类数称为准确数。准确数可以在它的小数部分右面增添"0"，以增加有效数字的个数。

（4）若某数字的第一位有效数字大于或等于 8，则有效数字的位数可多算一位，如 8.37 可看作四位有效数字。

（5）数字凑整（取整）规则。由数字的取舍而引起的误差称为"凑整误差"或"取舍误差"。为避免"取舍误差"的迅速积累而影响测量结果的精度，在计算中通常采用"四舍六入五留双"的凑整规则：

1）若拟舍去的第一位数字是 0 至 4 中的数，则被保留的末位数不变；

2）若拟舍去的第一位数字是 6 至 9 中的数，则被保留的末位数加 1；

3）若拟舍去的第一位数字是 5，其右边的数字皆为 0，则被保留的末位数是奇数时就加 1，是偶数时就不变。

在化学管理的分析工作中，分析结果是否准确、可靠是至关重要的问题。错误的分析结果往往会造成严重的后果。因此，在测定时要实事求是地记录原始数据，在测定工作结束后，还要对测得的各项数据进行处理。如发现分析所得结论与实际情况不相符，要以这些原始数据为根据仔细检查，找出错误的原因，而决不允许通过改动数据的方法达到所谓的"一致"。

1.2.3.2　分析的精密度与准确度

分析工作的主要目的是测量样品中某一组成分的含量，同时希望对同一个样品的多次测定都能得到一个相同的结果。但在实际的测定中，虽然真值是客观存在的，但在有限次的测定中，由于受到仪器及其他因素的影响，测定对象的真实值是无法确定的，多次测定结果之间也会有差别。测定值与真实值越接近，多次重复测定结果之间的差别越小，则分析结果的质量就越高，这就是分析工作中的准确度与精密度。

1. 准确度
准确度是测定值与真实值符合的程度，说明测定值的正确性。

2. 精密度
精密度是指在相同的条件下，一组平行测定结果之间相互接近的程度，说明测定数据的再现性。

3. 准确度与精密度的关系
（1）精密度是保证准确度的先决条件，精密度不符合要求，表示所测结果不可靠，失去衡量准确度的前提。

（2）精密度高不能保证准确度高。

换言之，准确的实验一定是精密的，精密的实验不一定是准确的。

准确度与精密度的概念可以用打靶的弹着点分布来说明。图 1-1 说明乙的准确度较高,图1-2说明甲的精密度较高。

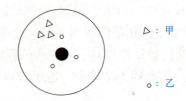

图1-1　乙有较高的准确度　　图1-2　甲有较高的精密度

1.2.3.3　分析过程的误差

1. 误差的种类与来源

（1）系统误差。系统误差是由于在测定过程中某些经常性原因造成的。它对分析结果的影响比较固定,误差常有重现性,使测定结果经常偏高或经常偏低。系统误差在找出产生误差的原因后,可以通过校正的手段将其消除。系统误差的来源主要有以下几方面。

定量分析中的误差

1）分析方法不够完善。例如：重量分析中沉淀有一定的溶解度,滴定时等当点和终点不十分相符；或测试方法本身不完善,如反应不完全或者有副反应、沉淀的溶解、共沉淀和络合物解离；使用的经验公式不完全符合真实情况,只是近似的,或者使用理论公式时,实验条件不完全符合建立理论公式所要求的条件；背景或空白值校正不正确等都会引起误差。

2）仪器的各部件制造不够精密、长期使用造成磨损引起精度下降、仪器未调整到最佳状态和器皿未经过校正、仪器本身的缺陷或使用了未经校正的仪器等。例如,天平的两臂不等长,滴定管等量器的真实值与标示值不完全相等。

3）使用的试剂或蒸馏水中含有杂质。

4）操作者个人的生理特点。系统误差常常是与操作人员相联系的,一是操作人员个人生理机能上的缺陷,如眼睛的辨别能力差,不能正确地读取刻度值和分辨颜色的色调和深浅,例如在滴定分析中对指示剂颜色的变化感觉迟钝等。二是操作人员的主观偏见和固有习惯,如后一次读数受到前一次读数的影响,不自觉地希望两次测定获得相同或相近的结果；观察终点变化的超前或滞后等。三是操作不合理,如取样的代表性不佳、灼烧沉淀温度选择不合适和反应条件控制不当等。

5）由于环境不完全符合测定所要求的条件而引起的误差。例如,环境温度的变化引起测定仪器和器皿精度的改变；大气污染对痕量组分测定空白值的严重影响；湿度、振动对准确称量的影响；照明情况变化引起视差对读数的影响等。

（2）偶然误差。由许多可变原因造成的误差。虽然操作者仔细操作,外界重要条件尽量控制一致,但测得的一系列数据也难免存在差别,并且所测得数据误差的大小和正负不一定,这类误差属偶然误差。偶然误差的发生是没有规律的,因此无法用校正的方法将其消除,只能通过规范操作,将其产生的影响降低到最低程度。产生偶然误差的原因有以下几种。

1）个人一时辨别的差异使读数不一致。例如,在读取滴管数据时,估计小数点后第二位的数值不准，会产生读数误差。

2）测定环境的温度、湿度、压力等发生微小变化。

3）操作中因疏忽产生的丢失、沾污、看错砝码、记录与计算错误等。

例如，在容量分析时用酸碱滴定法测定碱液的浓度时，所采用的碱液体积 $V_{碱}$ 与标准酸液的浓度 $N_{酸}$ 都是定值，当分析原理不当（如指示剂选择错误）或操作错误时引起的误差最终表现在滴定时所消耗的 $V_{酸}$ 的偏高或偏低。几种可能发生误差的情况见表 1-4。

表 1-4 用标准酸溶液滴定碱溶液时的误差

误差原因	滴定所耗用 $V_{酸}$	$N_{碱}$ 结果偏差 ($N_{碱}=N_{酸}V_{酸}/V_{碱}$)
酸式滴定管末端气泡未排出	偏高	偏高
酸式滴定管未用标准酸液润洗	偏高（由于实际应用的酸的浓度偏低）	偏高
滴定用的锥形瓶误用碱液润洗	偏高（由于实际应用的碱的体积偏高）	偏高
移液管内有水珠，未用未知碱液润洗	偏低（由于实际应用的碱液浓度偏低）	偏低

对于初学者来说，常因粗心、不遵守操作规程等，而引发许多操作上的错误，如果养成不良习惯，会对分析结果产生严重的影响。因此，严格按照操作规程，一丝不苟地认真操作，是操作者必须遵守的、基本的工作纪律。

在实际测定过程中，多种因素同时起作用，而且因素之间存在着交互效应，情况比较复杂。上面列举产生误差的可能原因，只是为找寻误差来源指出可能的、大致的方向，不能代替针对具体场合寻找产生误差真实原因。分析工作者的任务是要尽量减少误差，不断地提高测定结果的精密度和准确度。

2. 减少误差的方法

（1）正确使用仪器设备。按规程正确使用仪器或对仪器进行校正。例如，质量分析中称样品和沉淀、容量分析中称基准物质和试样，前后应尽可能使用同一台天平和同一盒砝码中相同的砝码（不要用 2 g、2 g、1 g 3 个砝码代替 5 g 的砝码。只用一个 2 g 砝码时，用无星号砝码），以抵消天平和砝码大部分的误差。再如，在标定标准溶液和滴定被测溶液时，尽量使用同一滴定管的相同间隔，也可以抵消因滴定管刻度不准引起的误差。

误差分类及减免误差的方法

（2）按规程正确进行分析操作。严格遵守操作规程、认真做好原始记录、细心审核计算等办法可以减小误差。试样成分的均匀性、代表性，化验操作环境，温度和湿度等对分析结果的准确性有一定的影响，也必须引起足够的重视。例如，标定标准溶液和滴定被测溶液时，使用同一指示剂，可以消除指示剂变色点与等当点的误差；条件允许时，标定和测定应由同一人操作，个人对结果掌握不同所引起的误差也可以消除。

（3）进行空白试验。空白试验是在不加试样或用蒸馏水代替试样的情况下，按照与试样分析同样的操作手法和条件进行分析试验。这样得到的结果称为空白值，从试样分析结果中扣除空白值，就可以校正由于试剂或水不纯等原因所引起的误差。

（4）进行对照试验。选用与试样成分相近的标准样品与试样在相同的条件下进行操作，把标准样品的分析结果与真实值加以比较，就可以测出方法不准、试剂不纯等引起的误差，从而在试样的分析结果中加以校正。

（5）增加平行测定的次数。在消除系统误差的前提下，测量的次数越多，其结果的平均值就越接近真实值，提高结果的准确度和精密度。

模块 2

原料及在制品日常分析

任务 2.1　密度法、折光法测在制品锤度

● 企业案例

主任从生产线上取回部分中间蔗汁，然后安排小明对以上样品的固溶物浓度进行测定，小明应该选择什么仪器和使用什么方法来测定呢？

● 任务目标

通过本任务的学习，学生达到以下目标：
（1）了解、掌握固溶物、干固物、视固溶物、锤度、折光锤度的概念。
（2）掌握锤度的测定方法：密度法和折光法。

● 素质目标

养成对比学习的好习惯。

● 任务描述

使用重力锤度计和折光锤度计分别测定样品的锤度。

● 实施条件

（1）样品：糖浆。
（2）设备及仪器：糖锤度计、阿贝折光仪、温度计（0～100 ℃）、量筒、烧杯。

● 程序与方法

1. 密度法测定锤度
（1）样品处理。

（2）锤度的测定。

2. 折光法测定折光锤度

（1）样品处理。

（2）折光锤度的测定。

详细内容见本模块任务 2.1 的相关内容。

3. 数据处理及计算

（1）计算。

1）密度法。锤度计是以 20 ℃为标准，如读数时温度不是 20 ℃，须查观测锤度温度改正表进行校正。温度低于 20 ℃时，则观测锤度减去查表数值；温度高于 20 ℃时，则观测锤度加上查表数值，即更正锤度。

对于稀释样品，则稀释样品锤度乘以稀释倍数即原样品的锤度。

2）折光法。阿贝折光仪是以 20 ℃为标准，如读数时温度不是 20 ℃，须查糖液折光锤度温度改正表进行校正。温度低于 20 ℃时，则测定值减去查表数值；温度高于 20 ℃时，则测定值加上查表数值，即更正折光锤度值。

对于稀释样品，则稀释样品折光锤度乘以稀释倍数即原样品的折光锤度。

（2）数据及结果处理。

1）密度法。

项目	数据及结果
样液观测锤度 /°Bx	
样液温度 /℃	
锤度的温度校正数	
样液锤度 /°Bx	
稀释倍数	
原样品锤度 /°Bx	

2）折光法。

项目	数据及结果
样品折光率	
样品观测折光锤度 /°Bx	
样品温度 /℃	
温度校正数	
折光锤度 /°Bx	

注意事项

（1）使用锤度计必须垂直拿取，以防折断。放入量筒时等锤度计浮定后方可放手，以免锤度计突然落下与量筒底撞击而损坏。观察读数时，锤度计不应附着于容器边缘，同时注意保持清洁，使与液面接触处有良好的弯月面。

（2）折光仪棱镜是软质玻璃，切勿用硬质物料触及，以防损伤。

（3）用折光计测定颜色很深的样品（如糖蜜）可用反射光，方法是调整反光镜，使无光线从进光棱镜射入，同时揭开折射棱镜的旁盖，使光线由折射棱镜的侧孔射入。

思考

折光锤度计所需样品量与重力锤度计相比，哪个多？

2.1.1 任务相关知识——样品的采集与处理

2.1.1.1 甘蔗制糖工业化学管理有关术语

1. 糖　Sugar

（1）糖类的统称术语。

（2）糖厂生产的基本由蔗糖组成的产品的统称。习惯上称为食糖。

2. 砂糖　Granulated sugar

砂粒状蔗糖。因颗粒大小不同，有幼砂、中砂和粗砂之分。因质量和色泽不同，有白砂糖和赤砂糖等。

专业术语

3. 白砂糖　White granulated sugar

甘蔗汁、甜菜汁或粗糖液用亚硫酸法或碳酸法等清净处理后，经浓缩、结晶、分蜜及干燥所得的洁白砂糖。

4. 赤砂糖　Brown granulated sugar

棕红色或黄褐色的带蜜砂糖。

5. 蔗料　Chopped or（and）shredded cane

甘蔗经切蔗机（或撕裂机）破碎后的碎料。

6. 蔗汁　Cane juice

从甘蔗中提取出来的汁液。

7. 糖汁　Juice

糖厂未经蒸发的各种稀糖液的统称。如混合汁、中和汁、澄清汁等。

8. 初压汁　First expressed juice

压榨机组的最初两个辊子榨出的蔗汁。

9. 末压汁（末辊汁）　Last expressed juice

压榨机组的最后两个辊子榨出的蔗汁。

10. 混合汁　Mixed juice

用压榨法或渗出法提取，送往清净处理的蔗汁。

11. 蔗渣　Bagasse

蔗料经压榨或渗出提汁后残余的物料。

12. 预灰汁　Preliming juice

经预灰处理后的糖汁。

13. 主灰汁（加灰汁）　Limed juice

经主灰（加灰）处理后的糖汁。

14. 中和汁　Neutralized juice

经硫熏和加灰中和后所得的糖汁。

15. 一碳汁　1st carbonatation juice

经第一次碳酸饱充处理后的糖汁。

16. 一清汁　1st carbonatation clear juice

一碳汁经固－液分离后所得的清糖汁。

17. 二碳汁　2nd carbonatation juice

经第二次碳酸饱充处理后的糖汁。

18. 二清汁　2nd carbonatation clear juice

二碳汁经过滤后所得的清糖汁。

19. 硫熏汁（硫漂汁）　Sulfitated juice

经硫熏处理后的糖汁。

20. 澄清汁　Clarified juice

经澄清处理后所得的清糖汁。

21. 滤清汁（滤汁）　Filtered juice

从过滤机滤出的清糖汁。

22. 泥汁　Mud juice

从澄清器（或增浓过滤设备）排出的含浓稠悬浮物的糖汁。

23. 滤泥　Filter cake，Mud

过滤机卸出的滤渣。

24. 糖浆　Syrup

糖汁经浓缩后所得的浓度较高的糖液。

25. 清净糖浆　Purified syrup

经清净处理后的糖浆。

26. 凝结水（汽凝水）Condensate

蒸汽或汁汽经热交换后，凝结而成的水。

27.（蔗糖）结晶　Crystallization of sucrose

糖浆或糖蜜在浓缩或冷却过程中，维持一定过饱和系数而析出晶体和养大晶体的过程。

28. 溶解度系数（饱和系数）　Solubility coefficient

不纯糖溶液的蔗糖溶解度对同温度下纯蔗糖溶液的溶解度的比值。

29. 过饱和系数（过饱和度）　Coefficient of supersaturation

过饱和糖液中每份水溶解蔗糖份数与同温度、同纯度下饱和糖液中每份水溶解蔗糖份数的比值。

30. 晶核　Nucleus

过饱和糖液中作为晶体长大的核心的微小晶粒。

31. 晶种　Seed

在晶种罐中起晶煮成或由低纯度砂糖与糖蜜混合，含一定数量晶粒，供养晶用料，习惯上称为种子。

32. 叠晶（梅花晶）　Conglomerate

由两颗或两颗以上晶体，面对面以不规则的角度相叠而成的晶体群。

33. 聚晶　Aggregate，Mounted grain

多个晶体聚集而成的晶体群。

34. 并晶　Twin crystals，Married grain

由两个晶体以一个公共面对称双生构成的晶体。

35. 伪晶　False grain

蔗糖在结晶过程中，新产生的、不必要的微小晶粒。

36. 糖糊　Magma

用机械方法将砂糖与糖液充分混合后所得的糊状混合物。有晶种糖糊和蜜洗糖糊两种。

37. 糖膏　Massecuite，Strike

通过煮糖而得到的晶体与母液的混合物。按煮糖顺序可分为甲、乙、丙糖膏等。

38. 糖蜜　Molasses

从糖膏或蜜洗糖糊中分离出来的母液的通称。有原蜜、洗蜜、废蜜等。

39. 原蜜　Green molasses

分蜜过程中从糖膏直接分离出来的母液。

40. 洗蜜（稀蜜）　White molasses

分蜜过程中汽洗或水洗后所得的稀糖蜜。

41. 废蜜（最终糖蜜）　Final molasses，Blackstrap molasses

从末号糖膏分离出来的母液。

42. 在制品（半成品）　Semi-product，Stock in process

糖厂生产过程中，除糖料、成品糖、废蜜、蔗渣和滤泥外，各工序正在处理或待处理的物料（或制品）统称为在制品。如各种糖汁、糖浆、糖膏和糖蜜等。

43. 糖品　Sugar products

制糖过程中含蔗糖较多的物料。通常指糖料、在制品和制品。

44. 规定量　Normal weight

根据国际统一规定，在 20.00 ℃时把一定质量的纯蔗糖配成 100.000 cm^3 水溶液，用 200.000 mm 观测管和带有国际糖刻度尺的旋光检糖仪测量其旋光度，读数为 100 °S（国际糖度），这一质量称为规定量。

在空气中用黄铜砝码称取 26.000 g（在真空下为 26.016 0 g）的质量为一规定量。

45. 固溶物　Soluble solids
溶于糖汁中的固体物质（包括蔗糖和非蔗糖物）。

46. 干固物（干物质）　Solids by drying，Dry substance，Total solids
糖品经干燥后所得的残留物。

47. 视固溶物　Apparent solids
用锤度计或糖用折射仪测定糖液中的可溶性固体物质（只是近似的固溶物）。

48. 锤度　Brix（简写 °Bx）
在 20 ℃时用锤度计测得的读数。对纯蔗糖溶液为蔗糖质量百分数；对不纯蔗糖溶液则表示溶液中视固溶物的质量百分数。

49. 折光锤度　Refractometer Brix
在 20 ℃时用糖用折射仪测得的读数。对纯蔗糖溶液为蔗糖质量百分数，对不纯蔗糖溶液则表示溶液中视固溶物的质量百分数。

50. 糖度（转光度）　Pol
用一次旋光法测得糖品中蔗糖质量百分数的近似值。

51. 纯度　Purity
糖品的固溶物中蔗糖所占的百分数。

52. 真纯度　True purity（or T.P.）
糖品的干固物中蔗糖所占的百分数。

$$真纯度（\%）= \frac{蔗糖分}{干固物} \times 100\%$$

53. 重力纯度　Gravity purity（or G.P.）
糖品的视固溶物（用锤度计测得）中蔗糖所占的百分数。

$$重力纯度（\%）= \frac{蔗糖分}{锤度} \times 100\%$$

54. 视纯度　Apparent purity（or A.P.）
糖品的视固溶物中所含糖度的百分数。

$$视纯度（\%）= \frac{糖度}{锤度} \times 100\%$$

注：式中锤度可由以下方式测得。
（1）甘蔗糖厂用锤度计测得。
（2）甘蔗糖厂用糖用折射仪测得，其计算结果称为折光视纯度（Refractometer purity）。

55. 电导灰分　Conductivity ash
用电导法测得糖品中灰分的质量百分数。

56. 色值（色度）　Colour
按规定方法测得的表示糖品颜色深浅的数值（以国际糖色值表示）。

57. 国际糖色值　ICUMSA colour

国际上统一规定的糖品色值单位。在一定的 pH 值范围内，采用适当的液层厚度和浓度，在规定的波长下（白砂糖及浅色糖品用 420 nm，深色糖品用 560 nm）溶液的吸光系数乘以 1 000，即国际糖色值。符号为 IU_x（x 为所用的波长，当用 420 nm 时，x 可不标出）。

$$IU_x = \frac{-\log T}{bc} \times 1\,000$$

式中　$-\log T$——在规定波长下的吸光度；

b——比色皿厚度（cm）；

c——糖液浓度（g/cm^3）。

58. 混浊度（浊度）　Turbidity

糖汁中由于胶体微粒存在而引起光的散射程度的数值。

59. 钙盐含量　Lime content，Calcium salt content

糖品中钙盐的质量百分数。用 100 g 样品含氧化钙质量表示。

60. 甘蔗纤维分　Fibercane

甘蔗组织中不溶于水的物质对甘蔗的质量百分数。

61. 蔗糖分　Sucrose

用规定的方法测得糖品中含蔗糖的质量百分数。

62. 还原糖分　Reducing sugar

用规定的方法测得糖品中具有还原性物质（以转化糖表示）的质量百分数。

63. 总糖分　Total sugar

用规定的方法测得糖品中含蔗糖分与还原糖分的总量。

64. 甘蔗糖分细胞破碎度（破碎度）　Cane sugar cell fragmentation（degree of fragmentation）

甘蔗在处理（斩切、压碎、撕裂、压榨等）过程中，由于受机械作用，甘蔗组织中含有蔗糖的细胞大部分破裂，已破裂的细胞数对处理前未破裂的全部细胞数的百分率。测定时以细胞所含的蔗糖分（或糖度）代表细胞数，即

$$破碎度（\%） = \frac{处理后破裂的细胞所含的蔗糖分（糖度）}{处理前全部未破裂的细胞所含的蔗糖分（糖度）} \times 100\%$$

65. 清净效率　Purification effect

在清净过程中，除去的非蔗糖物质对糖汁原来所含非蔗糖物质的百分数。

$$清净效率（\%） = \frac{100(P_2 - P_1)}{P_2(100 - P_1)} \times 100\%$$

式中　P_1——清净前糖汁的纯度（%）；

P_2——清净后糖汁的纯度（%）。

66. 二氧化硫吸收率　Absorption rate of sulfur dioxide

硫熏过程中糖汁吸收二氧化硫量对进入硫熏器的二氧化硫总量的百分数。

67. 硫熏强度　Intensity of sulfitation

糖汁吸收 SO_2 的数量。以滴定 10 mL 硫熏汁所耗用的 1/64 mol/L 碘液体积来表示。

68. 自然磷酸值　Phosphate in juice

蔗汁本身含可溶性磷酸盐的量，通常用每升含五氧化二磷毫质量表示。

69. 碱度　Alkalinity

糖汁呈碱性反应时所含的碱量。用 100 mL 糖汁中含氧化钙质量表示。

70. 自然碱度　Natural alkalinity

糖汁本身所含碱金属氢氧化物，以二碳饱充沉淀钙盐后，残留在糖汁中碱金属碳酸盐的量，用相当的氧化钙质量表示。

71. 全钙量（全氧化钙）　Total calcium

100 cm³ 主灰汁或一碳汁中所含未反应的氧化钙及已反应形成碳酸钙的氧化钙和部分碱金属相当于氧化钙的质量。

72. 纯度差　Purity difference

在制糖过程中处理前后两种物料之间的纯度差值。

73. 脱色率　Percent of decolorization，Decolorization ratio

糖汁脱色前后色值减少的百分数。

74. 提汁率　Juice extraction rate (Draft)

从渗出器或压榨机中提出的蔗汁质量与处理糖料质量的百分数。

75. 抽出率　Sucrose (or Pol) extraction

在提汁过程中从糖料提取蔗糖（或糖度）的百分数。

$$抽出率（\%）=\frac{提取糖汁中的蔗糖（或糖度）质量}{糖料中的蔗糖（或糖度）质量}\times100\%$$

76. 对比抽出率　Reduced sucrose (or pol) extraction

把实际抽出率换算为以标准甘蔗纤维分（12.5%）为基准的抽出率。

$$E_{12.5}=100-\frac{(100-E)(100-F)}{7F}$$

式中　$E_{12.5}$——对比抽出率（%）；

　　　E——实际抽出率（%）；

　　　F——甘蔗纤维分（%）。

77. 煮炼收回率　Boiling house recovery

成品白砂糖（包括赤砂糖等折的及在制品可制得的白砂糖）的蔗糖（或糖度）对混合汁中蔗糖（或糖度）的质量百分数。它表示制炼过程从混合汁（或渗出汁）的蔗糖（或糖度）中实际收回蔗糖（或糖度）质量的百分数。

$$煮炼收回率（\%）=\frac{已成及未成白砂糖的蔗糖（或糖度）质量}{混合汁（或渗出汁）中蔗糖（或糖度）质量}\times100\%$$

78. 对比煮炼收回率　Reduced boiling house recovery

煮炼收回率的高低影响因素较多，客观上受糖料纯度高低的影响。国际上采用对比煮炼收回率进行工艺技术的比较，即利用实际的混合汁（或渗出汁）纯度和煮炼收回率推算出理论废蜜纯度，然后换算至纯度为 85 的混合汁（或渗出汁）煮炼收回率。它不受混合汁（或渗出汁）纯度高低的影响，可作为各厂收回率直接比较的指标。

1965 年，国际甘蔗糖业技术师学会（ISSCT）正式采用 Deerr 的计算公式：

$$M_v = \frac{J(100-R)}{10\,000-JR} \times 100\%$$

$$R_{85} = \frac{100(85-M_v)}{85(100-M_v)} \times 100\%$$

1971 年，ISSCT 建议采用 Gundu-Rao 的公式代替原正式采用的 Deerr 公式。Gundu-Rao 计算对比煮炼收回率的公式为

$$R_{85} = R + \frac{M}{1-M} \times \frac{0.85-J}{0.85J}$$

式中　M_v——理论最终废蜜纯度（%）；

M——最终废蜜纯度，比率；

J——混合汁（或渗出汁）纯度（Deerr 公式为百分数，Gundu-Rao 公式为比率）；

R——煮炼收回率（同上）；

R_{85}——对比煮炼收回率（同上）。

79. 总收回率（蔗糖收回率）　Overall recovery，Total recovery

成品糖及在制品中可制成糖的蔗糖（或糖度）对糖料中的蔗糖（或糖度）的质量百分数。

$$总收回率（\%）= \frac{成品糖及在制品中可制成糖的蔗糖（或糖度）质量}{糖料中的蔗糖（或糖度）质量} \times 100\%$$

80. 对比总收回率　Reduced overall recovery

对比抽出率和对比煮炼收回率的积除以 100%。

$$对比总收回率（\%）= \frac{E_{12.5} \times R_{85}}{100} \times 100\%$$

式中　$E_{12.5}$——对比抽出率（%）；

R_{85}——对比煮炼收回率（%）。

81. 结晶率　Crystallization percent in massecuite

糖膏中结晶蔗糖对糖膏固溶物的质量百分数。

$$结晶率（\%）= \frac{糖膏纯度 - 糖蜜纯度}{100 - 糖蜜纯度} \times 100\%$$

82. 产糖率（等折白砂糖产率）　Sugar yield

本期已成及未成白砂糖质量对糖料的质量百分数。

$$结晶率（\%）= \frac{本期已成及未成白砂糖质量}{实际处理糖料质量} \times 100\%$$

83. 未测定损失　Undetermined losses

未测定损失等于蔗糖（或糖度）平衡中总蔗糖（或糖度）损失减去已测定损失，包括采样称重、分析上的误差、跑漏糖及化学转化和分解等的损失。甘蔗糖厂用对甘蔗中蔗糖

（或糖度）质量的百分数表示。

2.1.1.2 任务相关知识——采样目的与截样周期

1. 采样的目的

在糖厂化学管理工作中，待分析的对象是各式各样的，其中有些数量很大而组成部分又很不均匀。如甘蔗、白糖、蔗渣、石灰石、煤等，而实验室的分析不可能将所有的物料全部进行分析测定，只能处理少量的样品。在这种情况下，要为生产提供可靠而有效的分析数据，样品就必须能代表大量物料的平均组成。采样是分析工作的第一步，少量样品是否能代表物料的平均组成，决定了分析的结果是否具有代表性。在通常的情况下，采样造成的误差往往要大于分析过程产生的误差。所以加强样品采集管理，用科学的方法去采样，是提高分析准确度的重要手段。

制糖工业中的物料种类繁多，从物料的形态上分类主要有液体、固体及气体三类，其中液体物料占多数。

2. 截样周期

糖厂各种物料采样方法及截样周期见表 2-1。

表 2-1 采样方法及截样周期汇总

样品名称	采样方法	截取样品周期	使用的防腐剂	备注
甘蔗	2 h 一次	2h 或 4h		（1）甲醛 2/1 000 对样品量； （2）二氯化汞溶液 1/1 000 对样品量； （3）混合汁供测定泥砂、悬浮物样品，用甲醛作为防腐剂每 1 h 采样一次。测定钙盐样品不加防腐剂； （4）澄清汁供测定色值、酸值、钙盐的样品不加防腐剂。滤清汁测定色值、酸值的样品不加防腐剂
初压汁	连续	2h 或 4h	甲醛	
蔗渣	15 min 一次	2h 或 4h	甲醛	
初压汁	连续	2h 或 4h	甲醛	
混合汁	连续	2h 或 4h	二氯化汞溶液	
末压汁	连续或 15 min 一次	2h 或 4h	甲醛	
预灰汁	连续	1h 或 2h		
一碳汁 一碳清汁 二碳清汁	连续	2 h		
澄清汁	连续	2h 或 4h	二氯化汞溶液	
滤清汁	连续	2h 或 4h	甲醛	
清净糖浆	连续	1h 或 2h		
滤泥	15 min 一次	2h 或 4h		

续表

样品名称	采样方法	截取样品周期	使用的防腐剂	备注
糖膏		每煮成一罐时		
原蜜及洗蜜	15 min 一次	每编号糖膏分蜜完毕时		
废蜜	连续	每编号糖膏分蜜完毕时		
白砂糖	连续或 15 min 一次	每编号糖膏分蜜完毕时		
赤砂糖	15 min 一次	每编号糖膏分蜜完毕时		

2.1.1.3 采样方法

1. 液体及气体样品的采集

液体及气体物料的组成较为均匀，流动性好，一般可用连续采样法进行采样。糖厂中的糖汁、糖浆等，浓度不高且含悬浮物不多的液体物料都可以使用这种采样方法。对于浓度高的物料（如废蜜、糖膏）或浓度虽然不高但悬浮物多的物料（如石灰乳；一碳、二碳饱充汁）等，则可用间接采样法，也就是用人工（或自动）定时采集定量的样品。在规定的时间内所采集的样品，经充分混合后，取其中一部分进行分析。

样品的采集

至于气体，用小管从储藏箱或管道引入气体分析仪即可。

2. 固体样品的采集

由于固体物料的组成没有液体或气体均匀，取得有代表性的固体样品一直是较难解决的问题，尤其是原料甘蔗的采集。蔗渣、滤泥采用五点或七点采样法。蔗渣是按压榨辊长度均分的五点采样。滤泥可在滤泥斗车的车角及中央五点采样。如果使用真空吸滤机，则可按转鼓长度分为五点或七点采样。至于成品糖，则是在包装称量时每包取定量样品。所采集的样品（除甘蔗外）用"四分法"缩减其数量，取适量的样品进行分析。

3. 各工序样品采集的方法

（1）压榨工序。

1）甘蔗。甘蔗在卸入输蔗带前，每 2 h 采取有代表性的全茎甘蔗 2 条，放在不被日晒雨淋的阴凉地方，积集至所规定的时间截样，用于测定甘蔗夹杂物或其他项目。

注意：采取甘蔗样品必须小心轻放，防止夹杂物脱落。

2）蔗渣。在最后一座压榨机蔗渣出口处，每 15 min 采样一次（如遇塞辘或暂停注加渗浸水的情况，仍须按时采样）。采样时按压榨辊的长度均分为五点（在辊子的中点和中点至两端处等分三段，等分点共五点）为采样处。用双手（一上一下）或用夹子采取整层蔗渣样品放入内置防腐剂（以脱脂棉花湿润 10 mL 甲醛，装入有小孔的铁盒内，放于桶底）的蔗渣样品储桶中，并严密加盖，积集至所规定的时间截样。截样时，须于储桶内将样品混合均匀，然后取一部分（不少于 2 kg）置于样品桶中，用于测定糖度和水分。

3）初压汁。在第一座压榨机前辊蔗汁流出处装一斜槽，槽上盖一半圆形钢丝网以除去蔗屑，槽口装一引汁槽，以铜线引出蔗汁，调节流速，使其均匀不断地滴入样品桶中，多余蔗汁回流至原来蔗汁槽中。样品桶应先放入 4 mL 甲醛作为防腐剂。截样时样品容量应在 2 000～3 000 mL。

4）混合汁。在混合汁泵后管道适当位置处接一小管，管口接两条铜线，使混合汁分别均匀流入两个玻璃瓶中，其中一瓶内加入 3 mL 二氯化汞酒精饱和溶液作为防腐剂，另一瓶则不加防腐剂（用于钙盐测定）。瓶口上装有玻璃漏斗以减少蔗汁水分蒸发，所采样品每瓶每小时应不少于 2 000 mL（注意：每 1 000 mL 样品，防腐剂用量为 1 mL）。每 2 h 截样一次，分别将瓶内蔗汁摇匀，用滤网过滤，以除去蔗屑，分别按下述第 1、2 点留取样品。另外，供测定泥砂悬浮物的样品按下述第 3 点采集。

①供测定锤度、糖度、蔗糖分、还原糖分、磷酸值的样品：每 2 h 用玻璃瓶留取已加防腐剂的等量样品，积集至所规定的时间截样，样品不少于 2 000 mL。

②供测定钙盐的样品：每 2 h 用玻璃瓶留取不加防腐剂的等量样品，积集至所规定的时间截样，样品不少于 1 000 mL。

③供测定泥砂悬浮物的样品：在预灰前每 1 h 用勺采样一次，约 200 mL，放入内置 10 mL 甲醛作防腐剂的样品桶中，以后每 3 h 再加入 5～10 mL 甲醛以加强防腐，积集至所规定的时间截样。

5）末压汁。在最后一座压榨机后辊底装一斜槽，引出顶辊与后辊所榨出的蔗汁，连续或间歇采样，储入内置 10 mL 甲醛的样品桶中。如间歇采样，则每隔 15 min 于采完蔗渣样品后随即采样一次，积集至所规定的时间截样，样品应在 2 000～3 000 mL 间。

6）预灰汁。在混合汁预灰处理后，在中和之前的适当位置上连续采样，积集至所规定时间截样，样品应不少于 1 000 mL。

（2）提净及蒸发工序。

1）硫熏汁。在硫熏或硫熏管道的适当位置上，用间歇或连续采样法，积集至所规定的时间截样，样品应不少于 200 mL。

2）一碳饱充汁。在一碳饱充汁出口管道上，按分析时间不定时采样 200 mL。

3）一碳清汁。在一碳清汁进入二碳饱充前管道的适当位置上连续采样，积集至所规定时间截样，样品应不少于 1 000 mL。

4）二碳饱充汁。在二碳饱充汁的出口管道上装一小旋塞，由生产岗位操作人员根据需要不定期采样检查。

5）二碳清汁。在二碳清汁管道的适当位置上连续采样，积集至所规定的时间截样，样品应不少于 1 000 mL。

6）澄清汁或硫漂清汁。在澄清汁或硫漂清汁进入清汁储箱前管道的适当位置上连续采样，每 2 h 截取样品一次。不少于 2 000 mL，将样品搅拌均匀，按下列方法留样。

①供测定色值、酸值、钙盐的样品：每 2 h 用玻璃瓶留取等量样品，积集至所规定的时间截样，样品应不少于 1 000 mL。

②供测定锤度、糖度、蔗糖分、还原糖分的样品：每 2 h 用玻璃瓶留取等量样品。瓶内预先加入 2 mL 氯化汞酒精饱和溶液作防腐剂，积集至所规定的时间截样，样品应不少于 2 000 mL。

7）滤清汁。在滤清汁管道上连续采样，供测定锤度、糖度的样品以甲醛作防腐剂。供测定色值、酸值的样品不加防腐剂。

8）粗糖浆。在未经硫熏的粗糖浆管道的适当位置上，连续或间接采样，每 2 h 截取样品一次，样品应不少于 2 000 mL，将样品搅拌均匀后，按下列方法留样。

①供测定锤度、色值、酸值、钙盐的样品：每 2 h 用玻璃瓶留取等量样品，积集至所规定的时间截样，样品应不少于 1 000 mL。

②供测定锤度、糖度、蔗糖分、还原糖分的样品：每 2 h 用玻璃瓶留取等量样品，积集至所规定的时间截样，样品应不少于 2 000 mL。

9）清净糖浆。粗糖浆经硫漂后，在清净糖浆管道的适当位置上连续采样，积集至所规定的时间截样，样品应不少于 1 000 mL。

10）滤泥。

①在滤泥斗车上采样方法：用一直径约 3 cm、长约 60 cm 的竹子或金属半圆筒作为采样工具。当滤泥卸落斗车后，在车斗的四角和中间部分（五点），将工具深深插入泥中采样。将采得的样品刮入样品桶中。每一斗车的滤泥依上述方法采取等量样品，积集至所规定的时间截样。

②在螺旋输送槽采样方法：若卸下的滤泥是经螺旋输送槽运送厂外的，可在螺旋槽出口处，每 15 min 用长柄木铲采取等量样品一次，放入样品桶中，积集至所规定的时间截样。

③真空吸滤机可在进行过滤操作时，按转鼓长度分三点（两边及中间）每 15 min 采样一次，放入样品桶中，积集至所规定的时间截样。

（3）结晶工序。

1）糖膏。

①煮成糖膏：每罐糖膏煮成后，在卸糖的过程中，于流槽中央用勺取样品约 1 kg，置入样品盅内。

②在煮糖膏：当罐内物料在充分对流的情况下，在煮糖罐采样器处，采取样品约 1 kg，置入样品盅内。

2）糖蜜。

①洗蜜：在分蜜机糖蜜流出的槽口或适当位置进行间歇采样，等量的样品从开始分蜜后 5 min 直至分蜜完毕，每 15 min 采取一次，放入样品桶中。

②原蜜：采样方法与洗蜜相同。

3）废蜜。每编号糖膏从分蜜开始直至完毕，在废蜜箱内，每满一箱采取等量样品一次，不满一箱的按比例采样。或在废蜜秤的废蜜出口处，每称一次采取等量样品，采样次数按每称重量多少而定，一般以 1～2 d 采等量样品一次，但每编号采集样品应不少于 6 次。所采取的样品，其浓度必须与计量或称重时的废蜜浓度一致。

（4）成品糖及原糖。

1）白砂糖。每分离一罐糖膏为一个编号，在称量包装时，连续采集样品约 5 kg，放在带盖的容器中，调匀后为编号样品，该样品除供编号分析之用外，另取 0.5 kg 放在带盖的容器中，积累 24 h 后为日集合样品。

2）精糖。同白砂糖。

3）赤（黄）砂糖。赤（黄）砂糖在称重包装时，连续采集等量样品，放入样品桶中。每班截取样品一次，充分调匀，取样约 1 kg，供分析用。另从三班样品中，留取等量样品，调匀作为日集合样品。

4）红糖。同赤（黄）砂糖。

5）冰糖。每班包装的产品为一个编号。在包装部随机每 2 h 取 0.5 kg 样品（包括盆冰、盆底和柱冰）稍加粉碎混匀，放入双层食品袋中，积累 8 h 作为班样本。

6）方糖。每班（或每罐）生产并包装的产品为一个编号。在方糖机运输带终端每 2 h 随机取 0.5 kg 样品一次，放入带盖的容器中，累积 8 h 后作为班样品。

7）液体糖。每班（或每罐）生产并包装的产品为一个编号。在包装机或输送管道适当位置上，每 2 h 随机采集样品约 0.5 kg，放入带盖的容器中，累积 8 h 作为班样品。

8）原糖。原糖到达糖厂后，在卸运称量时，按照《原糖》（GB/T 15108—2017）国家标准的规定采集样品，放在带盖的容器中，将样品混匀后用四分法取出约 1 kg 样品供分析用。

（5）辅助材料。

1）石灰石。每批石灰石进厂时，在船（或车）上进行采样。先观察石灰石的品质、色泽等，然后在船（或车）上各点选择代表性的石块，将之逐一打碎成直径约 5 cm 的小块。在每堆小石块中，取等量样品，总量应不少于 3 kg。取得的样品按四分法缩减成小样（应不少于 0.5 kg），粉碎后以供分析。

2）石灰。每批石灰进厂后，首先观察石灰的品质、碎屑及块状石灰的色泽，然后从各处取最具代表性的石灰样品约 3 kg。对于碳酸法厂，则在石灰窑卸灰时，于卸下石灰堆中，观察石灰的品质及碎屑，分五点取最大代表性的石灰样品 3 kg。

将采集到的石灰打碎成均匀小块，用四分法反复处理，至留下约 60 g，打碎成粉末，并全部通过 0.15 mm 孔径筛网；注意最后过筛的细粒是最难打碎的，但决不可丢弃，以免影响样品的代表性。

3）硫黄（孔径 0.28 mm）。每批硫黄进厂时，在硫黄的总袋（或箱）数中，选取 5% 袋（或箱）于其 2/3 深处取样。如果是散装硫黄，注意在不同位置选取有代表性的试样，混合并按四分法缩分成 2 kg 左右。然后磨碎并通过 10 目筛（孔径 2 mm），以四分法分成两份：一份供测定水分用；另一份磨碎又分成两份，分别通过 30 目筛（孔径 0.63 mm）和 60 目筛（孔径 0.28 mm），分别用四分法各分成两份：一份供测定用，另一份保存以备复检。

4）磷酸。在每批磷酸进厂时，用清洁干燥的玻璃管从每批件的 10% 中取出试样。取样时将玻璃管垂直地插入容器内部均匀取样，每批取样总量不少于 500 mL，将试样混合均匀后收集于清洁干燥有磨口塞的玻璃瓶中，贴上标签送化验室分析。

5）絮凝剂。每批絮凝剂进厂时，随机抽取 5 桶，每桶采集等量样品，调匀。供分析用。

（6）其他。

1）煤（或煤堆）。当每批煤由货车或船进厂（或在煤堆处）卸下时，在煤堆的顶部、中央及底部采集有代表性样品约 5 kg（大块、小块应兼顾）放入密封的样品桶中，大块煤样稍经锤碎至约 13 mm 的粒度，迅速用四分法缩分至约 1 kg，供分析用。

2）润滑油。当每批润滑油进厂时，随机抽取 5 桶，用清洁干燥的玻璃管插入油层 10 cm 左右，采集等量样品，混匀，供分析用。

3）锅炉水。在锅炉水鼓旁的采样处，采取经过冷却装置的锅炉水鼓内的水。采样前必须先将管内积存的水排空，并将盛器冲洗两次，然后采集一小杯。

4）入炉水。在入炉热水鼓的采样处采样，采样前须先将内积存的水排空，并将盛器冲洗两次，然后采集一小杯。

5）凝结水。在蒸发罐或煮糖罐、加热器的汽水分离器采取。采样前须先将其内积存的水排空，并将盛器冲洗两次，然后采集一小杯。

6）冷凝水。在蒸发罐及煮糖罐的冷凝器排水管上离地面约 1 m 高处，装置两小管连续采样，务必使采样瓶内压力与排水管内的压力相等，才能将水引出。同时必须注意调节水的流量，使整个采样时间内的样品均衡，积集至所规定的时间截样。

7）窑气。在窑气储箱的上部，用一小管将窑气引入化验室，经一大玻璃瓶以分离水分，再经 U 形管，内部以玻璃丝滤除灰尘，然后与气体分析器连接，并装有排气管，在分析前，应先将上次残留在管内的窑气排除。

2.1.1.4 样品预处理

1. 物料的预处理

（1）组分比较均匀的样品。如成品糖、滤泥、糖液等，对样品稍加混合取其中一部分，即可成为具有代表性的分析试样。

（2）组分很不均匀的样品。对一些颗粒大小不均匀、成分混杂不齐、组成极不均匀的试样，选取具有代表性的均匀试样，是一项较为复杂的操作。为了使所采的试样具有代表性，必须按一定的程序，从物料的不同部位，取出一定数量、大小不同的颗粒。取出的数量越多，试样的组成与被分析物料的平均组成越接近。根据经验，平均试样选取量与试样的均匀度、粒度、易破碎度有关，通常可按以下公式估算：

样品的处理

$$Q=Kd^{\alpha}$$

式中：Q——采取平均试样的最小质量（kg）；

d——试样中最大颗粒的直径（mm）；

K，α——经验常数，可由物料的均匀度和破碎度等决定，通过试验求得，K 值为 0.02～0.15，α 值通常为 1.8～2.5。

例如，取 $K=0.1$，$\alpha=2.1$

若 $d=1.2$ mm

则 $Q=0.1\times1.2^{2.1}=0.15$（kg）

若 $d=0.8$ mm

则 $Q=0.1\times0.8^{2.1}=0.063$（kg）

根据各种样品所需分析的项目而定，分别对其进行不同的处理。但一般的步骤可分为破碎、过筛、混匀、缩分四个步骤。

1) 破碎。用机械或人工方法把样品逐步破碎,以达到所需的碎粒为止。每次破碎后要进行筛分,筛分出的粗粒子要再次破碎,不能扔掉,以保证样品的代表性。

2) 缩分。每次破碎样品后,用机械或人工取出一部分有代表性的样品,继续加以破碎,样品量就逐渐缩小,便于处理。

常用的手工缩分方法是"四分法"。先将已破碎的样品充分混匀,堆成圆锥形,再将其压成圆饼状,通过中心按"十"字形切成四等份,弃去任意对角的两份。由于样品中不同粒度、不同密度的颗粒,在混合后的物料中大体上分布均匀,留下样品的量是原样品的一半,但仍能代表原样品的组成。反复操作,直至样品的量合适为止。

2. 样品的稀释

(1) 倍数稀释。通常用于浓度较高或固体样品的稀释,如糖浆、糖膏、糖蜜等。这里的倍数是质量倍数,即稀释后样液的质量与原样品质量之比:

$$稀释倍数 = 稀释后样液的质量 / 原样品质量$$

采用倍数稀释时,加水的量通常用量筒量取,此时将水的密度视为1,每毫升水重1 g。

(2) 规定量稀释。规定量稀释法主要用于旋光测定。糖厂中将26.000 g 称为一个规定量。规定量稀释就是将26.000 g 样品用容量瓶配制成100 mL,或按相同的比例扩大、缩小,如将52.000 g 样品配制为200 mL 也称为规定量稀释。稀释后样液的浓度称为一个规定量。也可以将13.000 g 样品配制为100 mL,称为半规定量稀释,稀释后样液的浓度称为半个规定量。同理,也可以称取1/3 个规定量进行稀释,稀释后样液的浓度就称为1/3 个规定量。

采用规定量稀释时应保证稀释后样液的体积必须达到规定的值,如称 26.000 g 样品配为浓度为 1 个规定量的样液,则样液的总体积必须恰好为 100 mL。对于完全能溶解的样品来说,这点是能保证的。但对于一些不能完全溶解的样品,由于样品不完全溶解,固体部分在容量瓶内必然会占据一定的容积,造成样液的真实体积小于容量瓶的标称容积,即稀释时加入的水量少于理论上的计算值,溶液的浓度高于理论计算值。在实际操作中,为了弥补这一误差,对于不能完全溶解的样品,在进行规定量稀释时,可适当减少样品的质量。例如,在糖厂中,滤泥的规定量就定为 25.0 g。

采用规定量稀释时也可以计算样品的稀释倍数,计算方法如下:

$$稀释倍数 = (稀释后样液的体积 \times 稀释后样液的密度) / 原样品质量$$

稀释后样液的密度可按稀释后样液的观测锤度查表或按以下公式计算:

$$稀释后样液的密度 = 0.998\,2 + 0.003\,7B + 0.000\,018\,16B^2$$

式中 B——样液的观测锤度(°Bx)。

(3) 任意稀释。在生产中,有时要求快速提供样品纯度的数据。例如,煮糖配料。任意稀释主要用于纯度的测定。因样品的纯度不受稀释的影响,当需要进行快速纯度测定时,可取适量样品,加水至合适测定的浓度就可以直接进行测定,稀释后样液的纯度就是原样品的纯度。

3. 样品的防腐与澄清

(1) 防腐剂。糖厂样品多含糖分,是微生物的良好营养物,在温度及 pH 适宜时,微生物就很容易繁殖。在微生物的作用下，样品中原来的成分的相互比例就会发生变化，同时产生了新的成分，这样就不能代表原来的情况了。因此，在糖厂中，对于容易繁殖微生物的样品，如果不能立刻进行测定，一般需要在集合的样品中加入防腐剂。糖厂中常用的防腐剂有以下几种。

1）二氯化汞（$HgCl_2$）溶液。二氯化汞是重金属盐，能凝固蛋白质，有剧毒，具有强烈的杀菌作用。常用于糖厂液态样品的防腐，用量为每升液体样品加入二氯化汞酒精溶液 0.5 mL。因二氯化汞为重金属盐，水解后溶液呈酸性，故测灰分和 pH 值的样品不能使用。

2）甲醛（工业用甲醛溶液含甲醛 30%～40%）。甲醛的渗透性强，易渗入菌体中使蛋白质变性，故有抑制微生物繁殖的作用，从而达到防腐的目的。一般使用工业甲醛（含甲醛 30%～40%），每升样品溶液用量为 0.5～1 mL。常用作蔗渣的防腐剂。由于甲醛溶液具有酸性与还原性，故对拟测定 pH 值和还原糖的样品不适用；

3）氨水–氯仿溶液。用作蔗渣样品防腐，通常用棉花吸附后，置于样品桶的底部。

防腐剂的用量：

1）二氯化汞（$HgCl_2$）溶液：每升糖汁样品加入 0.5 mL；

2）甲醛：每升糖汁样品加入 0.5～1 mL。

（2）澄清剂种类。糖厂中待分析的样液，大多数带有颜色，常呈混浊状态，而且含有不同程度的杂质影响测定。为使样液的分析能够顺利进行，有必要将之澄清。例如，大多数样品都要用旋光法测定蔗糖分（或转光度），而旋光观测要求样液色浅、透明，故在旋光观测之前要先行澄清处理。

在有关文献中介绍的澄清剂种类不少，但最常用的是铅盐。因铅离子可与很多阴离子（如 Cl^-、PO_4^{3-}、SO_4^{2-}、CO_3^{2-}、酒石酸根等）结合，生成难溶的沉淀物，同时吸附除去部分杂质。

1）碱性醋酸铅（又称低醋酸铅或盐基性醋酸铅）。它是 4 份 $3Pb(CH_3COO)_2 \cdot PbO$ 与 3 份 $Pb(CH_3COO)_2 \cdot PbO$ 的混合物，近似的分子式为 $3Pb(CH_3COO)_2 \cdot 2PbO$。它的澄清能力较强，所形成的沉淀物不仅能吸附除去部分还原糖分解产物、蔗糖焦化产物、还原糖等，在碱性情况下还能使部分胶体物质凝聚，故在测定样液蔗糖分时，常用作澄清剂。但因其会把部分还原糖也沉淀，测还原糖的样液中不能用其作为澄清剂。

2）中性醋酸铅 $[Pb(CH_3COO)_2 \cdot 3H_2O]$。相对碱性醋酸铅而言，其澄清、脱色能力较差。因醋酸铅澄清剂澄清能力的大小与分子中 PbO 的含量多少有关，它不能把深色糖液完全澄清，但在澄清时它不会或很少沉淀糖液中的还原糖，所以适用于待测还原糖样液的澄清。

3）碱性硝酸铅 $[Pb(NO_3)_2 \cdot Pb(OH)]$。其澄清能力最强，适用于废蜜及顽性糖汁的澄清。使用时，把等体积的 $Pb(NO_3)_2$ 溶液与 NaOH 溶液依次加入待澄清的样液中。

这种澄清剂的缺点是把相当数量的 $NaNO_3$ 带进样液中，影响了蔗糖的旋光度。同时，它也能沉淀部分还原糖，因此，非必要时一般不用于澄清。

铅盐澄清剂效果良好，长期以来广泛使用于糖厂的制糖分析中。但具有毒，对人健康有不良的影响，在使用时要特别注意。

（3）澄清剂的用量。澄清剂的用量必须适当，用量太少，达不到澄清的目的；用量太多，则会因生成铅糖化合物等而产生铅误差。通常每 100 mL 样液加固体澄清剂约为 1 g。

糖厂样品所用澄清剂及用量见表 2-2、表 2-3。

表 2-2　甘蔗糖厂样品所用澄清剂及用量

样品	分析项目	澄清剂	用量
初压汁	转光度	碱性醋酸铅粉	每 100 mL 约用 1 g
混合汁	转光度、蔗糖分	碱性醋酸铅粉	每 100 mL 约用 1 g
	还原糖	54 °Bx 中性醋酸铅液	每 100 mL 用约 2 mL
末压汁	转光度	碱性醋酸铅粉	每 100 mL 约用 1 g
蔗渣	转光度	碱性醋酸铅粉	每 100 mL 蔗渣浸出液 0.2～0.3 g
清汁	转光度	碱性醋酸铅粉	每 100 mL 约用 1 g
	还原糖	54 °Bx 中性醋酸铅液	每 100 mL 用约 2 mL
滤清汁	转光度	碱性醋酸铅粉	每 100 mL 约用 1 g
滤泥	转光度	54 °Bx 碱性醋酸铅液	25 g 样品用 2～7 mL
粗糖浆	转光度	碱性醋酸铅粉	每 100 mL 约用 1 g
	蔗糖分	碱性醋酸铅粉	4 倍稀释时每 100 mL 用 0.5 g
	还原糖	54 °Bx 中性醋酸铅液	4 倍稀释时每 100 mL 用 2 mL
回溶糖浆	转光度	碱性醋酸铅粉	4 倍稀释时每 100 mL 用 1 g
糖膏、糖蜜	转光度	碱性醋酸铅粉	6 倍稀释时每 100 mL 用 1 g
废蜜	转光度	碱性醋酸铅粉	6 倍稀释时每 100 mL 用 1 g
	蔗糖分	碱性醋酸铅粉	1/3 规定量稀释液每 200 mL 用 3.5 g
	还原糖	54 °Bx 中性醋酸铅液	1/3 规定量稀释液每 50 mL 用 1 mL
白砂糖	蔗糖分	不用	
赤砂糖	蔗糖分	碱性醋酸铅粉	规定量稀释每 100 mL 用 1 g
	还原糖	54 °Bx 中性醋酸铅液	规定量稀释每 50 mL 用 1 mL

表 2-3　甜菜糖厂样品所用澄清剂及其用量

样品	分析项目	澄清剂	用量
甜菜	蔗糖分	碱性醋酸铅液（$d=1.235\sim1.240$ g/mL）	甜菜糊 26 g 加 7 mL
甜菜丝	糖度	碱性醋酸铅液（$d=1.235\sim1.240$ g/mL）	甜菜糊 26 g 加 4～7 mL
原汁	糖度	碱性醋酸铅粉	适量
渗出液	糖度	碱性醋酸铅粉	适量

续表

样品	分析项目	澄清剂	用量
废粕	糖度	碱性醋酸铅粉	适量
废粕	糖度	碱性醋酸铅粉	适量
二清汁	糖度	碱性醋酸铅粉	适量
滤泥	蔗糖分	碱性醋酸铅液（d=1.235～1.240 g/mL）	样品 50 g 加 4 mL
糖膏、糖蜜	糖度	碱性醋酸铅粉	规定量稀释液加入适量
废蜜	糖度	碱性醋酸铅粉	适量
废蜜	还原糖	中性醋酸铅液（d=1.235～1.240 g/mL）	1/2 规定量稀释液每 100 mL 加 10 mL
二、三号砂糖、复筛糖	糖度	碱性醋酸铅粉	规定量用 0.5 g

4. 样品的除铅、除钙

用适当数量的醋酸铅澄清后的样液，其中残留的铅盐对旋光测定的影响甚微，所以不必除铅。但在测定还原糖时，澄清后的样液中残留的铅盐以及由样品本身带来的钙盐，对测定均有影响。其中铅盐可能与还原糖（特别是果糖）结合生成铅糖化合物，钙盐能与葡萄糖、果糖形成络合物，会使测得的样液的还原糖含量降低。因此，测还原糖时样液必须除铅、除钙。

在糖厂的化学管理中，通常用磷酸氢二钠（$Na_2HPO_4 \cdot 12H_2O$）和草酸钾（$K_2C_2O_4 \cdot H_2O$）作除铅剂，使用时以溶液的形式加入。1 L 除铅剂溶液中含有磷酸氢二钠 70 g、草酸钾 30 g。测还原糖时，如果是 50 mL 样液加除铅剂 1.5～3 mL；如果用奥夫奈尔法测定还原糖，因测定过程中样液呈酸性，草酸根在此时有还原性，所以只用磷酸氢二钠作除铅剂。在使用兰—艾农法或兰—艾农恒容法测还原糖时，还可用 EDTA（乙二胺四乙酸）二钠代替草酸钾除钙。具体的方法及用量在后面的章节中介绍。

2.1.2　任务相关知识——在制品的日常分析

糖液中蔗糖及非蔗糖物总量称为干固物（固溶物）。当糖液中的水分完全蒸发至干时，所得的残留物称为真固溶物。纯蔗糖溶液中的真固溶物为蔗糖；不纯糖液中的真固溶物包括蔗糖及所有溶解于水中的非蔗糖杂质。

在糖厂化学管理中，测定干固物的分析方法有干燥法、密度法、折光法三种。

1. 干燥法测定干固物

干燥法是通过加热把被测样品中的水分蒸发直至恒重，经过计算而得到真固溶物含量。由于这种分析方法的测定过程太长，且比较麻烦。因此，在糖厂不用此法测定含水分较多的样品，而是用具有快捷、简单、方便等特点的密度法和折光法来测定固溶物。密度法和折光法测定的样品中所含的固溶物，是真固溶物含量的近视值，称为视干固物含量。

锤度测定是糖厂化学管理重要的基础数据之一，可用密度糖锤度计和阿贝折射仪测定。

锤度是指温度为 20 ℃时用锤度计测得的数值。对纯蔗糖溶液是表示蔗糖质量百分率，对不纯蔗糖溶液则表示溶液中可溶性固溶物的近似质量百分率。

测定含蔗糖为主的溶液锤度常用的方法有密度法及折光法两种。适用于含蔗糖为主的溶液中可溶性固溶物含量的测定。在制品（如蔗汁、糖浆、糖糊及各种糖膏、糖蜜等）均可适用。对于含有结晶糖的样品必须溶解稀释后才能测定。

2. 密度法测定干固物

（1）密度法测定干固物原理。

1）密度。密度是单位体积物质的质量（习惯称为重量）。糖厂中将 20 ℃时 1 cm³ 糖液在空气中称得的质量称为糖液的视密度，通常称为密度。

2）密度与干固物含量的关系。蔗糖溶液的密度随浓度的增加而增加，测定溶液的密度，可以确定溶液的浓度。

比重法测定锤度

（2）锤度计的结构及刻度方法。

1）锤度计结构。锤度计是一根下部较粗，上部较细并带有刻度的玻璃管。根据"物体在液体中所失去的质量等于它在液体中所排开的液体质量"这一原理，当质量一定的锤度计浮在溶液中时，浸没的体积的变化可以通过锤度计上部的刻度变化来反映。当液体的密度变小时，浸没的体积应变大，锤度计就下沉，锤度计上部的刻度的读数就变小；反之则变大。因为锤度计的上部很细，单位长度的体积很小，所以能明显地反映出密度的微小变化。

2）锤度计的刻度方法。锤度计的刻度则是以 20℃ 纯蔗糖溶液的质量百分浓度作为其刻度的基准。例如，将锤度计置于 20 ℃、质量百分浓度为 5% 的纯蔗糖溶液中，其液面所在的位置就定为 5 °Bx，而将同一锤度计置于 20 ℃ 的纯水中，其液面所在的位置就定为 0 °Bx，在它们之间等分就可以划分出不同的刻度。所以，对于纯蔗糖溶液，锤度即表示溶液蔗糖含量的质量百分数。但只有对纯蔗糖溶液时这个关系才是正确的。当溶液中含有其他杂质时，因杂质在水溶液中的密度与蔗糖不一定相同，溶液密度与纯蔗糖溶液的密度之间存在着差异，用同样的锤度计进行锤度的测定就会产生误差。杂质越多，误差就越大。例如，用锤度计测量质量百分浓度为 10% 的食盐溶液时，所显示的结果不是 10 °Bx，而是 18 °Bx。这一例子就充分说明了杂质对锤度测定的影响。因为糖厂在制品中也含有非糖分，所以锤度的测定结果应为固溶物的近似质量百分数，以锤度表示。溶液含糖量越高，锤度与固溶物的含量越接近。

（3）锤度计的使用。

1）规格。统一规定采用 20 ℃ 为标准的锤度计，如图 2-1 所示。锤度计的一般要求：总长度最好不超过 30 cm，最小分度值为 0.05 °Bx，附有温度计，温度范围为 0~40 ℃，分度值为 1 ℃，常用的锤度计测量范围有：0~6 °Bx、5~11 °Bx、10~16 °Bx、15~21 °Bx、20~26 °Bx。

图 2-1 重力锤度计

2）使用方法和注意事项。

①生产期中锤度计必须放在盛有清水且底部放有软胶垫片的盛器内。

②使用时必须垂直取出，不可斜拿上段，以防折断。

③测量样液前，先用干燥洁净软布抹去附着水液，或以样液冲洗，然后垂直轻放至样液中，锤度计浮定即可，切勿让锤度计突然落下，使锤度计与量筒撞击而损坏。

④观察读数时，锤度计不得贴附量筒边缘，同时必须保持清洁，不黏附油污，使与液面接触处有良好的弯月面，否则测量不准确。

⑤读数时，如用国产锤度计，则按锤度计内所附说明读数（国产锤度计一般按液体弯月面上缘读数）；用进口的则按液面水平视线读数。

⑥使用后，垂直轻取，用水冲洗干净后放回原盛器中，注意不要与盛器内其他锤度计碰撞。

（4）影响锤度测定的因素。

1）温度。锤度计的刻度必须以 20 ℃为标准，否则要进行温度校正。测定时温度高于 20 ℃，则因糖液体积增大而使密度相应减小；相反，测定时温度低于 20 ℃，密度相应增大，因此，测定时应使糖液温度尽量接近 20 ℃。

2）杂质。非蔗糖杂质因非蔗糖物质与蔗糖密度不同，糖用锤度计是以纯蔗糖溶液为基础而划分其刻度的，因此杂质含量越少，测定结果越准确。非蔗糖杂质（特别是无机非糖物质）含量多时，读数一般偏高。

3）体积收缩现象。当蔗糖溶解于水中或浓糖液用水稀释时，稀释后溶液的体积要小于稀释前两种溶液的总体积，这种现象称为体积收缩。因在进行锤度计的刻度划分时也存在着体积收缩现象，所以对纯糖液而言，体积收缩对测定没有影响。但不纯糖液的情况就比较复杂，由于溶液中非蔗糖杂质收缩现象与蔗糖不同，一般非蔗糖杂质的收缩程度要比蔗糖大，因此高浓度、低纯度样品读数往往偏高。

4）表面张力。当样品含非蔗糖杂质较多时，因某些表面活性物质的杂质能使糖液与锤度计间的向下垂直分力减小，故使读数偏高。

5）稀释。稀释会扩大观察误差，不纯糖液中离子的水化作用能影响测定结果，故测定时应该按统一规定的稀释倍数进行稀释，使测定结果能相互比较。

（5）混合汁锤度测定。

1）所用仪器设备。

①糖锤度计。标准温度为 20 ℃，其刻度必须符合下列要求。

标尺范围（°Bx）：0～6 °Bx；5～11 °Bx；10～16 °Bx；15～21 °Bx；20～26 °Bx。

标尺刻度：0.1 °Bx。

②温度计：0～100 ℃，刻度 1 ℃。

2）样品处理与测定。从车间采集的混合汁样品应先用筛网滤去其中的蔗糠，否则蔗糠会浮在液面上，使读数困难。

用少许混匀后的样液洗涤锤度测定筒内壁，然后盛满样液，静置，待样液中气泡全部浮上液面后并除去，把经样液冲洗过的锤度计慢慢插入筒中（如果锤度计内不附温度计，则另行插入温度计），当温度计正确表示样液温度时，按锤度计规定的读数方法读取锤度，并记录测定时样液的温度。

注意，锤度计规定的读数方法有两种：一种是以水平视线按样液的真正液面高度读取；另一种是以液面与锤度计接触形成的弯月面的上缘读取。所以，在使用前应仔细阅读锤度计的说明书，按正确的方法读取数据。

3）温度更正。由于温度上升溶液的体积增大、锤度下降，所以测定锤度应有一个对应的温度。现行的锤度计以 20 ℃为标准，如读数时温度不是 20 ℃，则须查看观测锤度温度改正表（附表 1）进行相应加、减，即为更正锤度。如不加任何说明，锤度即是指更正锤度。

【例 2-1】设在 19 ℃时混合汁的观测锤度为 20.00 °Bx，求混合汁的锤度。

【解】查附表 1 得 19 ℃时温度更正数 =0.06 °Bx，则更正锤度为 20.00−0.06=19.94（°Bx）。

【例 2-2】设在 22 ℃时混合汁的观测锤度为 19.50 °Bx，求混合汁的锤度。

【解】查附表 1 得 22 ℃时温度更正数 =0.12 °Bx，则更正锤度为 19.50+0.12=19.62（°Bx）。

目前计算器及计算机已普遍了应用，更正锤度也可以用以下公式计算：

$$更正锤度 = B + 0.001 T^2 + 0.015 T - 0.0029\sqrt{B} - 0.17$$

式中　T——样液温度（℃）；

　　　B——观测锤度（°Bx）。

对于【例 2-1】，更正锤度的计算如下：

$$更正锤度 = 19.50 + 0.001 \times 22^2 + 0.015 \times 22 - 0.002\,9\sqrt{19.50} - 0.17 = 19.80$$

（6）糖浆、糖膏锤度测定。

1）所用仪器设备。与混合汁相同。

2）样品处理与测定。糖浆可稀释 4 倍，糖膏、糖蜜、糖糊可稀释 6 倍，并须全部溶解其晶体，加水后要充分搅拌均匀。

其余步骤与混合汁相同。

3）锤度计算。与混合汁相同。但计算出被测样液的锤度后，应乘以稀释倍数，得到原样品的锤度。

（7）温度计的使用。

温度计是化验室最常用的仪器之一。

1）规格。

①糖液转化用温度计：可用 0 ～ 100 ℃、分度为 1 ℃ 的普通水银温度计，长度约为 270 mm，直径应小于 10 mm，并要求在 60 ℃ 刻度处与水银球的距离不应少于 170 mm，以便插入 100 mL 容量瓶中，该刻度能露出瓶口而利于观察。

②测定旋光读数时的温度：可采用 0 ～ 40 ℃（或 0 ～ 60 ℃）、分度为 0.1 ℃ 的水银温度计，水银球以较短为宜（以利于测定有侧斗形观测管内的糖液温度）。

③烘箱用温度计：可采用 0 ～ 150 ℃、分度为 1 ℃、刻度与水银球距离较长的水银温度计。

2）使用方法和注意事项。

①测量溶液的温度时，将干燥洁净的温度计小心地插入，待温度恒定后读数。

②温度计毛细管中的水银柱容易脱开成两段或数段，致使所测得的温度不够准确，应特别注意。

③不能将温度计置入高于该温度计测量范围的温度区域内，以免温度计破裂。

3）检定方法。部分温度计出厂时其温度刻度可能会存在着一定的误差，所以新的温度计在使用前应统一进行检定。

①低于室温的温度检定：将标准温度计和被检定的温度计用棉线系在一起，使温度计下端的水银球齐平，上端固定悬挂于铁架所附的铁环上，下端放置于 250 mL 烧杯内，杯中盛水，加入冰或适量的冷却剂（如 100 g 10 ～ 15 ℃ 的水，加入 30 g 氯化铵可降至 -5.1 ℃，或加入 60 g 硝酸铵可降至 -13.6 ℃），使能其降到所需检定的温度，搅拌使水温均匀一致。当标准温度计达到检定的温度（必须为整数），立刻读出被检温度计的温度，两者之差，即为被检温度计的改正数。然后令水温上升，按照以上方法逐度检查（或根据所需要的度数进行检查）。

②高于室温的温度检定：将上述烧杯加热，使温度逐渐上升，逐度检定，若检定 100 ～ 200 ℃ 之间的温度，可用甘油代替水，其检定方法同上。

3. 折光法测定干固物

（1）折光法测定干固物原理。

1）光的反射现象与反射定律。一束光线在两种介质的分界面上时，要改变它的传播方向，但仍在原介质中传播，这种现象叫称为的反射。

例如，暗室里有一束光 AO 射到平面镜 MM 上（图 2-2），就有一束光 OB 从镜面的 O 点反射出来。AO 称为入射线，O 点称为入射点，OB 称

折光法测定锤度

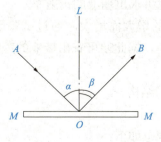

图 2-2 光的反射

为反射线。从入射点所作的一条垂直镜面的直线 LO 称为法线，入射线与法线的夹角 α 称为入射角，反射线与法线的夹角 β 称为反射角。

如果转动镜面，改变入射角的大小，则反射角的大小也会随之改变，经试验证明，光的反射符合以下定律：

①入射线、反射线及法线总是在同一平面内，入射线和反射线分居法线的两侧；

②入射角等于反射角。

2）光的折射现象与折射定律。光线从一种介质（如空气）射到另一种介质（如水）时，除了一部分光线反射回第一介质外，另一部分进入第二种介质中并改变它的传播方向，这种现象称为光的折射。

如图 2-3 所示，OD 为折射线，折射线与法线的夹角 γ 为折射角。当入射角改变时折射角也随之改变。试验证明，光的折射符合下述的折射定律：

①入射线、法线及折射线在同一平面内，入射线与折射线分居法线的两侧；

②无论入射角怎样改变，入射角与折射角的正弦之比，恒等于光在两种介质中的传播速度之比。

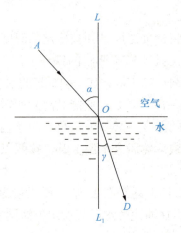

图 2-3　光的折射

3）全反射。光线从光密介质射入光疏介质时，折射线偏离法线。当入射角 $α_2$ 逐渐增大至某一角度，如图 2-4 中入射线 4 的位置，此时折射线 4′ 恰好与 OM 重合，折射光线将沿两介质的接触面 OM 平行射出，不再进入光疏介质而产生全反射。将折射角成 90° 时所对应的入射角为临界角。若光线从 1′～4′ 的范围内反向进行（从光疏到光密），经折射后，产生 OU 直线、左边明亮、右边黑暗的现象。通过试验可测出临界角 $α_2$。设光线从棱晶射入糖液，当发生全反射时可求出糖液的折射率。

纯蔗糖溶液折射率随浓度的增大而增大，通过试验已制定出纯蔗糖溶液折射率与浓度的对照表，测出糖液的折射率，据此便可查出糖液的蔗糖质量百分数或者视干固物的百分数。

（2）常用折光锤度计类型与结构。

1）折光锤度计类型。

糖厂使用的折光锤度计的规范如下：

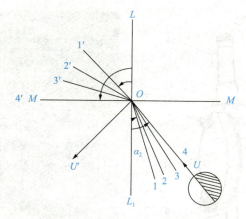

图 2-4 光的全反射

①标准温度：20 ℃；
②测量范围：0～30 °Bx、0～85 °Bx 或 0～95 °Bx；
③标尺刻度最小分度值：0.1 °Bx，0.25 °Bx 或 0.5 °Bx（误差不超过半分格）。

化验室中使用的折光仪主要有以下几种类型：
①双筒阿贝折光仪；
②单筒阿贝折光仪；
③自动阿贝折光仪；
④数字式全自动阿贝折光仪。

2）折光锤度计结构。各类折光锤度计的光学部分的结构基本相同，主要区别在于读数和方式。在原理上都是基于光线在糖液和棱镜的分界面上产生折射，并在一定条件下能产生全反射的特性，测出其糖液的折射率，同时使其在刻度盘上反映出读数。

下面以双筒阿贝折光仪为例进行说明。

阿贝折射仪的光学系统由观察系统与读数系统两部分组成。光线从反射镜反射，经过进光棱镜、折射棱镜及其间的糖液薄层折射后的光线，由阿米西棱镜消除折射棱镜及糖液所产生的色散，由物镜将所产生的明暗分界线成像于分划板上，通过目镜放大后，成像于观测者视线中。在测定时，可根据视野的情况，判别明暗分界线是否刚好通过十字线的交点。旋钮用于调节物镜与折射棱镜之间的消色棱镜，使明暗分界线清晰，调节旋钮，棱镜摆动，使明暗分界线通过十字线的交点，此时，可在读数筒中读取折射率或糖液浓度。在浓度的光学系统中，光线由小反光镜反射，经毛玻璃射到刻度盘，经转向棱镜及物镜，将刻度成像于分划板上，通过目镜放大后，成像于观测者眼中。

阿贝折射仪的构造如图 2-5 所示。

当旋动旋钮使棱镜摆动，视野内明暗分界线通过十字线交点时，表示光线从棱镜射入糖液的入射角达到了临界角。由于测定的糖液浓度不同，折射率不同，所以临界角的数值也有所不同，因糖液折射率与临界角的正弦成正比，故在折光计的刻度盘上，直接刻上折射率或糖液的锤度。

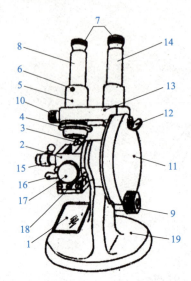
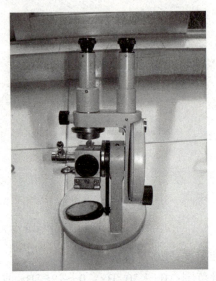

图 2-5 阿贝折射仪

1—反光镜；2—棱镜组（包括进光棱镜和反射镜）；3—棱镜反手；4—色散值刻度（阿米西棱镜组）；5—物镜；6—明暗分界划分线调节旋钮；7—目镜；8—观测镜筒；9—明暗分界线调节旋钮；10—色散的调节旋钮；11—圆盘组（内有刻度板）；12—小反光镜（读数透光用）；13—支架；14—读数镜筒；15—温度计插座；16—恒温水浴接头；17—保护罩；18—轴；19—底座

（3）折光锤度计的使用。根据纯蔗糖溶液具有折光性，其折光率随浓度的增加而加大的规律，只要测出糖液的折光率，便可求得蔗糖的质量百分率。不纯糖液中含有非糖杂质，所得出的结果为折光固溶物的质量百分率。该数值为真固溶物的质量百分率的近似值。折光法测定时，非糖杂质对测定结果的影响较密度法小，且比较准确、测定快捷，在生产及科研中广泛应用。

折光法测定糖液的浓度，可在仪器刻度尺上直接读数，或由折光率查附表进行温度校正，由此得出的糖液中的质量百分数为折光锤度，对不纯糖液则称为真固溶物质量百分数的近似值。

随着电子工业的不断发展，现在的折光仪都是电子数字显示，数据可直接读取，正在被越来越广泛地使用。WAY-2S 数字型阿贝折射仪如图 2-6 所示。

图 2-6 WAY-2S 数字型阿贝折射仪

WAY-2S 数字型阿贝折射仪具有精度高、自动校正温度对锤度值的影响、显示样液温度等优点。

1）WAY-2S 数字型阿贝折射仪的结构如图 2-7 所示。

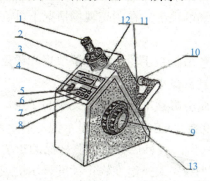

图 2-7 WAY-2S 数字型阿贝折射仪的结构图
1—目镜；2—色散校正手轮；3—显示窗；4—"POWER"电源开关；5—"READ"读数显示键；
6—"Bx—TC"经温度修正锤度显示键；7—"nD"折射率显示键；8—"Bx"未经温度修正锤度显示键；
9—调节手轮；10—聚光照明部件；11—折射棱镜部件；12—"TEMP"温度显示键；13—RS232 插口

其规格如下：

测量范围：折射率 nD　　　　　$1.300\,0 \sim 1.700\,0$
　　　　　锤度 Bx-TC　　　　　$0 \sim 95\%$
　　　　　锤度 Bx　　　　　　　$0 \sim 95\%$
测量精度：折射率 nD　　　　　$\pm 0.000\,2$
　　　　　锤度 Bx-TC　　　　　$\pm 0.1\%$
　　　　　锤度 Bx　　　　　　　$\pm 0.1\%$
　　　　　温度：显示范围　　　$0 \sim 50\,℃$
　　　　　修正 Bx 的温度范围　$15 \sim 45\,℃$

2）操作步骤及使用方法。

①按下"POWER"波形电源开关，聚光照明部件灯亮，同时显示窗显示"00000"（有时显示窗先显示"——"，数秒后显示"00000"）。

②打开折射棱镜部件，移去擦镜纸（擦镜纸只需用单层，不使用仪器时放在两棱镜之间，防止关上棱镜时，留有细小硬粒损坏棱镜工作表面）。

③检查上、下棱镜表面，并用水或酒精小心清洁其表面。每一个样品测定后都要仔细清洁两块棱镜表面，因为留在棱镜上少量的原来样品将影响下一个样品的测量准确度。

④将被测样品放在下面的折射棱镜的工作表面上。如样品为液体，可用干净滴管吸 1~2 滴液体样品放到棱镜工作表面上，然后将上面的进光棱镜盖上。如样品为固体，则固体样品必须有一个经过抛光加工的平整表面。测量前需将抛光表面擦清，并在下面的折射棱镜工作表面上滴 1~2 滴比固体样品折射率高的透明的液体（如溴代萘），然后将固体样品抛光面放在折射棱镜工作表面上，使其接触良好。测固体样品时无须将上面的进光棱镜盖上。

⑤旋转聚光照明部件的转臂和聚光镜筒，使上面的进光棱镜的进光表面（测液体样品）

或固体样品前面的进光表面（测固体样品）得到均匀照明。

⑥通过目镜观察视场，同时旋转调节手轮，使明暗分界线落在交叉线视场中。如从目镜中看到视场是暗的，可将调节手轮逆时针旋转。看到视场是明亮的，则将调节手轮顺时针旋转。明亮区域是在视场的顶部。在明亮视场情况下可旋转目镜，调节视度看清晰交叉线。

⑦旋转目镜方缺口里的色散校正手轮，同时调节聚光镜位置，使视场中明暗两部分具有良好的反差和明暗分界线具有最小的色散。

⑧旋转调节手轮，使明暗分界线准确对准交叉线的交点，如图 2-8 所示。

图 2-8　明暗分界线准确对准交叉线的交点

⑨按"READ"读数显示键，显示窗中"00000"消失，并显示"——"，数秒后显示被测样品的折射率。获取该样品的锤度值，可按"Bx"未经温度修正锤度显示键或按"Bx-TC"经温度修正锤度（按 ICUMSA）显示键。"nD""Bx-TC"及"Bx"三个键是用于选定测量方式。经选定后，再按"READ"键，显示窗按预先选定的测量方式显示。有时按"READ"键，显示"——"，数秒后显示窗全暗，无其他显示，反映该仪器可能存在故障，此时仪器不能正常工作，需进行检查修理。当选定测量方式为"Bx-TC"或"Bx"时，如果调节手轮旋转超出锤度测量范围（0～95%），按"READ"键后，显示窗将显示"·"。

⑩检测样品温度，可按"TEMP"温度显示键，显示窗将显示样品温度。如果按"READ"键后，显示窗显示"——"时，按"TEMP"键无效，则无法检测，在其他情况下都可以对样品进行温度检测。显示温度时，按"nD""Bx-TC"或"Bx"键，将显示原来的折射率或锤度。为了区分显示值是温度还是锤度，在温度前加"┝"符号；在"Bx-TC"锤度前加"┌"符号；在"Bx"锤度前加"┣"符号。

⑪样品（糖溶液）测量结束后，必须用酒精或水进行小心清洁。

⑫仪器折射棱镜部件中有恒温器结构，如需测定样品在某一特定温度下的折射率，仪器可外接恒温器，将温度调节到所需温度再进行测量。

⑬计算机可用 RS232 连接线与仪器连接。首先，送出一个任意的字符，然后等待接收信息（参数：波特率 2 400 bit/s，数据位 8 位，停止位 1 位，字节总长 18）。

3）仪器校正。仪器定期进行校准，或对测量数据有怀疑时，也可以对仪器进行校准。校准用蒸馏水或玻璃标准块。如测量数据与标准有误差，可用内六角扳手通过色散校正手轮中的小孔，小心旋转里面的螺钉，使分划板上交叉线上下移动，然后再进行测量，直到测数符合要求为止。样品为标准块时，测数要符合标准块上所标定的数据。如样品为蒸馏水时测数要符合表 2-4。

表 2-4　蒸馏水折光率表

温度 /℃	折射率 /nD	温度 /℃	折射率 /nD
18	1.33316	25	1.33250
19	1.33308	26	1.33239

续表

温度 /℃	折射率 /nD	温度 /℃	折射率 /nD
20	1.33299	27	1.33228
21	1.33289	28	1.33217
22	1.33280	29	1.33205
23	1.33270	30	1.33193
24	1.33260		

【例 2-3】设在 18 ℃时观测样品锤度为 19.5 °Bx，求样品的更正折光锤度。

【解】查折光锤度更正表得 18 ℃时温度更正数 =0.14 °Bx，

则更正折光锤度 =19.5-0.14=19.36（°Bx）

【例 2-4】设经四倍稀释后的糖浆在 23 ℃时观测锤度为 16.30 °Bx，求糖浆的更正折光锤度。

【解】查折光锤度更正表得 23 ℃时温度更正数 =0.22 °Bx，

则更正折光锤度 =（16.30+0.22）×4 =66.08（°Bx）。

4. 波美计的使用

糖厂中除使用锤度计来测量锤度外，还使用波美计来测量石灰乳的浓度。波美计的形状也如同锤度计一样，是一根下部较粗、上部较细并带有刻度的玻璃管。波美计以波美度（简写为 °Bé）为单位。波美计的刻度是蒸馏水作为 0 °Bé，以食盐的质量分数（重量百分浓度）为 15% 作为刻度 15 °Bé，然后在 0 与 15 间等分并向下延伸而成。如果测定的结果为 5 °Bé，则表示溶液的密度相当于 5% 的食盐溶液的密度。因为没有测量石灰乳浓度的专用设备，而波美计是通用设备，波美度与石灰乳中的氧化钙的质量百分率近似相等（10 °Bé 的石灰乳中氧化钙的含量为 8.74%），所以糖厂中就习惯用它来测定石灰乳的浓度。

石灰乳中氧化钙的含量与波美度间的关系可用以下公式表示：

$$石灰乳中氧化钙的含量（\%）=0.891\ 8Be-0.114\ 1$$

式中 Be——石灰乳中的波美度；

波美计的使用与锤度计相同。在一般情况下，1 °Bé 近似地等于 1.8 °Bx。

任务 2.2 在制品的糖度及简纯度测定

● 企业案例

主任从生产线上取回部分中间蔗汁，然后安排小明对以上样品的糖度和简纯度进行测定，小明应该选择什么仪器和使用什么方法来测定呢？

● 任务目标

通过本任务的学习，学生达到以下目标：
(1) 理解、掌握糖度、蔗糖分、简纯度和重力纯度等概念。
(2) 掌握旋光检糖仪的使用方法。
(3) 掌握测定糖度、简纯度的方法及其计算。

● 素质目标

养成对比分析学习的好习惯。

● 任务描述

使用旋光仪对在制品进行糖度和简纯度的测定。

糖度及简纯度
的测定

● 实施条件

(1) 原料：糖汁。
(2) 设备及仪器：锤度计、旋光仪、200 mm 观测管、温度计、250 mL 锥形瓶、量筒、烧杯、滤纸。
(3) 试剂：碱性醋酸铅。

● 程序与方法

(1) 样品处理：糖蜜、糖膏、糖糊等含有晶体的样品采用 6 倍稀释，必须保证使晶体全部溶解。
(2) 测定样液的观测锤度及温度，记录数据。
(3) 糖度的测定：倾取样品约 100 mL 于干燥洁净 250 mL 锥形瓶内（如不干燥，则须先用样液洗涤），加入适量碱性醋酸铅粉（以最少量而又收到澄清效果为宜），摇匀，过滤，用最初滤液洗涤盛器（小烧杯），倾取后，收集滤液，用 200 mm 观测管测其旋光读数。观测管须先用滤液洗涤 2～3 次。

(4)数据处理及计算:
1)计算。

$$样液的糖度（\%）= 观测旋光度 \times 糖度因数$$

式中　糖度因数——以观测锤度查糖度因数检索表而得。

$$简纯度（\%）= \frac{糖度}{锤度} \times 100\%$$

2)数据及结果处理。

项目	数据及结果
样液观测锤度 /°Bx	
样液温度 /℃	
样液锤度 /°Bx	
样液观测旋光度 /°Z	
样液糖度 /%	
样液简纯度 /%	
样品糖度 /%	
样品简纯度 /%	

● 思考

近年来,为了保护环境,促进生态文明的发展,近红外旋光仪逐渐替代普通光旋光仪,请问近红外旋光仪在环保方面有何优势?请同学们广泛查阅资料,给出答案。

任务相关知识——旋光法测定糖度、蔗糖分

1. 旋光法基本原理

（1）基本知识。旋光法是测定蔗糖分的最简单、最快速的方法之一,它基于蔗糖的旋光性和蔗糖溶液的旋光度与浓度成正比的性质。

旋光法可分为一次旋光法和二次旋光法。二次旋光法测定的结果比较准确,但测定过程较复杂,时间较长。有些特殊样品(纯度很高或蔗糖含量很少),两种方法的测定结果相差不大或生产上对结果的准确度要求不高,此时可用一次旋光法代替二次旋光法进行测定。

旋光检糖仪(简称旋光仪)是糖厂中测定蔗糖溶液蔗糖分的主要仪器。旋光检糖仪是检测旋光度的一种仪器,它的工作原理建立在糖液具旋光性的基础上,利用旋转偏振光的振动方向来检测旋光度的大小。在介绍旋光仪之前,必须对蔗糖分的概念、偏振光及旋光现象等基本知识有一定的了解。

1)糖度(转光度)。用一次旋光法测得的样品中蔗糖含量的质量百分数。

$$糖度（转光度）（\%） = \frac{蔗糖质量}{样品质量} \times 100\%$$

2）蔗糖分。用规定的方法测得的样品中蔗糖含量的质量百分数。一般纯度很高（如白砂糖）或蔗糖含量很少的样品（如滤泥）采用一次旋光法测定，其他的中间制品采用二次旋光法测定。

$$蔗糖分（\%） = \frac{蔗糖质量}{样品质量} \times 100\%$$

3）纯度。是指样品干固物中蔗糖含量的质量百分数。根据测定方法不同可分为重力纯度和简纯度。

①重力纯度（GP）。

$$重力纯度（\%） = \frac{蔗糖分}{锤度} \times 100\%$$

②简纯度（AP）。

$$简纯度（\%） = \frac{糖度（转光度）}{锤度} \times 100\%$$

4）自然光与偏振光（偏振面）。光是一种电磁波，一种横波，即光线前进的方向与光波振动的方向互相垂直。

自然光：光线具有无数个与光线前进方向互相垂直的振动平面。假如有一束光线由纸面射出，指向人眼，我们看到光线的振动平面如图2-9（a）所示，有无数个振动平面，每个双箭头表示一个振动平面。

偏振光：由于某些晶体只允许某一振动方向的光通过，当光线通过这些特殊的晶体后，会变为只有一个振动平面的光线。假如有一束光线由纸面射出，通过特殊的晶体后指向人眼，我们看到光线的振动平面如图2-9（b）所示，只有一个振动平面，这种单一振动平面的光线就称为"偏振光"。它的振动平面称为"偏振面"。这一类特殊晶体不仅能产生偏振光，而且还能检测偏振光旋转的角度，即测定旋光度的大小。

(a)　　　　　(b)

图2-9　自然光与偏振光
(a) 自然光；(b) 偏振光

5）旋光质。当光线通过某些物质后，光线的振动平面会转动一定的角度，这种现象称为"旋光性"，凡是能把偏振光的振动平面旋转一定角度的物质称为"旋光质"。糖类溶液都具有旋光性，其中蔗糖、葡萄糖能把偏振光的振动平面向右转动，称为右旋，以（+）号表示，这些物质称为右旋物质；果糖能把偏振光的振动平面向左转动，称为左旋，以（−）号表示，这些物质称为左旋物质。

6）旋光度。当温度、光源的波长及光在旋光性物质中通过的距离为一规定值时，光线通过旋光性物质后振动平面旋转的角度称为旋光度。

（2）旋光检糖仪。测定光线通过旋光性物质后振动平面旋转的角度的仪器称为旋光检糖仪（旋光仪）。图 2-10 所示为 WZZ-2SS 数字式糖度旋光仪。

图 2-10　WZZ-2SS 数字式糖度旋光仪

为了比较光线振动平面旋转的角度，旋光检糖仪必须使用只有一个振动平面的偏振光。

旋光检糖仪的种类很多，但基本原理是相同的。它利用蔗糖溶液的旋光性进行测定。当蔗糖溶液的液层厚度、温度、光源波长为一定值时，蔗糖溶液将偏振平面旋转的角度与蔗糖的浓度成正比，通过测定糖液将偏振平面旋转的角度，就可以求出溶液中蔗糖的浓度，即

$$\alpha = K \cdot C \cdot L$$

式中　α——偏振平面旋转的角度；

　　　C——糖液浓度；

　　　L——光线在溶液中通过的距离，即糖液的厚度（观测管长度）；

　　　K——系数。

K 可以通过测定已知浓度为 C_0 糖液的旋光度 α_0 用 $K=\alpha_0/(C_0 \cdot L)$ 的关系计算求出。

1）旋光检糖仪的基本部件。不管是那一种旋光检糖仪，都有以下几个组成部分：

①光源。要求能产生波长稳定的光。因为同一旋光性物质，在不同波长的光线下其旋光度也不同，所以要求光源的波长稳定（也称色纯度高）。常用钠光灯泡作光源，这种灯泡能产生黄色的单色光。例如，WZZ-2SS 型自动旋光仪采用 20 W 钠光灯作光源，由小孔光栅和物镜组成一个简单的点光源。

②起偏器。其作用是将光源产生的单色光变为只有一个振动平面的偏振光。一般用由两块方解石晶体组成的"尼科尔"棱镜作为起偏器。方解石晶体具有双折射作用，当光线通过晶体时，能将光线分解为两束振动方向相互垂直、前进方向各不相同的光线。通过适当地选择两块晶体黏合面的角度及黏合材料，可以使其中一束光线在黏合面上被全反射至棱镜的黑色侧面上，并在那里被吸收。而另一束光线可以几乎无偏折地透过棱镜，从棱镜中出来的这束光线就成为只有一个振动平面的偏振光。

③观测管。用来盛放被测样液。这是一根细长的玻璃管，两端有由螺旋帽固定的玻璃盖片，螺旋帽中间开孔以允许光线通过。形状如图 2-11 所示。测定时，将样品装入观测管后将其置于偏振光的光路中。如果样品具有旋光性，偏振光通过观测管后，偏振光的

振动平面就会转动一定的角度。因样品将振动平面旋转的角度与光线通过旋光性物质的路程成正比,所以观测管的长度必须准确。观测管的规定长度为 200 mm,半规定长度为 100 mm,双倍长度为 400 mm。

图 2-11 观测管

④检偏器。检偏器的结构与起偏器相同,用来测定偏振光振动平面旋转的角度。当两块晶体的方向相同时(即产生偏振光的振动平面相同),从起偏器产生的偏振光可以完全通过检偏器,从检偏器的后面看光线最亮(即检偏器的"视野"最亮)。当两块晶体的方向相互垂直在仪器的零点时,从起偏器产生的偏振光完全不能通过检偏器(即检偏器的"视野"最暗)。当在光路中放入旋光性物质后,由于偏振光的振动平面被旋转了一定角度,部分偏振光能通过检偏器,检偏器的"视野"应变亮;如果转动起偏器或检偏器,使进入检偏器偏振光的振动与检偏器晶体的方向相互垂直,检偏器的"视野"又会重新变为黑暗。起偏器或检偏器在这一过程中转动的角度就是被测物质的旋光度。

2) 旋光检糖仪工作原理。

①零点。如图 2-12 所示,自然光通过起偏器产生竖直方向的偏振光,洁净的观测管装蒸馏后放入旋光检糖仪,将检偏器的方向调为与起偏器垂直,即只有水平方向的偏振光能通过检偏器,光线不能同时通过起偏器和检偏器晶体,此时,在检偏器后观察无光线透过,观察镜筒中视野黑暗。

②放入样品后。如图 2-13 所示,将装有糖液的观测管置于旋光检糖仪中,即起偏器和检偏器晶体之间,自然光通过起偏器产生竖直方向的偏振光,偏振光通过装有糖液的观测管后振动平面被糖液旋转一定角度,旋转后的光可分解为水平部分和垂直部分,其中水平方向的光能通过检偏器晶体,此时,从观测镜筒看到的光线比原来的减弱了,即观测镜筒中视野稍变亮。

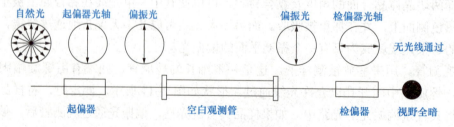

图 2-12 观测管未装糖液时的情况

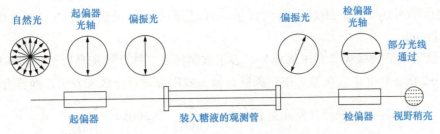

图 2-13　观测管装入糖液后的情况

③旋光角度的测定。如图 2-14 所示，转动检偏器晶体，使检偏器的轴面与从观测管出来的偏振光的振动平面相垂直，即使偏振光无法通过检偏器，镜筒中视野重新变黑暗，此时，检偏器晶体所转动的角度即为样品使偏振平面旋转的角度。

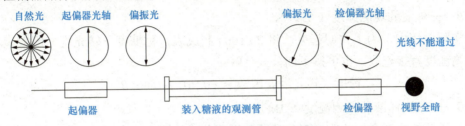

图 2-14　检测器光轴调整后的情况

3）旋光检糖仪的刻度划分方法。用旋光检糖仪可以测出样品的旋光角度，为了方便起见，作为制糖行业专用的设备，可以将偏振平面旋转的角度直接转变为样液的蔗糖浓度。旋光检糖仪的读数统一采用国际糖度（°Z）作为单位。

①旋光检糖仪的刻度划分。

a. 以 20 ℃为被测样液的标准温度，观测管的长度为 200 mm。

b. 将蒸馏水装入洁净的观测管内，置于旋光仪上进行测定，并将读数设为仪器的"0"点。表示样液中蔗糖的浓度为"0"，即样品中蔗糖的含量的质量百分数为"0"。

c. 将 26.000 g（一个规定量）纯蔗糖溶解后移入 100 mL 容量瓶中，加水至标线（规定量稀释），摇匀后将糖液装入 200 mm 观测管（已用测定糖液多次洗涤）并置于旋光仪上进行测定，所得的读数定为仪器的"100"。表示原样品（注意：不是被测样液）中蔗糖含量的质量百分数为"100"。

d. 将"0"至"100"间进行等分，就可以得到旋光仪的刻度。

②国际糖度（°Z）。在制糖专用的旋光检糖仪上测得的读数称为"国际糖度"，用符号"°Z"表示。

从旋光仪刻度的划分方法可以看出，当样品采用规定量稀释时，若在旋光仪上测得的读数为 100 °Z，表示原样品中蔗糖的含量的质量百分数为"100"。而被测样液的蔗糖含量为 26 g/100 mL。因糖液旋光角度与样液中蔗糖的浓度成正比，1 °Z 就表示被测样液中蔗糖的含量为 0.26 g/100 mL。

4）旋光读数与蔗糖关系。假设某样品测得旋光读数为 P，并不一定就直接表示样品的糖度（或蔗糖分），样品采用不同的稀释方法，旋光测定的结果表示的含意也不同。

①样品采用规定量稀释。表示原样品（稀释前）中蔗糖含量的重量百分数为 P（即

原样品的蔗糖分为 P）；或表示被测样液（样品经稀释后得到的样液）的蔗糖含量为 $0.26P$ g/100 mL。

②样品采用倍数稀释（含未稀释）。表示被测样液（样品经稀释后得到的样液）的蔗糖含量为 $0.26P$ g/100mL。在糖厂中，蔗糖含量一般用质量百分数来表示，换算方法如下：

$$被测样液的蔗糖含量质量百分数 = \frac{0.26P}{100d} \times 100\% = \frac{26P}{100d}（\%）$$

式中　P——旋光读数（°Z）；
　　　d——被测样液的密度（g/mL）。

注：d 可以根据被测样液的观测锤度查相关的表或按下面公式计算：

$$被测样液的视密度 = 0.998\,2 + 0.003\,7B + 0.000\,018\,16B^2$$

式中　B——样液的观测锤度（°Bx）。

在实际应用中，为了方便起见，将 $26/100d$ 制成表，称糖度（转光度）因数表，由样液的观测锤度直接查取。也可按下面公式计算：

$$糖度（转光度）因数 = (0.510\,7 - 0.001B)^2$$

式中　B——被测样液的观测锤度（°Bx）。

被测样液的蔗糖含量的质量百分数可按下面公式计算：

$$被测样液的蔗糖含量的质量百分数 = P \times 糖度（转光度）因数$$

原样品的蔗糖含量质量百分数可以按下面公式计算：

$$原样品的蔗糖含量的质量百分数 = 被测样液的蔗糖含量的质量百分数 \times 稀释倍数$$

【例 2-5】接近 20 ℃ 的条件下，测得糖液的观测锤度为 20.00 °Bx，用 200 mm 观测管测得旋光度读数为 71.20 °Z，求糖液的糖度。

【解】解法一：
d 可以根据被测样液的观测锤度查相关的表或按下面公式计算：
被测糖液视密度 $d = 0.998\,2 + 0.003\,7B + 0.000\,018\,16B^2$
$\qquad = 0.998\,2 + 0.003\,7 \times 20 + 0.000\,018\,16 \times 20^2 = 1.079\,6$

$$被测糖液的糖度（\%）= \frac{0.26P}{100d} \times 100\% = \frac{26P}{100d}\% = 17.14\%$$

解法二：
由样液的观测锤度直接查取糖度（转光度）因数表，得糖度（转光度）因数或按下式计算：

$$糖度（转光度）因数 = (0.510\,7 - 0.001B)^2 = (0.510\,7 - 0.001 \times 20)^2 = 0.240\,76$$

被测糖液的糖度（%）= $P \times$ 糖度（转光度）因数 $\times 100\%$ = $71.2 \times 0.240\,76 \times 100\%$ = 17.14%

5）影响旋光测定的因素。

①光的波长。同一种物质，在不同波长下测得的旋光角度也不同。对于糖类物质，波长减小，测得的旋光角度增大。光的波长与光源的温度有关，所以在使用旋光仪前，一定要先预热，待灯泡发出的光的波长稳定后开始测定。

②温度。在温度较高时，温度对旋光测定的影响较小；而温度较低时，温度对旋光

测定的影响较大。蔗糖的旋光角度随温度升高而减小,但还原糖的旋光角度随温度升高而增大(还原糖的旋光性与蔗糖相反,为左旋,它的旋光角度的读数为负值,左旋角度减小即为旋光角度的数值增大)。在对结果要求较高的情况下,应对旋光测定的结果进行温度校正。

③杂质。糖厂样品中对旋光测定产生影响的杂质主要有两类:一类是具有旋光性的物质,主要是还原糖等;另一类是对糖类旋光测定有影响的物质,主要是碱金属、碱土金属及各种盐类。

a. 还原糖。糖厂中的还原糖主要是葡萄糖(+右旋糖)和果糖(-左旋糖)。这两种糖的综合旋光特性为左旋,即旋光度为负值,可以使蔗糖样品的旋光测定结果减小。这也是糖厂中通常测定同一样品的糖度要比蔗糖分小的主要原因之一。

b. 碱金属、碱土金属。这两类金属的氢氧化物及碱金属的碳酸盐能与蔗糖生成右旋度较小的蔗糖盐,减小蔗糖的旋光角度。当溶液较稀时,由于蔗糖盐的水解度较大,由碱引起的测定误差会小一些。

碱金属的氯化物、硝酸盐、硫酸盐、醋酸盐和柠檬酸盐,碱土金属的氯化物、硫酸镁及其他多种类盐,均能减小蔗糖的旋光度。减小程度随盐类量的增加和盐类分子量的减小而增大。

④澄清剂、酸类。澄清剂能沉淀果糖,使旋光读数增大。在测定方法不能避免还原糖对结果的影响情况下,为了减小误差,使同一样品不同批次的测定结果具有可比性,澄清剂的用量必须严格符合规定要求。

无机酸能影响还原糖的旋光度,当样品中含有较多还原糖时,应设法校正无机酸对旋光测定的影响。有机酸对旋光测定的影响较小,可以忽略不计。

⑤变旋作用。还原糖类具有光学活性。这类物质溶解后,其旋光度会发生变化。最初变化比较迅速,然后逐渐变得缓慢,经数小时后,旋光度基本达到稳定。将糖液加热,可使变旋作用迅速达到平衡。

2. 旋光仪的使用

(1)使用方法。

1)将仪器电源插头插入220 V交流电源,打开电源开关,这时钠光灯应启亮,需经5 min钠光灯预热,使之发光稳定。

2)将光源开关切换至直流电源,使钠光灯在直流电源下点亮。如果切换为直流电源后钠光灯不亮,则应切换回交流电源,增加预热时间后再切换为直流电源。

3)打开测量开关,这时数码管应有数字显示。

4)将装有蒸馏水或其他空白溶剂的试管放入样品室,待读数稳定后,按清零按钮或调节零点旋钮,使仪器的计数为零。

5)取出空白溶剂的试管,将待测样品注入观测管,放入样品室内,仪器数显窗将显示出该样品的旋光度。

6)逐次按下复测按钮,重复读几次,取平均值作为样品的测定结果。

7)仪器使用完毕后,应依次关闭测量、光源、电源开关。

(2) 注意问题。

1) 仪器应放在干燥通风处，防止潮气侵蚀，尽可能在 20 ℃，相对湿度不大于 85% 的工作环境中使用仪器，搬动仪器应小心轻放，避免震动。

2) 在调零或测量时，观测管中不能有气泡，如果有气泡，应先将气泡从中间漏斗处排除。如果通光面两端有雾状水滴，应用软布擦干。

3) 钠灯在直流电源供电系统出现故障不能使用时，仪器也可在钠灯交流电源供电的情况下测试，但仪器的性能可能略有降低。

(3) 旋光观测管。

观测管用于盛放样液，是旋光仪中试验人员最常接触的部件之一。偏振光就是在通过观测管内样液的过程中而发生旋光现象，所以它的质量直接影响旋光测定的准确度。

1) 规格。检糖用的观测管以具有斗形测管的为宜。长度分为 400 mm、200 mm、100 mm 三种，其长度必须经严格检定。

200 mm 观测管供测定各种糖液糖度（转光度）用，是糖厂中最常用的观测管，100 mm 管供测定颜色较深的糖液糖度（转光度）用，400 mm 管一般为测定蔗渣糖度（转光度）用。

大部分国产的自动旋光仪的观测管槽只能容纳 200 mm 观测管，这种情况下，采用常规方法测定颜色较深的糖液糖度（转光度）和蔗渣糖度（转光度），要注意测定读数的处理。

2) 使用方法和注意事项。

①观测管必须是洁净的，如果管内附有醋酸铅粉或积垢，可以用 10%～20% 醋酸（必要时可用稀盐酸）洗涤，再用蒸馏水冲洗洁净。

②测定时用蒸馏水将观测管冲洗干净，再以样液冲洗 2～3 次，然后盛满样液。

③观测管内样液，不能混有气泡，以免引起测定误差。

④观测管的螺母不可旋得过紧，以免在盖玻片上产生应力，影响测量结果。松紧以不漏液为宜。

⑤必须将管身及两端旋帽和帽内盖玻片黏附的样液（或水滴）用软布拭干，然后进行测定，以免引起观察视野模糊不清，造成误差。

⑥不得把持玻璃管（应把持螺母），以防样液受热，温度升高而引起误差。

⑦使用后将样液倒出，即以蒸馏水冲洗洁净，晾干，置于铺有软棉布的小盒内。

⑧必须定时以蒸馏水检查观测管的盖玻片是否因黏上污物而具有旋光性，并对盖玻片进行清洁或予以校正。

3) 检定方法。观测管长度检查：除去观测管两端螺母，用游标卡尺夹着玻璃管的两端量度（必须变换不同位置量度数次），记其长度，其长度允许误差见表 2-5。

表 2-5 观测管长度误差

公称长度 /mm	100	200	400
A 级 /μm	±10	±20	±40
B 级 /μm	±200	±400	±800

如果观测管长度误差超过以上数字，最好不用，必要时可用长度改正数，算出正确的旋光读数。

例：200 mm 观测管实测长度为 199.8 mm。

则长度改正数 =200/199.8=1.001。

设：以该管测得某白砂糖旋光度数为 99.60 °Z，则该白糖正确的旋光度数为

$$99.60 \times 1.001 = 99.70$$

3. 一次旋光法的应用

通过测定一次样品的旋光读数，从而求出样品中蔗糖分含量的方法，称为一次旋光法。由于糖厂在制品中除主要成分蔗糖外还含有少量的其他旋光性物质（如葡萄糖、果糖等），这就影响了旋光测定的准确性，所以一次旋光法测得的结果对于不纯糖液来说只是蔗糖质量百分率的近似值，称为"糖度"（也称"转光度"）。但白砂糖中所含的杂质很少，对旋光测定的干扰甚微，故一次旋光法测定的结果可作为样品的蔗糖分。而对含糖量甚少的蔗渣、滤泥，因其中蔗糖含量的绝对值很小，可以允许较大的测定误差，也可以将一次旋光法测定的结果视为样品的蔗糖分。

（1）混合汁、糖浆、糖膏糖度测定。

1）主要设备及试剂。

①设备及仪器：旋光仪、200 mm 观测管、工业天平或台秤。

②试剂：蒸馏水或去离子水、无水乙醇、碱性醋酸铅。

2）测定步骤。

①样品处理。对于浓度在 25 °Bx 以下的样品可不用稀释。浓度较高的样品可加蒸馏水稀释。糖浆采用 4 倍稀释，糖蜜、糖膏、糖糊等含有晶体的样品采用 6 倍稀释，必须保证使晶体全部溶解。

②锤度测定。测定样品（或稀释后样液）的观测锤度。

③澄清过滤。取样液约 100 mL，放入干燥洁净的 250 mL 锥形瓶内（如不干燥，则须先用样液洗涤 2～3 次）。加入适量碱式醋酸铅粉（以最少量而又收到澄清效果为宜），摇匀，过滤；用最初的滤液洗涤盛器（一般为小烧杯）后倾取，收集滤液于盛器中。

注意：澄清剂的用量应根据样品的实际情况确定使用量，过少，滤液混浊，旋光读数不稳定，甚至不能进行测定，同时过滤速度明显降低；过多，会增大旋光读数，使测定结果误差偏大。

④旋光测定。取 200 mm 观测管，先用滤液洗涤 2～3 次，然后注满滤液，放入旋光仪中，测定其旋光度读数 P（°Z）。

3）计算。

$$\text{被测样液的糖度（\%）} = \frac{26P}{100d} \times 100\%$$

式中　P——被测样液旋光度读数（°Z）；

d——被测样液视密度（按观测锤度查表或按下面公式计算）。

被测样液视密度（d）$=0.998\,2+0.037B+0.000\,018\,16B^2$

式中　B——样液的观测锤度（°Bx）。

为了使用方便起见，可将 $26/(100d)$ 按不同的观测锤度制成表，称为糖度（转光度）因数表。

也可以根据观测锤度按以下公式计算：

$$糖度（转光度）因数 = (0.5107-0.001B)^2$$

式中　B——被测样液的观测锤度（°Bx）。

$$样液糖度（\%）= P \times 转光度因数 \times 100\%$$

或

$$样液糖度（\%）= P(0.5107-0.001B)^2 \times 100\%$$

注意：若样品在测定过程进行倍数稀释，则：

$$原样品糖度 = 被测样液糖度 \times 稀释倍数。$$

【例 2-6】糖浆采用四倍稀释后，测得糖液的观测锤度为 15.35 °Bx，用 200 mm 观测管测得旋光度读数为 54.48 °Z，求糖浆的糖度。

【解】根据观测锤度按公式计算糖度（转光度）因数。

糖度（转光度）因数 = $(0.5107-0.001 \times 15.35)^2$ = 0.2454

稀释后糖液的糖度（%）= $54.48 \times 0.2454 \times 100\%$ = 13.37%

因样品采用 4 倍稀释，所以：

糖浆的糖度（%）= $13.37 \times 4 \times 100\%$ = 53.48%

(2) 蔗渣糖度测定。蔗渣糖度测定是利用蔗渣加水蒸煮使其中的糖分渗出，最终达到蔗渣内部溶液的糖分与外部渗出液的糖分基本相等。由于糖厂将蔗料视为由蔗汁与纤维分两部分组成，蔗糖和所有可溶性物质均包括在蔗汁内，其他不溶部分则视为纤维分。只要预先求出样品的纤维分，就可以算出其中蔗汁的含量。再根据蔗汁的糖度，就可以算出蔗渣的糖度。

1）主要设备及试剂。

①检糖仪（旋光仪）。检糖仪应是根据国际糖度标尺，按糖度（°Z）刻度的，测量范围为 −30 ~ +120 °Z

②蔗渣蒸煮器。蔗渣蒸煮器包括水浴锅、蔗渣盅及盅压三部分。

③ 1 000 W 电炉。

④ 12.5 °Bx 碳酸钠溶液。

⑤碱性醋酸铅。

2）测定步骤。

①测定样品纤维分。样品纤维分可以直接测定，也可以按规定估算。

蔗渣纤维分 F（%）规定如下：

$R \leqslant 1.0$ 时，　　　　　$F=60$

$R=1.05 \sim 2.0$ 时，　　　$F=55$

$R>2.0$ 时，　　　　　　$F=50$

R 为蒸煮液用 200 mm 观测管测得的旋光读数。

②蒸煮。先称蔗渣盅与盅压的质量，然后迅速称取蔗渣 100.0 g，加入含 5 mL 12.5 °Bx 碳酸钠溶液的 70 ℃ 500 mL 水，以盅压作盖，将蔗渣轻轻压平，置入沸水浴中蒸煮。0.5 h 后用盅压作第一次加压，以后每 15 min 加压一次，共压三次，使糖分充分渗出。蒸煮时间共 1 h。蒸煮完毕，将盅置入冷水浴中冷却至室温，擦干称重，记录蔗渣连同溶液的质量。

③澄清过滤。用盅压将溶液尽量挤出，注入 200 mL 锥形瓶中，加入适量碱性醋酸铅粉，摇匀后过滤，最初部分滤液洗涤盛器后弃去。收集滤液。

④旋光测定。200 mm 观测管用滤液洗涤 2～3 次后将滤液装入，置于旋光仪上测定旋光读数。

3) 计算。

$$蔗渣糖度（\%）=\frac{蔗渣中蔗糖质量}{蔗渣质量}\times 100\%$$

$$蔗渣中蔗糖质量 = 渗出液质量 \times 渗出液糖度$$

$$渗出液质量 = 蔗渣连同溶液的质量 - 纤维质量$$

$$纤维质量 = 样品质量 \times 样品纤维分$$

因渗出液的浓度很低，所以其密度可以取为 1。

$$渗出液糖度（\%）=\frac{26P}{100d}\times 100\%=0.26R\times 100\%$$

所以

$$蔗渣糖度（\%）=\frac{0.26R（W-F）}{100}\times 100\%$$

式中　R——用 200 mm 观测管测得的旋光读数（°Z）；
　　　W——蔗渣连同溶液的重量（g）；
　　　F——样品纤维分（根据 R 按规定选取）。

【例 2-7】设蔗渣盅及盅压共重 855.5 g，蒸煮后蔗渣盅、盅压及溶液的总质量为 1 301.5 g，用 200 mm 观测管测得的旋光读数为 2.2（°Z），求蔗渣的糖度。

【解】蔗渣连同溶液的质量 =1 301.5−855.5=446（g）

按 R 选择样品纤维分 F=50

则

$$蔗渣糖度（\%）=\frac{0.26\times 2.2\times（446-50）}{100}\times 100\%=2.27\%$$

（3）滤泥糖度（蔗糖分）测定。滤泥糖度采用一次旋光法测定，虽然滤泥中含有较多的非糖分，对旋光测定有一定的影响，但因滤泥含的蔗糖分较少，蔗糖分与糖度相差的绝对值不大，对测定结果的要求也较低，在计算时也可将滤泥的糖度视为蔗糖分。

滤泥糖度测定采用规定量稀释。因滤泥不是全部可以溶于水中，有一部分不溶物占去容量瓶的部分体积，导致稀释后容量瓶内的液体不足 100 mL，而按规定量稀释的要求，稀释后样液的体积应为 100 mL。所以应对这一体积误差进行校正。

假设滤泥的水分含量为 50%，则干固物含量也为 50%，滤泥中干固物的密度约为 2.9 g/cm³。当称取 26 g 样品时，其中干固物为 13 g，干固物所占的体积应为 13/2.9=4.86（mL）。按规定量稀释的要求，应使用 104.86 mL 的容量瓶，若用 100 mL 容量瓶，样品的量就应适当减少。

$$26 : 104.86 = X : 100$$
$$X = 26 \times 100 / 104.86 = 24.79（g）$$

因这只是近似估算，为了方便称量，就将滤泥的规定量定为 25 g。

1）主要设备及试剂。

①检糖仪（旋光仪）。检糖仪应是根据国际糖度标尺，按糖度（°Z）刻度的，测量范围为 −30 ～ +120 °Z。

②瓷蒸发皿。

③阔口容量瓶 100 mL±0.08 mL。

④碱性醋酸铅溶液（54 °Bx）。

⑤对于碳酸法糖厂，须增加滴定管，分度为 0.1 mL，20 ℃标准，最大允许误差为 0.05 mL。

2）测定步骤。

①亚硫酸法糖厂。称取样品 25.0 g 于瓷蒸发皿中，加入 2 ～ 7 mL 碱性醋酸铅溶液（54 °Bx）及少量蒸馏水，用圆头玻璃棒或瓷棒将其小心研磨成完全不含粒子的均匀糊状，然后用蒸馏水洗入 100 mL 阔口容量瓶内，加水至标线，充分摇匀后过滤。最初部分滤液洗涤盛器后弃去。收集滤液。200 mm 观测管用滤液洗涤 2 ～ 3 次后将滤液装入，置于旋光仪上测定旋光读数。

②碳酸法糖厂。称取样品 25.0 g 于瓷蒸发皿中，加入 2 ～ 7 mL 碱性醋酸铅溶液（54 °Bx）及少量蒸馏水，用圆头玻璃棒或瓷棒将其小心研磨成完全不含粒子的均匀糊状，然后用蒸馏水洗入 100 mL 阔口容量瓶内，滴入酚酞指示剂 2 滴，摇匀后用稀醋酸慢慢滴至红色消失为止。加水至标线，充分摇匀后过滤。最初部分滤液洗涤盛器后弃去。收集滤液。200 mm 观测管用滤液洗涤 2 ～ 3 次后将滤液装入，置于旋光仪上测定旋光读数。

注：因碳酸法糖厂的滤泥中含有部分化合糖分，即蔗糖与石灰生成的不溶性钙盐，测定前需要用稀醋酸将样品中和，使蔗糖钙盐分解，释放出这一部分蔗糖。

3）计算。因样品已采用规定量稀释，所以旋光仪的读数就是滤泥的糖度（蔗糖分）。

任务 2.3　蔗糖分及重力纯度的测定

● 企业案例

主任从生产线上取回部分中间蔗汁，然后安排小明对以上样品的蔗糖分和重力纯度进行测定，小明应该选择什么仪器和使用什么方法来测定呢？

● 任务目标

通过本任务的学习，学生达到以下目标：
（1）进一步理解蔗糖分及重力纯度等概念。
（2）了解一次旋光法与二次旋光法的区别。
（3）掌握测定蔗糖分、重力纯度的方法及其计算。

**蔗糖分及重力
纯度的测定实验**

● 素质目标

养成对比分析学习的好习惯。

● 任务描述

使用旋光仪对在制品进行蔗糖分和重力纯度的测定。

**蔗糖分及重力
纯度的测定原理**

● 程序与方法

（1）原料：糖汁。
（2）设备及仪器：锤度计，旋光仪，200 mm 观测管、恒温水浴锅、分析天平、温度计（0～100 ℃）、水银温度计（0～50 ℃，分度为0.1 ℃）、200 mL 锥形瓶、100 mL 容量瓶、50 mL 移液管、量筒、烧杯、滤纸。
（3）试剂：碱性醋酸铅粉，24.85 °Bx 盐酸溶液［比重 d（20 ℃）= 1.102 9］、氯化钠溶液（231.5 g/L）、锌粉。
（4）步骤。

1）样品制备。各种糖汁可用原样品进行测定，其他浓度较高的样品可用稀释法或规定量溶液法配制样液。

2）测定样液的观测锤度及温度，记录数据。

3）蔗糖分的测定。倾取样液约 200 mL 于锥形瓶中，加入碱性醋酸铅粉约 2 g，迅速摇匀，过滤，用吸管吸取两份 50 mL 滤液，分别移入两个 100 mL 容量瓶中。其中一瓶加入 10 mL 氯化钠（231.5 g/L）溶液，然后加水至刻度，摇匀，如发现混浊，则过滤，滤液用 200 mm 观测管测其观测转光度，以此数乘 2 即得直接旋光读数 P，并记录读数时糖液的温度。在另一瓶先加入 20 mL 蒸馏水，再加入 10 mL 24.85 °Bx 的盐酸，插入温度计，在水浴中准确加热至 60 ℃，并在此温度保持 10 min（在最初 3 min 内应不断摇荡），取出浸入冷水中，迅速冷却至接近读取直接转光度时的温度。以洗瓶喷少量水，将附着于温度计上的糖液洗入瓶内，取出温度计，加水至刻度（如溶液色较深，可加入少量锌粉），充分摇匀，如发现混浊，则过滤。用 200 mm 观测管测其观测旋光度，以此数乘 2 得转化旋光读数 P'，并用 0.1 ℃ 刻度温度计测出读数时糖液的温度 t（测 P 及 P' 时的糖液温度，二者相差不得超过 1 ℃）。

（5）记录数据处理及计算结果。

项目	数据及结果
样液的观测锤度 /°Bx	
样液温度 /℃	
样液锤度 /°Bx	
直接旋光读数 P /°Z	
测定直接旋光度时液温 /℃	
转化旋光读数 P′/°Z	
测转化旋光度时液温 /℃	
样液蔗糖分 /%	
样液的重力纯度 /%	
样品蔗糖分 /%	
样品的重力纯度 /%	

● 思考

糖度和蔗糖分、简纯度和重力纯度有什么关联？又是分别在什么场景下使用？请同学们用"具体情况具体分析"的哲学思想去思考。

任务相关知识——二次旋光法

旋光法是测定蔗糖分的最常用的一种方法。它是利用蔗糖溶液具有旋光性，能将通过溶液的光线的振动平面旋转一定角度，旋转的角度与糖液的浓度成正比的原理进行测定。但对于含有杂质的样品，由于一些杂质也具有旋光性，这些杂质也会对旋光的角度各自产生影响。对这类样品进行旋光测定，实际上得到的是一个各种旋光性物质对旋光度的综合影响。蔗糖产生的旋光度只是测定结果中的一部分。为了从含有杂质的样品中单独将蔗糖产生的旋光度分离，就需要采用一种新的测定方法，就是二次旋光法。

1. 二次旋光法基本原理

二次旋光法除了利用糖液的旋光性外，还做了如下假设：蔗糖在适当的条件下可以完全转化为还原糖，同时样品中所有杂质的旋光性在蔗糖转化的过程中不发生变化。如果假设成立，用浓度为一个规定量（26.000 g/100 mL）的纯蔗糖溶液做试验，则

$$C_{12}H_{22}O_{11}+H_2O \rightarrow C_6H_{12}O_6 + C_6H_{12}O_6$$
蔗糖　　　　葡萄糖　　果糖

即蔗糖在转化剂（盐酸或转化酶）的作用下，水解成等量的葡萄糖及果糖，这两种糖的混合物称为转化糖。

如果将转化前后的糖液放在旋光仪上分别进行测定，其旋光读数应为：

转化前　$P=100\ °Z$；

转化后　$P'=-32.1\ °Z$。

则转化前后旋光度的变化与蔗糖的旋光度之比为：

$$\frac{P-P'}{P} = \frac{100-(-32.1)}{100} = 1.321$$

以上是用浓度为一个规定量的纯糖液测定得到的结果，因旋光度与糖液的浓度成正比，糖液的浓度发生变化，P 与 P' 也会以相同的比例同时变化，所以上面的公式适用于任何浓度的纯糖液，即对于纯蔗糖溶液，蔗糖转化前后旋光度的变化与蔗糖旋光度之比为固定值 1.321。

对于不纯糖液，旋光测定得到的 P 与 P' 是样品中各种旋光性物质旋光度的综合结果。对任意一个样品，若样品中纯蔗糖部分产生的旋光度是 α_s。将样品在适当的条件下进行转化，蔗糖转化后生成的转化糖的旋光度是 α_r。样品中各种杂质的旋光度分别为 α_1，α_2，α_3，α_4，…，α_n，因已假设杂质的旋光度在蔗糖转化的过程中不发生变化，所以有

$$P=\alpha_s+\alpha_1+\alpha_2+\alpha_3+\alpha_4+\cdots+\alpha_n$$
$$P'=\alpha_r+\alpha_1+\alpha_2+\alpha_3+\alpha_4+\cdots+\alpha_n,$$

将 P 与 P' 相减，因其他杂质的旋光度没有发生变化，相减时完全相互抵消。所以样品转化前后的旋光度变化也就是样品中纯蔗糖部分旋光度发生的变化，即

$$P-P'=\alpha_s-\alpha_r$$

而 α_s 及 α_r 是样品中纯蔗糖部分和以上部分蔗糖转化后产生的旋光度，应有以下关系：

$$\frac{\alpha_s-\alpha_r}{\alpha_s} = \frac{100-(-32.1)}{100} = 1.321$$

因为 $P-P'=\alpha_s-\alpha_r$

所以

$$\frac{P-P'}{P} = \frac{\alpha_s-\alpha_r}{\alpha_s} = \frac{100-(-32.1)}{100} = 1.321$$

即任何浓度的样品转化前后旋光度的变化（$P-P'$）与样品中的纯蔗糖产生的旋光度 α_s 之比也是固定值（1.321）。式中的 P 与 P' 是旋光仪的测定结果，而 α_s 是样品中纯蔗糖产生的旋光读数是旋光测定结果中的一部分，无法直接用旋光仪测定，但可以通过计算间接得出。

根据这一原理，首先测定样品的直接旋光度（转化前的旋光度），其次用适当的方法将样品转化，再次测定转化后样品的旋光度，最后根据以上公式推算出样品中纯蔗糖产生的旋光度。即样品中纯蔗糖所产生的旋光度不是直接在旋光仪上测出，而是用计算的方法

间接求得。因为要进行两次旋光测定，所以称为二次旋光法。

$$\alpha_s = \frac{P-P'}{1.321} = \frac{P-P'}{132.1} \times 100$$

式中　α_s——样品中纯蔗糖产生的旋光度（°Z）；
　　　P——样品直接旋光度（转化前的旋光度）（°Z）；
　　　P'——样品转化后旋光度（转化后的旋光度）（°Z）。

2. 克来杰除数

蔗糖转化前后旋光度的变化与纯蔗糖旋光度之比称为克来杰除数，即上面公式中的分母项。克来杰除数的大小受蔗糖转化方式（如酸法或酶法转化）、样品干固物含量及温度的影响。糖厂一般采用酸法转化，克来杰除数的计算公式如下：

$$\text{克来杰除数} = 132.56 - 0.0794(13-g) - 0.53(t-20)$$

式中　g——100 mL 被测定样液中干固物的克数（$g = \frac{\text{观测锤度}}{2} \times$ 对应观测锤度的密度）（g）；
　　　t——测定转化后样液旋光度的温度（℃）。

注：因为锤度的测定是在旋光测定前进行，而旋光测定时由于需要加入盐酸使蔗糖转化，通常是取 50 mL 测定锤度后的样液，加入盐酸后定容至 100 mL，即进行了二倍稀释，所以在计算被测定样液中干固物的质量时应将观测锤度除以 2。

3. 二次旋光法的基本操作

（1）适用范围。在糖厂中，除了杂质很少的白砂糖及糖分很少的蔗渣、滤泥等样品的蔗糖分用一次旋光法测定外，所有中间制品的蔗糖分均采用二次旋光法进行测定。

（2）主要仪器及试剂。

1）主要仪器。

①旋光检糖仪（旋光仪），国际糖度单位为 °Z。

②标准石英管。

③200 mm 观测管。

④容量瓶。标准温度 20 ℃，容积为（200±0.10）mL，（100±0.08）mL。

⑤移液管。标准温度 20 ℃，容积为（50±0.05）mL。

⑥精密温度计。0～50 ℃，分度为 0.1 ℃ 的水银温度计。

⑦普通温度计。0～100 ℃，分度为 1 ℃。

⑧恒温水浴锅。分度为 ±1 ℃。

⑨分析天平。感量 0.001 g。

2）主要试剂。

①碱性醋酸铅。化学纯，主要质量指标如下：

总铅（以 PbO 计）> 76%；

碱性铅（以 PbO 计）> 33%。

粒度全部可以通过孔径 0.42 mm，且至少有 70% 可以通过孔径 0.12 mm 的筛网。

②盐酸溶液（24.85 °Bx，密度 1.102 9 g/cm³）。

③氯化钠溶液（浓度为 231.5 g/L）。
④锌粉。
⑤氢氧化铝浆。

4. 测定步骤及计算

（1）样品制备

1）稀释。一般糖汁可用原样品直接测定，浓度较高的样品可用规定量稀释法或重量稀释法进行稀释。固体样品可考虑用规定量稀释，如赤砂糖、片糖可配制一个规定量的溶液，即称取样品 52.000 g，稀释至 200 mL 或称取样品 65.000 g，稀释至 250 mL。废蜜配制成 1/3 规定量的溶液，即称取样品 43.333 g，稀释至 500 mL，因为采用规定量稀释时，旋光仪的读数（或二次旋光计算结果）直接表示原样品的蔗糖分，可简化计算。糖浆采用 4 倍稀释，糖膏、糖蜜采用 6 倍稀释，有结晶的样品必须保证所有结晶完全溶解。

2）澄清过滤。样品稀释后加入适量的澄清剂进行澄清过滤。澄清剂的用量以能进行澄清为度，不要过量，否则在加入 NaCl 溶液后可能会生成沉淀，会对旋光测定的结果有影响。

倾取样液约 200 mL 于锥形瓶中，加入碱性醋酸铅粉约 2 g，迅速摇匀、过滤。样液过滤时应弃去最初的约 25 mL 滤液。由于测定要求用两份各 50 mL 的样液，所以滤液量超过 100 mL 后才能开始测定，以保证两份样品完全一致。

（2）测定。

1）锤度测定。测定制备后样液的观测锤度及温度。

注：温度是用于重力纯度的计算，如果不用计算重力纯度，可不测定温度。

2）直接旋光度测定。以吸管吸取 50 mL 滤液，移入 100 mL 容量瓶中。加入 10 mL 氯化钠（231.5 g/L）溶液，然后加水至刻度，摇匀，如发现混浊，则过滤，滤液用 200 mm 观测管测其观测转光度，以此数乘 2 即得直接旋光读数 P，并记录读数时糖液的温度。

3）转化旋光度测定。用吸管吸取 50 mL 滤液，移入 100 mL 容量瓶中，加入 20 mL 蒸馏水，再加入 10 mL 24.85 °Bx 的盐酸，插入温度计，在水浴中准确加热至 60 ℃，并在此温度保持 10 min（前 3 min 内应不断摇荡），取出浸入冷水中，迅速冷却至接近读取直接转光度时的温度。用洗瓶喷少量水，将附着于温度计上的糖液洗入瓶内，取出温度计。加水至刻度（如溶液色较深，可加入少量锌粉）充分摇匀，如发现混浊，则过滤。用 200 mm 观测管测其观测旋光度，以此数乘 2 得转化旋光读数 P'，并用 0.1 ℃ 刻度温度计测出读数时糖液的温度 t（测 P 及 P' 时的糖液温度，两者相差不得超过 1 ℃）。

4）数据记录。

①稀释后样液的观测锤度（采用规定量稀释时不需要测定这一数据）。

②直接旋光读数 P（°Z）。

③转化旋光读数 P'（°Z）。

④测定转化旋光读数 P' 时样液温度 t（要求准确至 1/10 ℃）。

（3）计算方法。

1）稀释后样液中纯蔗糖产生的旋光读数 S。通过测定样品转化前后的旋光读数，可以推算出样品中纯蔗糖产生的旋光度 S。

$$S(°Z) = \frac{100(P-P')}{132.56 - 0.0794(13-g) - 0.53(t-20)}$$

式中　P——直接旋光读数（°Z）；

　　　P'——转化旋光读数（°Z）；

　　　g——100 mL 被测定样液中干固物的质量 [$g = \dfrac{\text{观测锤度}}{2} \times$ 对应观测锤度的密度

　　　　或 $g = \dfrac{1}{2}(0.9982B + 0.0037B^2 + 0.00001816B^3)$]；

　　　B——样液观测锤度；

　　　t——测定转化旋光读数 P' 时样液温度。

2）被测样品的蔗糖分。以上计算结果 S 只相当于被测样液中纯蔗糖产生的旋光读数。其确切的含义，与样品的稀释方法有关。按规定量稀释的样品，S 就表示原样品（未经稀释的样品）的蔗糖分（质量百分数）。对未稀释或按倍数稀释样品，则 S 只是相当于纯糖液在旋光仪上产生的读数，它的单位是 °Z，表示稀释后样品的蔗糖含量为 $0.26S$ g/100 mL。这是一个体积质量百分浓度，应通过适当的方法换算为质量百分浓度，即将 100 mL 换算为质量。具体方法是按稀释后样液的观测锤度，查表得糖度（转光度）因数或密度，然后进行计算。

$$\text{稀释后样品蔗糖分}(\%) = S \times \text{糖度（转光度）因数} \times 100\%$$

糖度（转光度）因数可以按以下公式计算：

$$\text{糖度（转光度）因数} = (0.5107 - 0.001B)^2$$

式中　B——稀释后样液的观测锤度。

或：

$$\text{稀释后样品蔗糖分}(\%) = \frac{26S}{100 \times \text{稀释后样液密度}} \times 100\%$$

3）原样品蔗糖分。

$$\text{原样品蔗糖分}(\%) = \text{稀释后样品蔗糖分} \times \text{稀释倍数}$$

注：若样品未经稀释，则稀释倍数为 1。

当样品采用规定量稀释时，26.000 g 样品稀释为 100 mL，此时相应的稀释倍数为：

$$\text{按规定量稀释的稀释倍数} = \frac{100 \times \text{稀释后样液的密度}}{26.000}$$

$$\text{原样品蔗糖分} = \frac{26S}{100 \times \text{稀释后样液密度}} \times \frac{100 \times \text{稀释后样液的密度}}{26.000} = S$$

即当样品采用规定量稀释后，经计算得到的被测样液中纯蔗糖所产生的旋光读数就是原样品的蔗糖分，能简化计算。

（4）注意事项。

1）因为蔗糖转化需要加入 HCl，这就等于样品中加入了一种杂质，按二次旋光法原理，在测定直接旋光度时，加入一种对转化糖旋光度产生影响与 HCl 基本相同的杂质 NaCl，加入量与 HCl 相同。通过 $(P-P')$ 就可以将 HCl 对转化后糖旋光度的影响抵消。

注：在转化后的糖液中加入 m 克 HCl 或 NaCl 盐类后，对旋光度的影响为：NaCl，$0.540m$；HCl，$0.5407m$。两者很相近，所以可以抵消。

2）糖液中干固物的含量 g 的计算应使用观测锤度，由于 g 的单位是 100 mL 被测定样液中干固物的克数，而锤度的单位是 100 g 被测定样液中干固物的克数，将锤度换算为 g 就应将 100 g 被测定样液除以观测锤度对应的密度（换算为体积）。

3）因克来杰除数的温度校正只有一项，所以要求两次旋光测定的温度相差不要超过 1 ℃，校正时的温度使用测定转化后样液旋光度的温度。

4）因蔗糖在适当的条件下可以完全转化为还原糖只是一个假设，实际上很难实现完全转化，为了使各次测定结果能相互比较，要严格控制蔗糖转化的时间与温度。

5）因在进行旋光测定前，样品进行了二倍稀释（取 50 mL 定容为 100 mL），所以旋光读数应乘以 2 得到样品稀释前的旋光读数。

6）若澄清剂加入过量，加入 NaCl 溶液后会产生沉淀，应过滤后再进行旋光测定（必要时可加入氢氧化铝浆作为助滤剂）。

5. 测定实例

（1）混合汁蔗糖分测定。

1）锤度测定。将混合汁样品用筛网滤去蔗渣后测定观测锤度及温度。

注：如果不进行重力纯度计算，可以不测定温度。

2）样液澄清过滤。取滤去蔗渣后的混合汁约 200 mL，加入碱性醋酸铅粉约 2 g，迅速摇匀、过滤。样液过滤时应弃去最初的约 25 mL 滤液。由于测定要求用两份各 50 mL 的样液，所以滤液量超过 100 mL 后才能开始进行旋光测定，以保证两份样品完全一致。

3）直接旋光度测定。以吸管吸取 50 mL 滤液，移入 100 mL 容量瓶中。加入 10 mL 氯化钠（231.5 g/L）溶液，然后加水至刻度，摇匀，如发现混浊，则过滤，滤液用 200 mm 观测管测其观测转光度，以此数乘 2 即得直接旋光读数 P，并记录读数时糖液的温度。

4）转化旋光度测定。以吸管吸取 50 mL 滤液，移入 100 mL 容量瓶中，加入 20 mL 蒸馏水，再加入 10 mL 24.85 °Bx 的盐酸，插入温度计，在水浴中准确加热至 60 ℃，并在此温度保持 10 min（前 3 min 内应不断摇荡），取出浸入冷水中，迅速冷却至接近读取直接转光度时的温度。以洗瓶喷少量水，将附着于温度计上的糖液洗入瓶内，取出温度计。加水至刻度（如溶液色较深，可加入少量锌粉），充分摇匀，如发现混浊，则过滤。用 200 mm 观测管测其观测旋光度，以此数乘 2 得转化旋光读数 P'，并用 0.1 ℃刻度温度计测出读数时糖液的温度 t（测 P 及 P' 时的糖液温度，两者相差不得超过 1 ℃）。

5）数据记录。设测得某混合汁的数据如下：

混合汁的观测锤度为 16.10 °Bx，温度为 18.5 ℃。

直接旋光读数 $P=27.83\times2=55.66$（°Z）。

转化旋光读数 $P'=-9.84\times2=-19.68$（°Z），温度 $t=18.52$ ℃。

6）计算（以上面的数据为例）。

①计算 100 mL 被测定样液中干固物的克数 g。

根据混合汁的观测锤度为 16.10 °Bx，

则

$$g=\frac{1}{2}(0.998\,2B+0.003\,7B^2+0.000\,018\,16B^3)$$

$$=(0.998\,2\times16.10+0.003\,7\times16.10^2+0.000\,018\,16\times16.10^3)/2$$

$$=8.552\,9\,(\text{g}/100\,\text{mL})$$

②计算混合汁中纯蔗糖产生的旋光读数 S。

$$S=\frac{100(P-P')}{132.56-0.079\,4(13-g)-0.53(t-20)}$$

$$=\frac{100[55.66-(-19.68)]}{132.56-0.079\,4(13-8.552\,9)-0.53(18.52-20)}=56.650\,(°Z)$$

③计算混合汁蔗糖分。

混合汁蔗糖分（%）= 混合汁中纯蔗糖产生的旋光读数 $S\times$ 糖度因数 $\times100\%$。

糖度因数 $=(0.510\,7-0.001B)^2\times100\%$

则混合汁蔗糖分 $=56.650\times(0.510\,7-0.001\times16.10)^2\times100\%=13.86\%$

（2）糖浆蔗糖分测定。糖浆蔗糖分测定与混合汁蔗糖分测定基本相同，不同点在于样品的稀释及结果的计算。

1）样品稀释。取适量糖浆样品，四倍稀释后（例如，取糖浆 150 g，加水 450 g）测定稀释后样液的锤度。

余下操作与混合汁相同。

2）结果计算。与混合汁相同，但要注意所计算得到的结果是被测定样液的蔗糖分，而原样品是糖浆，所以要将计算结果乘以稀释倍数才能得到糖浆的蔗糖分。

（3）废蜜蔗糖分测定。

1）样液制备。将糖蜜样品搅拌均匀，称取样品 43.333 g，加入少许蒸馏水，用小玻璃棒混合均匀，将样品小心洗入 500 mL 容量瓶中，加水至标线，充分摇匀。

2）澄清过滤。从容量瓶中倒出制备后的样液测定锤度，然后取约 200 mL 于锥形瓶中，加入碱性醋酸铅粉约 3.5 g（因糖蜜含的杂质较多，所以澄清剂的用量要适当加大），摇匀后过滤。最初部分滤液洗涤盛器后弃去，待滤液的量足够吸取两份 50 mL 的样液后开始进行旋光度测定。

3）旋光度测定。参照混合汁蔗糖分测定的 3）、4）所示的方法进行，测定滤液的直接旋光度读数（P）及转化旋光度读数（P'），并用 1/10 分度的温度计测定转化旋光读数时糖液的温度（t）。

注：测定转化旋光读数时，若糖液的颜色较深，可在加水至标线后，加入少量锌粉，利用锌粉与糖液中的盐酸起反应，生成具有漂白作用的初生态氢，使颜色变浅，过滤后再进行测定。

4）数据记录。与混合汁蔗糖分相同。

5）糖蜜蔗糖分计算。按样液的制备方法，100 mL 样液中含样品的量为：

100 mL 样液中含样品的量 $=43.333/5\approx8.667$（g）

因为糖厂的规定量为 26.000 g，而 26.000 g/3 ≈ 8.667 g，因此按以上方法制备的样液的浓度为 1/3 个规定量。因旋光读数与浓度成正比，所以测得的读数也只有按一个规定量稀释的 1/3，只要将旋光测定的读数乘以 3，就可以得到浓度为一个规定量的溶液对应的旋光读数。按旋光仪的刻度方法，当样品按规定量稀释后，旋光仪的读数就表示原样品的蔗糖分（或糖度），即置于旋光仪上进行测定的是稀释后的样液，而仪器上显示的读数表示原样品（未稀释）的蔗糖分。

【例 2-8】取糖蜜 43.333 g 定容为 500 mL 后，测得观测锤度为 15.08 °Bx，直接旋光读数 $P=3.45×2=6.90$（°Z）；转化旋光读数 $P'=-3.23×2=-6.46$（°Z）；温度 $t=19.45$ ℃。计算糖蜜蔗糖分。

【解】稀释后每 100 mL 糖液中含糖蜜克数为 43.333/5 g。

假设稀释后糖液的浓度为 X（规定量浓度），则

（43.333/5）：X=26：100

X=33.33（即 1/3 个规定量）

计算 100 mL 被测定样液中干固物的质量 g：

根据配制糖液的观测锤度为 15.08 °Bx，则

$$g = \frac{1}{2}(0.998\,2B + 0.003\,7B^2 + 0.000\,018\,16B^3)$$

$$= (0.998\,2 × 15.08 + 0.003\,7 × 15.08^2 + 0.000\,018\,16 × 15.08^3)/2$$

$$= 7.98\ (\text{g}/100\ \text{mL})$$

因糖液的浓度为 1/3 规定量，而旋光度读数与浓度成正比，所以将测得的旋光度读数乘以 3 就可以换算为浓度为一个规定量时的读数。当被测样液的浓度为一个规定量时，旋光仪的读数就表示原样品的蔗糖分。

$$\text{糖蜜的蔗糖分} = \frac{3×100(P-P')}{132.56-0.079\,4(13-g)-0.53(t-20)} × 100\%$$

$$= \frac{3×100[6.90-(-6.46)]}{132.56-0.079\,4(13-7.98)-0.53(19.45-20)} × 100\% = 26.16\%$$

任务2.4　兰-艾农法测定蔗汁还原糖

● 企业案例

主任看了看刚刚出来的报表，发现清汁的还原糖特别高，他立即从生产线上取回部分清汁，安排小明对以上样品的还原糖进行测定，小明应该选择什么仪器和使用什么方法来测定呢？

● 任务目标

通过本任务的学习，学生达到以下目标：
（1）了解兰-艾农法的测定原理。
（2）掌握兰-艾农法的测定方法及计算。

● 素质目标

养成对比分析学习的好习惯。

● 任务描述

使用兰-艾农法对在制品进行还原糖分的测定。

● 程序与方法

（1）原料：蔗汁。
（2）设备及仪器：电炉、秒表、250 mL 容量瓶、50 mL 滴定管、5 mL 移液管。
（3）试剂：费林试剂甲、乙液，54 °Bx 中性醋酸铅溶液，脱铅剂（磷酸盐和草酸盐混合液），1% 四甲基蓝指示剂。
（4）步骤：
1）样液的制备。对浓度不高的样品（26 °Bx 以下）如各种糖汁可不必稀释，按样品原浓度进行配制。

取样品蔗汁 50 mL 移入 250 mL 容量瓶中，加入 1 mL 54 °Bx 中性醋酸铅溶液，充分摇匀，然后加入 3 mL 脱铅剂摇匀，加蒸馏水至刻度，充分摇匀，过滤，弃去最初滤液约 15 mL，该滤液即为配制糖液。配制糖液的还原糖浓度应保持为 2.50～4.00 g/L。

2）滴定。
①预检：用两支 5 mL 吸管，先吸取 5 mL 费林试剂乙液于 250 mL 锥形瓶中，然后吸取 5 mL 甲液于乙液中，混匀。从滴定管加入 15 mL 糖液于锥形瓶内，摇匀，放在铺有石棉网的电炉上加热，并准确煮沸 2 min（用秒表控制），加入 3～4 滴四甲基蓝指示剂，继续滴加糖液至蓝色消失为止，即为终点。此项操作不可超过 1 min，使整个沸腾和滴加

操作总时间控制在 3 min 以内。记录滴定耗用配制糖液体积。

②复检：按上述次序吸取费林试液乙、甲液各 5 mL 于 250 mL 锥形瓶内，从滴定管加入比预检时耗用量约少 1 mL 的配制糖液，摇匀。滴定步骤与预检相同，其沸腾时间亦应准确控制为 2 min，滴定至终点亦不可超过 1 min，滴定时须轻轻摇动锥形瓶，但不可离开热源，使溶液继续保持沸腾，以免空气进入瓶内使四甲基蓝再被氧化而发生误差。

（5）数据处理及计算。

样品的锤度：　　　　　　蔗糖分：

项目	数据及结果		
滴定耗用配制糖液体积			
还原糖 /%			

● 思考

蔗糖与还原糖有何联系？

还原糖分的测定

在糖厂的各种物料中，都含有一定量的还原糖（RS），糖料中的还原糖反映了糖料的成熟程度，即糖料的质量，中间制品中的还原糖含量则是制定工艺条件的重要依据，成品糖中的还原糖关系到产品的储存性能，所以还原糖的测定是糖厂化学管理中的一项重要项目。

糖厂的中间制品还原糖测定使用兰-艾农法或兰-艾农恒容法，白砂糖的还原糖使用奥夫纳尔法。这几种方法都是根据 Cu^{2+} 离子在一定的条件下能定量与还原糖反应生成 Cu^+ 离子，从而确定还原糖的量。但各种方法对还原糖的定量方法不同。

1. 兰-艾农法测定还原糖

兰-艾农恒容法是基于费林试剂的 Cu^{2+} 离子与还原糖发生氧化还原反应的一种容量分析方法。费林试剂是一种铜盐的碱性溶液，由硫酸铜、酒石酸钾钠和氢氧化钠组成。硫酸铜提供反应所需的 Cu^{2+} 离子，氢氧化钠提供反应所需的碱性条件。硫酸铜在碱性条件下会生成氢氧化铜沉淀，由于有酒石酸钾钠的存在，氢氧化铜能与酒石酸钾钠生成可溶性的络合物，从而使 Cu^{2+} 离子在碱性的情况下也能稳定地存在于溶液中。

还原糖分检测——兰-艾农法实验

（1）基本原理。在沸腾的状态下，用含有还原糖的样液滴定一定量的费林试剂，在定量的费林试剂中 Cu^{2+} 离子量一定，只能与相当量的还原糖反应。样液的还原糖含量高，滴定时消耗的体积就小，反之则大，所以根据样液的耗用量就可以计算出还原糖的量。为了判别终点，加入氧化能力较 Cu^{2+} 弱的四甲基蓝。滴定时还原糖先将氧化能力较强的 Cu^{2+} 全部还原后，过剩的还原糖才与四甲基蓝作用，使其由蓝色变为无色，指示终点到达。测定过程中主要的化学反应如下：

还原糖分检测——兰-艾农法测定

$$还原糖 + CuSO_4（蓝色）\rightarrow 有机酸 + Cu_2O（砖红色）$$

$$还原糖 + 四甲基蓝（蓝色）\rightarrow 四甲基蓝（无色）$$

上述反应是可逆的,当无色的四甲基蓝与空气中的氧气结合时又会变为蓝色。为了防止这种情况发生,在滴定时一定要使溶液保持沸腾状态,利用上升的蒸汽阻止空气进入锥形瓶中。

1) 费林试剂。滴定过程中的 Cu^{2+} 离子由费林试剂提供。费林试剂由分别存放的甲、乙两种溶液组成。

甲液：$CuSO_4$ 溶液,提供反应所需的 Cu^{2+} 离子。称取 69.278 g 结晶硫酸铜（$CuSO_4 \cdot 5H_2O$）,用水溶解后移入 1 000 mL 容量瓶中,加水至标线,摇匀后过滤即成。

乙液：含有酒石酸钾钠和氢氧化钠,提供反应所需的条件。称取 346 g 酒石酸钾钠（$NaKC_4H_4O_6 \cdot 4H_2O$）,溶于约 500 mL 水中,另称取氢氧化钠 100 g 溶于约 200 mL 水中,将两者混合后移入 1 000 mL 容量瓶,加水至标线。静置 2 d,如液面下降,须再加水至标线,摇匀后过滤即成。

Cu^{2+} 离子在碱性条件下才能与还原糖起反应,因为在碱性条件下,$CuSO_4$ 会缓慢地与酒石酸钾钠发生反应,生成 Cu_2O 沉淀,使 $CuSO_4$ 溶液的浓度发生变化。而 $CuSO_4$ 溶液的浓度是定量还原糖的关键,所以甲、乙液只能在使用前混合,混合后要立刻使用。

乙液中的酒石酸钾钠能与 Cu^{2+} 离子形成稳定的可溶性络合物,有利于反应的进行。向锥形瓶内移入费林试剂时,要先移乙液,后移甲液。先将乙液移入,锥形瓶内有大量的酒石酸钾钠,能保证在甲液移入时,Cu^{2+} 离子会立刻形成稳定的可溶性络合物,而不会生成氢氧化铜沉淀。如果先移甲液,当乙液移入与甲液接触的一瞬间,由于环境中是以甲液为主,酒石酸钾钠的量很少,Cu^{2+} 离子就有可能生成氢氧化铜沉淀,这种沉淀一旦形成,就不能保证在短时间内完全重新溶解,沉淀状态的 Cu^{2+} 离子与还原糖的反应速度大大地低于溶解状态的 Cu^{2+} 离子,在相同的反应时间内,有部分 Cu^{2+} 离子可能就来不及与还原糖反应,从而影响测定的结果。

2) 费林试剂的浓度。还原糖的测定是用样品去滴定一定量的费林试剂,所以费林试剂中 Cu^{2+} 离子量是定量还原糖的关键。Cu^{2+} 离子的量与费林试剂的浓度、用量及反应条件有关。一定量的 Cu^{2+} 离子能与多少还原糖起反应并不是一个固定不变的值,它与反应的条件有关,不能用简单的化学反应式来计算,只能用试验的方法来确定。所以在进行测定前要在一定的条件下,用标准的转化糖滴定一定量的费林试剂,然后根据标准转化糖液的耗用量,计算费林试剂相当的还原糖量。目前,统一规定在没有蔗糖存在的情况下,若 10 mL 费林试剂（甲、乙液各 5 mL）消耗 2.000 g/L 的转化糖液 25.64 mL（试验数据）,则将其浓度定为一个当量,相当的还原糖质量为 51.28 mg；否则应按费林试剂的消耗量计算其浓度校正系数。

$$费林试剂浓度校正系数 K=V/25.64$$

式中　V——标定时实际消耗的 2 g/L 的标准转化糖液量（mL）。

3) 还原糖因数。由于影响还原糖与 Cu^{2+} 离子反应的因素很多,包括使用的电炉及锥形瓶规格（与加热至沸腾所用的时间有关）、样品中蔗糖的含量（因蔗糖也能与 Cu^{2+} 离子起反应)、样品中还原糖的含量(因还原糖的含量不同,滴定时消耗的配制糖液也不同,与定量的费林试剂混合后,体积各不相同,锥形瓶内 Cu^{2+} 离子的浓度也不同,配制糖液消耗量大时,总体积就大,Cu^{2+} 离子的浓度就会下降,反之则上升,Cu^{2+} 离子的氧化能力也不同)等,所以费林试剂的浓度只能用标准转化糖溶液标定后确定其浓度。要求标定的条件与检测时的条件完全一致。10 mL 费林试剂(甲、乙液各 5 mL)相当的还原糖质量称为还原糖因数,它随样液中还原糖与蔗糖的含量改变而略有变化。通过试验可以得到在不同蔗糖浓度下、样品耗用量在 15～50 mL 时 10 mL 费林试剂所相当的还原糖,并制成"还原糖因数表"供计算使用。

还原糖因数可由 100 mL 配制糖液中蔗糖质量 G 和滴定消耗配制糖液体积 V 查附表 8 得到,也可以按以下公式计算:

$$F=(50.62-0.548\,8V+0.305\,6G)/(1-0.011\,33V+0.015\,98G-0.000\,1G^2)$$

式中　F——还原糖因数;

G——100 mL 配制糖液中蔗糖质量(g);

V——滴定消耗配制糖液体积(mL)。

(2)测定步骤。

主要设备及试剂:

①碱式滴定管分度 0.1 mL,20℃标准,最大允许误差 0.05 mL。

②250 mL 锥形瓶。

③5 mL 移液管。

④电炉、秒表等。

⑤标准转化糖液,质量浓度:2.000 g/L。

⑥54 °Bx 中性醋酸铅溶液。

⑦费林试剂。

⑧四甲基蓝指示剂。

⑨除铅剂(磷酸氢二钠与草酸钾混合液)。

(3)测定步骤。

1)样液制备。

①稀释。26 °Bx 以下的样品可以不稀释,糖浆 4 倍稀释,糖膏、糖蜜 6 倍稀释。兰－艾农法要求用于滴定的配制糖液的还原糖含量为每 100 mL 配制糖液中 250～400 mg。样品稀释的倍数还应按预检的结果进行调整,如果预检消耗的配制糖液过多(超过 50 mL),则应减小稀释倍数;过少会增大分析的误差,则应加大稀释倍数。

对于浓度较高的样品,也可以用直接称重方式制备配制糖液。例如,糖浆可称 12.500 g 左右;糖蜜、糖膏、糖糊可称 1.500～12.500 g(要求准确至小数点后三位),用蒸馏水溶解后移入 250 mL 容量瓶,再按容量稀释的方法进行操作。

②澄清、除铅。取稀释后样液 50 mL 以细密铜网过滤后移入 250 mL 容量瓶中,加入适量(约 1 mL)54 °Bx 中性醋酸铅溶液进行澄清,充分摇匀后加入除铅剂(约 3 mL),摇匀后加水至标线,过滤并收集滤液(称为配制糖液)。

2)预检。用移液管先吸取 5 mL 费林试剂乙液于 250 mL 锥形瓶中,然后用另一支移液管吸取 5 mL 甲液于乙液中，混匀。用滴定管加入 15 mL 糖液于锥形瓶内，摇匀，放在铺有石棉网的电炉上加热，并准确煮沸 2 min（用秒表计数），加入 3～4 滴四甲基蓝指示剂，在保持糖液沸腾的情况下，继续滴加配制糖液至蓝色消失即为终点。此项操作不可超过 1 min，使整个沸腾和滴定操作总时间控制在 3 min 以内。记录滴定耗用配制糖液体积 V_1。

3）复检。按上述次序吸取费林试液乙、甲液各 5 mL 于 250 mL 锥形瓶内，按预检时耗用的配制糖液 V_1，从滴定管加入较预检时耗用量 V_1 约少 1 mL 的配制糖液，摇匀。滴定手续与预检相同，其沸腾时间也应准确控制为 2 min，加入指示剂后滴定至终点的时间也不可超过 1 min，滴定时须轻轻摇动锥形瓶，但不可离开热源，使溶液继续保持沸腾，以免空气进入瓶内使四甲基蓝再被氧化而发生误差。记录滴定耗用配制糖液体积 T。

（4）计算方法。影响费林试剂与还原糖反应的因素很多，如沸腾时间、滴定结束时锥形瓶内溶液的体积及蔗糖含量等。在这些影响因素中，沸腾时间可以通过规范操作来控制，滴定结束时锥形瓶内溶液的体积受配制糖液中还原糖浓度的影响，当还原糖浓度高时，耗用的配制糖液量少，滴定结束时锥形瓶内溶液的体积就小，反之则大。因蔗糖也能消耗部分 Cu^{2+} 离子，使用的配制糖液量越大、配制糖液的蔗糖分越高，滴定结束时锥形瓶内溶液的蔗糖就越多，蔗糖消耗的 Cu^{2+} 离子也越多，能参与还原糖反应的 Cu^{2+} 离子就越少。所以不能直接按反应方程式进行计算，10 mL 费林试剂所相当的还原糖量，应根据实际情况，按相关数据查"还原糖因数表"而确定。

根据配制糖液耗用量及 100 mL 配制糖液中蔗糖的质量这两个参数来确定 10 mL 费林试剂所相当的还原糖量，即"还原糖因数"。其中配制糖液耗用量可从滴定结果中直接得出，而 100 mL 配制糖液中蔗糖的质量要由样品的蔗糖分（或糖度）计算而得。查表所得的还原糖因数，就是 10 mL 费林试剂所相当的还原糖量，即在锥形瓶内样品中所含的还原糖质量。

1）计算过程所需数据如下。
①样品的质量（用容量法配制样液无此项数据），g。
②样品的锤度 B（用直接称重法配制样液无此项数据），°Bx。
③样品蔗糖分 S。
④费林试剂浓度校正系数 K。
⑤滴定耗用配制糖液毫升数 T。
2）计算可按以下方法进行。
①计算 100 mL 配制糖液中含样品质量 W。
a. 用容量法配制样液：

$$W=Vd$$

式中　V——100 mL 配制糖液中含样品的体积（mL）；
　　　d——样液密度（由观测锤度 B 查表或按公式计算：$d=0.9982+0.0037B+0.00001816B^2$）。

例如：取 50 mL 样液配为 250 mL，样品的锤度为 15.32 °Bx，则 100 mL 配制糖液中含样品的体积 V 为

$$V=50\times100/250=20（mL）$$

$$W=20×(0.998\ 2+0.003\ 7×15.32+0.000\ 018\ 16×15.32^2)=21.18\ (g)$$

b. 用直接称重法配制糖液：

$$W=样品质量/2.5$$

②计算 100 mL 配制糖液中蔗糖质量 G。

$$G=WS/100$$

式中　G——100 mL 配制糖液中蔗糖质量（g）；

　　　W——100 mL 配制糖液中含样品质量（g）；

　　　S——样品蔗糖分（%）。

③计算 100 mL 配制糖液中还原糖质量 I。

$$I=100FK/T$$

式中　I——100 mL 配制糖液中还原糖质量（mg）；

　　　K——费林试剂浓度校正系数；

　　　T——滴定消耗配制糖液质量（mg）；

　　　F——还原糖因数（由 100 mL 配制糖液中蔗糖质量 G 和滴定消耗配制糖液体积 V 查附表 8 可得），也可以按以下公式计算：

$$F=(50.62-0.548\ 8V+0.305\ 6G)/(1-0.011\ 33V+0.015\ 98G-0.000\ 1G^2)$$

式中　G——100 mL 配制糖液中蔗糖质量；

　　　V——滴定消耗配制糖液体积。

④计算样品还原糖分 R。

$$R=100I/(1\ 000W)=I/(10W)$$

式中　R——样品还原糖分（%）。

其余符号同上。

【例 2-9】混合汁的观测锤度为 15.32 °Bx，蔗糖分为 13.58%。吸取样品 50 mL 稀释至 250 mL，费林试剂的浓度校正系数为 0.975，复检耗用配制糖液 35.00 mL，求混合汁的还原糖分。

【解】100 mL 配制糖液中含样品的体积 V_1 为

$$V_1=50×100/250=20\ (mL)$$

100 mL 配制糖液中含样品的质量 W 为

$$W=20×(0.998\ 2+0.003\ 7×15.32+0.000\ 018\ 16×15.32^2)=21.18\ (g)$$

100 mL 配制糖液中蔗糖质量 G 为

$$G=21.18×13.58\%=2.88\ (g)$$

还原糖因数 F 由 100 mL 配制糖液中蔗糖质量 G 和滴定消耗配制糖液体积 V 查附表 8 得到，也可以按以下公式计算：

$$F=(50.62-0.548\ 8V+0.305\ 6G)/(1-0.011\ 33V+0.015\ 98G-0.000\ 1G^2)$$

$$F=\frac{50.62-0.548\ 8×35+0.305\ 6×2.88}{1-0.011\ 33×35+0.015\ 98×2.88-0.000\ 1×2.88^2}=49.78$$

费林试剂的浓度校正系数为 0.975，10 mL 费林试剂相当于还原糖量为：

$$10 \text{ mL 费林试剂相当于还原糖} = 0.975 \times 49.78 = 48.54 \text{ (mg)}$$

即 10 mL 费林试剂相当于还原糖 48.54 mg；或滴定消耗的 35.00 mL 配制糖液中含有还原糖 48.54 mg。则 100 mL 配制糖液中含有还原糖为：

$$35 : 48.54 = 100 : X$$

$$X = 100 \times 48.54 / 35 = 138.69 \text{ (mg)}$$

$$\text{混合汁的还原糖分} = \frac{100 \text{ mL 配制糖液中还原糖质量}}{100 \text{ mL 配制糖液中样品质量}} \times 100\% = \frac{138.69}{21.18 \times 1000} \times 100\% = 0.65\%$$

【例 2-10】赤砂糖的蔗糖分为 88.45%，用规定量稀释后取 50 mL 配为 200 mL，费林试剂浓度校正系数为 1.028，复检耗用配制糖液 32.35 mL，求赤砂糖还原糖分。

【解】规定量浓度的样液每 100 mL 中含样品 26.000 g，取 50 mL 配为 200 mL 后，在 200 mL 配制糖液中含有原样品 13.000 g。

32.35 mL 配制糖液中含样品的质量 W 为

$$13 : 200 = W : 32.35$$

$$W = 13 \times 32.35 / 200 = 2.10 \text{ (g)}$$

100 mL 配制糖液中含蔗糖的质量 G 为

$$(13 \times 88.45\%) : 200 = G : 100$$

$$G = 13 \times 88.45\% \times 100 / 200 = 5.75 \text{ (g)}$$

按 G 及配制糖液耗用量 32.35 mL 计算还原糖因数 F。

$$F = (50.62 - 0.548\,8V + 0.305\,6G) / (1 - 0.011\,33V + 0.015\,98G - 0.000\,1G^2)$$

$$F = \frac{50.62 - 0.548\,8 \times 32.35 + 0.305\,6 \times 5.75}{1 - 0.011\,33 \times 32.35 + 0.015\,98 \times 5.75 - 0.000\,1 \times 5.75^2} = 47.95$$

10 mL 费林试剂相当于还原糖 $= 1.028 \times 47.95 = 49.29$ （mg）

$$\text{赤砂糖的还原糖分} = \frac{49.29}{2.10 \times 1\,000} \times 100\% = 2.35\%$$

（5）注意事项。

1）费林试剂甲、乙液要分开存放。如果将它们预先混合，则酒石酸钾钠会缓慢地将 Cu^{2+} 离子还原为 Cu^+ 离子，造成 Cu^{2+} 离子浓度发生变化。使用时用移液管准确吸取，要按先移乙液后移甲液的顺序移入锥形瓶。

2）煮沸时间准确控制为 2 min，当糖液全面沸腾时才开始计算时间。

3）所用热源及锥形瓶规格，对分析结果有一定影响。故在标定费林试剂和测定样品时，最好能使用同一套仪器用具。

4）滴定时应使锥形瓶内溶液保持沸腾状态，以免空气进入瓶内使四甲基蓝被氧化而呈蓝色，使测定产生误差。

5）四甲基蓝指示剂本身能消耗一定量的转化糖，故必须按规定量加入。

6）预检的目的。一是了解滴定用的配制糖液的稀释倍数是否合适。配制糖液的耗用量应在 20～40 mL。若加入配制糖液 15 mL 于费林试剂中，煮沸后蓝色全部褪去（所

有 Cu^{2+} 离子均被还原为 Cu^+ 离子），说明配制糖液过浓；若滴定耗用的配制糖液超过 50 mL，则说明配制糖液过稀。这两种情况都要求重新调整稀释倍数。二是了解滴定耗用配制糖液的大概数量，并按这个数量算出应预先加入配制糖液量，以便复检时能在溶液保持沸腾状态下，在 1 min 内完成最后的滴定操作，提高测定的准确性。

7）因还原糖能与钙与铅生成络合物，干扰测定，所以必须进行除钙、除铅处理。

2. 兰-艾农恒容法测定还原糖

兰–艾农恒容法通常也称恒容法，它是兰–艾农法的一种改进。兰–艾农法曾为 ICUMSA（国际糖品分析统一方法委员会）建议的正式方法，国家赤砂糖新标准中还原糖分测定以兰–艾农恒容法代替兰–艾农法，1978 年 ICUMSA 第 17 届会议已撤销兰–艾农法的正式方法资格，而采用兰–艾农恒容法为正式方法。用这种方法可以克服因样品的还原糖含量不同，滴定结束时锥形瓶内溶液的总体积也有所变化，导致 Cu^{2+} 离子的氧化能力产生差异、计算较繁杂的缺点。

还原糖分检测–（兰–艾农恒容法）实验

（1）基本原理。兰–艾农恒容法是传统的兰–艾农法的改进，它的测定原理及主要的化学反应与兰–艾农法完全相同，但使用的费林试剂量为 20 mL（甲、乙液各 10 mL）。

试验证明，$CuSO_4$ 的氧化能力受很多因素影响，如 Cu^{2+} 的浓度、温度、时间、样品中蔗糖的量等。为了减少这些因素的影响，兰–艾农恒容法除了规定费林试剂的用量、沸腾时间外，还要求滴定结束时锥形瓶内溶液的体积恰好为 75 mL（通过预先加入适量的水来控制），以减少体积变化对测定的影响，简化计算工作。所以这种方法称为恒容法。

还原糖分检测–（兰–艾农恒容法）原理

恒容法是通过预先往锥形瓶内加入一定量的水来达到滴定结束时锥形瓶内溶液的体积恰好为 75 mL 这一目的。但由于事先并不知道滴定要消耗多少样品，所以首先要进行预检，初步确定样品的耗用量，根据样品的耗用量，算出应加入水量，并将水预先加入锥形瓶中。滴定结束时，锥形瓶内溶液的体积就基本上为 75 mL 了。

（2）费林试剂的浓度。兰–艾农恒容法所用的费林试剂与兰–艾农法相同，但标定方法略有区别。

由于影响还原糖与 Cu^{2+} 离子反应的因素很多，所以费林试剂的浓度只能用标准转化糖溶液标定后确定其浓度，要求标定的条件与检测时的条件完全一致。如果费林试剂的浓度恰当，在没有蔗糖存在的情况下，20 mL 费林试剂（甲、乙液各 10 mL）应消耗 2.5 g/L 的标准转化糖液 40 mL，即相当于还原糖 100 mg。若浓度不恰当，应根据标定时所相当的还原糖量，计算校正系数（K）。

$$费林试剂浓度校正系数\ K=V/40$$

式中 V——标定时实际消耗的 2.5 g/L 的标准转化糖液量（mL）。

若样品中有蔗糖存在，蔗糖会消耗部分 Cu^{2+} 离子。锥形瓶内蔗糖量越大，消耗的 Cu^{2+} 离子也越多。此时应根据锥形瓶内蔗糖的质量，查相关的表（附表 9）得校正系数 f，则 20 mL 费林试剂（甲、乙液各 10 mL）相当的还原糖质量应按下面公式计算。

$$20\ mL\ 费林试剂相当的还原糖质量 =100fK$$

式中　K——费林试剂浓度校正系数；

　　　f——蔗糖校正系数。

(3) 主要仪器与试剂。

1) 主要仪器。碱式滴定管、300 mL 锥形瓶、10 mL 移液管、电炉、秒表等。

2) 试剂。EDTA 溶液（40 g/L）、四甲基蓝、费林试剂等。

(4) 测定步骤。

1) 样液的制备。对浓度不高的样品（26 °Bx 以下）如各种糖汁可不必稀释，按样品原浓度进行配制。对浓度较高的样品（26 °Bx 以上）可加蒸馏水稀释（糖浆稀释 4 倍，糖蜜稀释 6 倍，糖膏、糖糊等含有结晶的样品稀释 6 倍，且必须全部溶解其晶体）。

取上述稀释后的糖液（糖汁不用稀释）100 mL 移入 300 mL 容量瓶中，对每克干固物加入 1～2.4 mL 草酸钾溶液（50 g/L），充分摇匀，加蒸馏水至刻度，充分摇匀后过滤。加入草酸钾溶液是为了除去对还原糖测定有干扰的钙离子。弃去最初滤液约 15 mL，该滤液即为配制糖液。

也可在溶液中对每克干固物加入 4 mL EDTA 溶液（40 g/L），充分摇匀，加蒸馏水至刻度。因 EDTA 与钙离子生成的络合物比还原糖与钙离子生成的络合物更稳定，所以不用过滤。

对于浓度较高的样品，也可以用直接称重方式制备配制糖液。例如，糖浆可称 12.500 g 左右；糖蜜、糖膏、糖糊可称 1.500～12.500 g（要求准确至小数点后三位），以蒸馏水溶解后移入 250 mL 容量瓶，再按容量稀释的方法进行操作。

配制糖液的还原糖浓度应保持在 2.50～4.00 g/L 之间。如果不在这一范围，滴定时消耗的配制糖液就会过多或过少，此时应通过改变稀释倍数来调整。

2) 测定方法。

①预检。用移液管先吸取 10 mL 费林试剂乙液于 300 mL 锥形瓶中，然后用另一支移液管吸取 10 mL 甲液放于乙液中，混匀。从滴定管加入 25 mL 糖液于锥形瓶内，用量筒加入 15 mL 蒸馏水，摇匀，放在铺有石棉网的电炉上加热，并准确煮沸 2 min（用秒表控制），加入 3～4 滴四甲基蓝指示剂，继续滴加配制糖液至蓝色消失为止，即为终点。此项操作不可超过 1 min，将整个沸腾和滴加操作总时间控制在 3 min 以内。记录滴定耗用配制糖液体积 V_1。

②复检。按上述次序吸取费林试液甲、乙液各 10 mL 于 300 mL 锥形瓶内，按预检时耗用的配制糖液体积 V_1，用量筒加入蒸馏水（75-20-V_1）mL，从滴定管加入较预检时耗用量 V_1 约少 1 mL 的配制糖液，摇匀。滴定手续与预检相同，其沸腾时间也应准确控制为 2 min，加入指示剂后滴定至终点的时间也不可超过 1 min，滴定时须轻轻摇动锥形瓶，但不可离开热源，使溶液继续保持沸腾，以免空气进入瓶内使四甲基蓝被氧化而发生误差。记录滴定耗用配制糖液体积 T，并将这一数据用于还原糖的计算。

如果由于操作不熟练，导致 V_1 与 T 相差较远，可将复检的结果计为 V_1，重复测定一次。

（5）计算方法。

1）所需数据。

①滴定耗用配制糖液体积 T。

②费林试剂浓度校正系数 K。

③稀释后样液蔗糖分 S。

2）计算方法。兰－艾农恒容法与兰－艾农法的原理相同，费林试剂与还原糖反应的能力也不能直接按反应方程式进行计算，只能用试验的方法，测定在不同条件下，20 mL 费林试剂所相当的还原糖量，并将这些数据制成"校正系数表"供实际测定使用。但兰－艾农恒容法将滴定结束时锥形瓶内溶液的量控制为一个固定的值，费林试剂与还原糖反应的能力就不受配制糖液耗用量的影响，在计算时只需考虑蔗糖对费林试剂氧化能力的影响，所以只要计算出锥形瓶内蔗糖的质量，就可以查出费林试剂的校正系数，与兰－艾农法相比，较为简单。

在没有蔗糖存在的情况下，20 mL 费林试剂相当于还原糖 100 mg，若有蔗糖存在，则乘以相对应的费林试剂的校正系数。计算的方法如下所述。

①计算滴定结束后，锥形瓶内溶液含蔗糖克数 A：

采用容量稀释法时，首先计算 100 mL 配制糖液含样品体积 V_0：

$$V_0=100\times 100/250=40（\text{mL}）$$

锥形瓶内溶液含蔗糖的质量 $A=\dfrac{V_0\times d_{(20℃)}\times T\times S}{100\times 100}$

采用质量稀释法时，可直接按样品的质量 W 进行计算。

锥形瓶内溶液含蔗糖的克数 $A=\dfrac{W\times T\times S}{100\times 250}$

式中　W——采用质量稀释法时，称取样品的质量（g）。

②根据锥形瓶内溶液含蔗糖克数 A 用下面的公式计算或查附表 9 得蔗糖校正系数 f：

$$f=\dfrac{0.994\,7+0.041\,8A}{1+0.067\,2A}$$

式中　A——锥形瓶内溶液含蔗糖质量（g）。

③样品还原糖分：

$$\text{还原糖分}（\%）=\dfrac{f\times K\times 100}{T\times \dfrac{V_0}{100}\times d_{(20℃)}\times 1\,000}\times 100\%=\dfrac{10\times f\times K}{T\times V_0\times d_{(20℃)}}\times 100\%$$

式中　f——蔗糖校正系数；

V_0——100 mL 配制糖液含样品体积（mL）；

T——滴定耗用配制糖液体积（mL）；

K——费林试剂浓度校正系数；

$d_{(20℃)}$——样品的视密度（20℃）。

样品的视密度 $d_{(20℃)}$ 由样品的观测锤度查表或用以下公式计算：

样品的视密度 $d_{(20℃)}=0.998\ 2+0.003\ 7B+0.000\ 018\ 16B^2$

式中　B——样品的观测锤度。

注：对于糖浆、糖膏等要进行稀释后才能测定的样品，应将计算结果乘以稀释倍数后才得到原样品的还原糖分。

【例2-11】混合汁蔗糖分为13.98%，观测锤度为16.55 °Bx，吸取样品100 mL配为250 mL，费林试剂浓度校正系数为1.028，复检耗用配制糖液33.35 mL，求混合汁还原糖分。

【解】100 mL配制糖液中含样品体积V为：

$V=100×100/250=40$（mL）

100 mL配制糖液中含样品质量W为：

$W=40(0.998\ 2+0.003\ 7×16.55+0.000\ 018\ 16×16.55^2)$
　　$=42.58$（g）

锥形瓶内溶液（33.35 mL）含蔗糖克数A为：

$A=33.35×42.58×13.98/(100×100)=1.99$（g）

按以下公式计算费林试剂校正系数：

$$f=\frac{0.994\ 7+0.041\ 8A}{1+0.067\ 2A}$$

$$f=\frac{0.994\ 7+0.041\ 8×1.99}{1+0.067\ 2×1.99}=0.950\ 7$$

20 mL费林试剂相当的还原糖量R：

$R=100×1.028×0.950\ 7=97.73$（mg）

滴定结束后锥形瓶内样品的质量W为：

$W=33.35×42.64/100=14.22$（g）

$$混合汁的还原糖分=\frac{R}{W}×100\%=\frac{97.73}{14.22×1\ 000}×100\%=0.69\%$$

【例2-12】赤砂糖的蔗糖分为88.45%，费林试剂浓度校正系数为0.986，将赤砂糖按1/2规定量进行稀释测定还原糖，滴定耗用量为31.30 mL，求赤砂糖的还原糖。

【解】因样品按半规定量进行稀释，每100 mL含样品13 g。

锥形瓶内溶液（31.30 mL）含蔗糖质量A为：

$A=31.3×13×88.45/(100×100)=3.60$（g）

按公式计算费林试剂校正系数：

$$f=\frac{0.994\ 7+0.041\ 8A}{1+0.067\ 2A}$$

$$f=\frac{0.994\ 7+0.041\ 8×3.60}{1+0.067\ 2×3.60}=0.922$$

按锥形瓶内溶液含蔗糖质量A计算样品质量W：

$W=3.6/0.884\ 5=4.07$（g）

20 mL 费林试剂相当的还原糖量 R：
$R=100\times0.986\times0.922=90.91$（mg）

$$\text{赤砂糖的还原糖分} = \frac{R}{W}\times100\% = \frac{90.91}{4.07\times1\,000}\times100\% = 2.23\%$$

【例 2-13】废蜜蔗糖分为 34.32%，每 100 mL 配制糖液含废蜜为 1.733 g，费林试剂浓度校正系数为 0.988 5，滴定耗用配制糖液为 28.65 mL，求废蜜还原糖分。

【解】锥形瓶内溶液（28.65 mL）含蔗糖克数 A 为：
$A=28.65\times1.733\times34.32/(100\times100)=0.170\,4$（g）

按公式计算费林试剂校正系数：

$$f=\frac{0.994\,7+0.041\,8A}{1+0.067\,2A}$$

$$f=\frac{0.994\,7+0.041\,8\times0.170\,4}{1+0.067\,2\times0.170\,4}=0.990$$

20 mL 费林试剂相当的还原糖量 R：
$R=100\times0.988\,5\times0.990=97.86$（mg）

滴定结束后锥形瓶内样品的质量 W 为：
$W=28.65\times1.733/100=0.497$（g）

$$\text{废蜜还原糖分} = \frac{R}{W}\times100\% = \frac{97.86}{0.497\times1\,000}\times100\% = 19.69\%$$

（6）注意事项。滴定终点时溶液总体积一定要保持恒定（75 mL）。其他与兰-艾农法相同。

任务 2.5　中和汁硫熏强度的测定

● 企业案例

厂长发现今天清汁的质量很不好，相当混浊，他估计与最近甘蔗质量和清净过程的硫熏强度有关，他让化验室主任和小明认真地去采集中和汁，然后分析硫熏强度，用来判断他的想法是否准确。

● 任务目标

通过本任务的学习，学生达到以下目标：
（1）了解硫熏强度的定义。
（2）掌握硫熏强度的测定方法。

● 素质目标

养成全面考虑问题的习惯。

● 任务描述

测定中和汁硫熏强度。

● 程序与方法

（1）滴定管：50 mL。
（2）锥形瓶：150 mL。
（3）量筒：10 mL。
（4）1/64 mol/L I_2 标准溶液。
（5）10 g/L 淀粉溶液。
（6）步骤。用量筒量取 10 mL 硫熏汁，倒入 150 mL 锥形瓶中，再以该量筒量取 30 mL 蒸馏水，倒入锥形瓶中将样品稀释，使用 pH 试纸测试；若为碱性，则加入适量的稀醋酸（或稀盐酸）中和，使其呈酸性。然后加入 10 g/L 淀粉溶液数滴，随即用 1/64 mol/L I_2 标准溶液滴定至溶液呈蓝色为止，记录滴定所耗用碘标准溶液的体积。

1/64 mol/L I_2 标准溶液 1 mL 相当于 1 mg SO_2。

（7）数据处理及计算。

样品	消耗 1/64 mol/L I_2 标准溶液体积
1	
2	
3	

> 思考
>
> 硫熏强度是亚硫酸法重要的控制指标，请同学们思考为何这个指标如此重要。

2.5.1 任务相关知识——硫熏强度的测定

硫熏强度是表示蔗汁中吸收二氧化硫的数量，是亚硫酸法糖厂直接影响提净效果的重要工艺条件，通过硫熏强度的测定，可以了解硫熏中和工艺情况并加以控制，以获得良好的提净效果。

1. 基本原理

本方法基于碘量法的原理，其基本反应为：

$$I_2 + 2e^- = 2I^-$$

I_2 是较弱的氧化剂，利用 I_2 标准溶液可以直接滴定有较强还原性的 SO_2，其反应化学方程式如下：

$$I_2 + SO_2 + 2H_2O = 2I^- + SO_4^{2-} + 4H^+$$

此反应可用淀粉作为指示剂，淀粉与 I_2 反应形成蓝色络合物，根据蓝色的出现指示终点。

2. 主要仪器设备及试剂

（1）滴定管：50 mL。
（2）锥形瓶：150 mL。
（3）量筒：10 mL。
（4）1/64 mol/L I_2 标准溶液。
（5）10 g/L 淀粉溶液。

3. 测定步骤

用量筒量取 10 mL 硫熏汁，倒入 150 mL 锥形瓶中，再以该量筒量取 30 mL 蒸馏水，倒入锥形瓶中将样品稀释，使用 pH 试纸测试；若为碱性，则加入适量的稀醋酸（或稀盐酸）中和，使其呈酸性，然后加入 10 g/L 淀粉溶液数滴，随即用 1/64 mol/L I_2 标准溶液滴定至溶液呈蓝色为止，记录滴定所耗用碘标准溶液的体积。

1/64 mol/L I_2 标准溶液 1 mL 相当于 1 mg SO_2。

4. 计算方法

硫熏强度 = 滴定用去 1/64 mol/L I_2 标准溶液体积。

如滴定耗用的碘液为 18 mL，即硫熏强度为 18 mL，则表明中和汁中 SO_2 的含量为 1.8 g/L。

2.5.2 任务拓展知识——碱度和全钙量的测定

碱度是表示糖汁呈碱性反应时所含的碱量。碱度一般以 100 mL 溶液中含有氧化钙的质量来表示，也可用每升样品中含 CaO 的质量来表示。

测定预灰汁、一碳汁、一碳清汁的碱度，了解石灰的使用情况以及一碳饱充的效果，从而改变饱充的控制条件，以获得良好的清净效果。

全钙量是表示 100 mL 糖汁中所含未反应的氧化钙及已形成碳酸钙的氧化钙和部分碱金属相当氧化钙的质量。全钙量用于碳酸法糖厂一碳饱充汁的分析。分析方法与碱度基本相同。

碱度测定基于酸碱滴定反应原理，用已知浓度的标准酸溶液滴定待测定的样液，直到酸碱中和反应进行完全为止，然后根据所消耗标准溶液的浓度和体积，按照化学反应的计量关系，计算出待测组分的含量。

1. 主要设备与试剂

（1）滴定管：50 mL，分度 0.1 mL。

（2）锥形瓶：150 mL、250 mL。

（3）量筒：10 mL。

（4）硫酸溶液：1/56 mol/L，1/5.6 mol/L。

（5）酚酞指示液：1%。

（6）甲基橙指示液：0.1%。

2. 碱度测定步骤

（1）预灰汁、一碳汁、一碳清汁碱度测定。将样品搅匀，过滤。用移液管取 10 mL 滤液放入 150 mL 锥形瓶中。在锥形瓶内加入 20 mL 蒸馏水稀释，加酚酞指示剂 1～2 滴，摇匀，溶液呈红色。用酸式滴定管装 1/56 mol/L 的 H_2SO_4 标准溶液，调好零点，在不断摇动下滴定锥形瓶内溶液，直至锥形瓶内溶液红色刚好消失为止，记录所消耗硫酸标准的体积 V_0（重复测定三次，V 取平均值）。

（2）碱度计算。

1）每毫升 1/56 mol/L 的 H_2SO_4 标准溶液相当于 0.001 g CaO。

2）样品碱度以 100 mL 糖汁中含相当的氧化钙质量表示。

$$碱度 = \frac{0.001 \times V}{10} \times 100$$

式中　V——滴定所耗用 1/56 mol/L 的 H_2SO_4 标准溶液体积（mL）。

3. 全钙量的测定

（1）一碳汁全钙量测定步骤。量筒量取 10 mL 未经过滤的样品，移入 250 mL 锥形瓶中，加入 50 mL 蒸馏水稀释，加甲基橙指示剂 2～3 滴，摇匀后用 1/5.6 mol/L 的 H_2SO_4 标准溶液滴定，并不断摇荡，直至溶液中出现微红色并在 30 s 内不消失为止，记录所消耗的标准溶液的体积 V_0（重复测定三次，V 取平均值）。

（2）全钙量计算。每毫升 1/5.6 mol/L 的 H_2SO_4 标准溶液相当于 0.01 g CaO。

$$全钙量 = \frac{0.01 \times V}{10} \times 100$$

式中　V——滴定所耗用 1/5.6 mol/L 的 H_2SO_4 标准溶液体积（mL）。

注意：测定全钙量的样品不得过滤，否则影响分析结果的准确性。

任务 2.6　糖汁 pH 的测定

● 企业案例

今天的生产报表反映蔗糖损失很大,化验室主任马上安排小明去采集各种中间产品进行 pH 的测定,这时他应该选择什么仪器?又该如何检测呢?

● 任务目标

通过本任务的学习,学生达到以下目标:
理解、掌握糖汁 pH 的测定原理及方法。

● 素质目标

掌握对比分析学习法的习惯。

● 任务描述

使用 pH 计对在制品进行酸度的测定。

● 程序与方法

(1)原料:糖汁。

(2)设备及仪器:酸度计(测量范围 pH=0~14,精度范围 pH=0.01~0.1)、玻璃电极、甘汞电极、电磁搅拌器。

(3)试剂:苯二钾酸氢钾缓冲溶液(pH = 4.00)、硼砂缓冲溶液(pH = 9.22)、混合磷酸盐缓冲溶液(pH = 6.88)。

上述三种缓冲溶液在不同温度下 pH 的变化见表 2-6。

表 2-6　三种缓冲溶液在不同温度下 pH

温度/℃	苯二钾酸氢钾缓冲溶液 0.05 mol/L	硼砂缓冲溶液 0.01 mol/L	混合磷酸盐缓冲溶液 0.025 mol/L
5	4.00	9.39	6.95
10	4.00	9.33	6.92
15	4.00	9.27	6.90
20	4.00	9.22	6.88
25	4.01	9.18	6.86
30	4.01	9.14	6.85

续表

温度 /℃	苯二钾酸氢钾缓冲溶液 0.05 mol/L	硼砂缓冲溶液 0.01 mol/L	混合磷酸盐缓冲溶液 0.025 mol/L
35	4.02	9.10	6.84
40	4.03	9.07	6.84
45	4.04	9.04	6.83
50	4.06	9.01	6.83
55	4.08	8.99	6.84
60	4.10	8.96	6.84

(4）步骤。

1）样液的配制。在测定固体（白砂糖除外）及黏性大的样品的 pH 时，须将样品 1∶1 稀释，糖厂的其他在制品可在原浓度测定 pH。

2）酸度计的校正。在进行测量前，先以合适的缓冲溶液校正酸度计，使仪器标度所示的 pH 恰好是缓冲溶液的 pH。

3）测定。调节仪器的温度调节器，使所指示的温度与待测溶液的温度相同。将用待测溶液冲洗过的电极浸入溶液中，开动搅拌器搅拌溶液，按下测量开关，待仪器稳定地显示出溶液的 pH 时，即可记录读数。具体操作步骤按仪器使用说明书进行。

(5）注意事项。

1）初次使用或久置重用的玻璃电极应先在蒸馏水中浸泡 24 h 以上，以稳定其不对称电位。用时球泡部分应全部浸在被测溶液中；用完或短期不用可放回蒸馏水中浸泡。

2）甘汞电极不要经常浸于蒸馏水中，应保留少许氯化钾晶体，经常保持氯化钾溶液饱和，电极内溶液中不能有气泡。

3）校正酸度计用的缓冲溶液 pH 最好与被测溶液的 pH 相近。缓冲溶液如有沉淀物产生，则不能使用。

4）使用甘汞电极时，将上端加氯化钾溶液处的小橡皮塞拔去，使毛细管保持足够的液位差，防止样液进入毛细管而影响测定结果。

5）玻璃电极玻璃膜脆、薄，极易损坏，使用时应小心。安装玻璃电极时，其下端球泡应比甘汞电极陶瓷芯端稍高些，以免碰坏。球泡玻璃膜如果黏有油污，可先浸入乙醇内，再浸入乙醚或四氯化碳中，最后浸入乙醇后，用蒸馏水冲洗干净。

◎ 思考

pH 计用毕或短期不用是否可放回蒸馏水中浸泡？请同学们养成爱护仪器设备、精心维护仪器设备的好习惯。

任务 2.7　蔗汁磷酸值的测定

● 企业案例

今天的生产报表反映在产品稳定的情况下，最近消耗了很多磷酸，厂长判断有可能磷酸过多使用，随后立即让化验室主任安排小明去采集蔗汁进行磷酸值的测定，这时他该怎么做？

● 任务目标

通过本任务的学习，学生达到以下目标：
（1）了解蔗汁磷酸值的测定原理。
（2）掌握蔗汁磷酸值的测定方法及计算。

● 素质目标

养成透过现象看本质的习惯。

● 任务描述

测定蔗汁的自然磷酸值。

● 程序与方法

（1）原料：蔗汁。
（2）设备及仪器：光电比色计或分光光度计一台、比色皿一套、25 mL 比色管、移液管、100 mL 容量瓶。
（3）试剂：磷酸二氢钾标准溶液、25 g/L 钼酸铵硫酸混合溶液、50 g/L 氯化亚锡甘油溶液。
（4）步骤。

1）蔗汁配制液：用吸管吸取 5 mL 蔗汁，移入 100 mL 容量瓶中，加水至标线，摇匀、过滤，弃去最初的滤液，吸取 5 mL 滤液，移入另一个 100 mL 容量瓶中，加水至约 95 mL，准确加入 25 g/L 钼酸铵硫酸混合溶液 2 mL，随即加入 50 g/L 氯化亚锡甘油溶液 5 滴，然后加水至标线，充分摇匀。静置 5 min，该配制液为蔗汁的 400 倍稀释液。

2）磷酸值标准色液的配制：用吸管分别吸取磷酸二氢钾标准溶液 0.5 mL、1.0 mL、1.5 mL、2.0 mL、2.5 mL、3.0 mL、3.5 mL、4.0 mL 于 8 个 100 mL 容量瓶中，加入蒸馏水稀释后定容，则各个容量瓶中 P_2O_5 含量分别为 50 mg/L、100 mg/L、150 mg/L、200 mg/L、250 mg/L、300 mg/L、350 mg/L、400 mg/L。将上述溶液同样稀释 400 倍，并加入钼酸铵硫酸混合溶液及氯化亚锡甘油溶液，静置 5 min 后便成为颜色由浅到深的蔚蓝色标准色液。

3）比色法：使用分光光度计或光电比色计，能更迅速、准确地测定蔗汁磷酸值。但须先行测定并绘制磷酸值曲线。由于色值与吸光度成正比，因此该曲线为一直线。

①磷酸值曲线的绘制：在比色皿中盛入蒸馏水，用红色滤光片（波长约为 660 nm）校正零点。然后用上述已知磷酸值的标准色液分别测出其吸光度。用数理统计的回归法，根据回归方程式作出较准确的磷酸值曲线。

②蔗汁磷酸值的测定：在比色皿中盛入未加任何试剂的空白样液，用红色滤光片（波长约为 660 nm）校正零点。然后测定蔗汁配制液的吸光度，从磷酸值曲线中查取对应的磷酸值，或代入回归方程式计算出磷酸值。

4）作曲线图。

任务相关知识——蔗汁成分分析

蔗汁中的各种成分对制糖工艺过程、节能、糖分回收及成品糖的质量等都有重要的影响。通过对蔗汁成分的分析，为制订最佳的澄清工艺条件提供依据。

1. 磷酸值的测定

（1）目的。蔗汁中含可溶性磷酸盐的量称为磷酸值，以 P_2O_5 表示。糖厂中一般只分析蔗汁的自然磷酸值，即能迅速生成磷酸根离子并在提净过程中生成磷酸钙沉淀的磷酸盐。自然磷酸值是蔗汁有效澄清的一个重要因素，一般要求蔗汁中 P_2O_5 的含量应不少于 300 mg/L，若蔗汁中 P_2O_5 的含量不足此数，则可假设适当数量的磷酸或过磷酸钙清液，以帮助澄清的正常进行。所以，测定蔗汁酸磷值的目的在于为制订最佳提净工艺条件提供依据。

（2）常用方法。测定磷酸值的常用方法为钼酸铵比色法。

（3）钼酸铵比色法。

1）基本原理。在一定的酸度下，磷酸盐与所加入的钼酸铵作用，生成黄色晶状的磷钼酸铵沉淀。反应式为：

$$PO_4^{3-}+3NH_4^++12MoO_4^{2-}+24H^+ = (NH_4)_3PO_4 \cdot 12MoO_3 \cdot 6H_2O+6H_2O$$

磷钼铵很容易被适当的还原剂（如氯化亚锡等）所还原，生成深蓝色的磷钼蓝——$(MoO_2 \cdot 4MoO_3) \cdot H_3PO_4$。其蓝色深浅与磷酸盐的含量成正比，故可用比色法测出蔗汁中可溶性磷酸盐的含量。反应式为：

$$(NH_4)_3PO_4 \cdot 12MoO_3+11H^++4Sn^{2+} = (MoO_2 \cdot 4MoNO_3)_2 \cdot H_3PO_4+2MoO_2+4Sn^{4+}+3NH_4^++4H_2O$$

2）主要仪器、设备及试剂。

①设备及仪器：光电比色计或分光光度计一台、比色皿一套、25 mL 比色管、移液管、100 mL 容量瓶。

②试剂：磷酸二氢钾标准溶液、25 g/L 钼酸铵硫酸混合溶液、50 g/L 氯化亚锡甘油溶液。

3）测定方法。

①蔗汁配制液。用吸管吸取 5 mL 蔗汁，移入 100 mL 容量瓶中，加水至标线，摇匀、过滤，弃去最初的滤液，吸取 5 mL 滤液，移入另一个 100 mL 容量瓶中，加水至约 95 mL，准确加入 25 g/L 钼酸铵溶液 2 mL，随即加入 50 g/L 氯化亚锡甘油溶液 5 滴，然后加水至标线，充分摇匀。静置 5 min，该配制液为蔗汁的 400 倍稀释液。

②磷酸值标准色液的配制。用吸管分别吸取磷酸二氢钾标准溶液 0.5 mL、1.0 mL、1.5 mL、2.0 mL、2.5 mL、3.0 mL、3.5 mL、4.0 mL 于 8 个 100 mL 容量瓶中，加入蒸馏

水稀释后定容,则各个容量瓶中 P_2O_5 含量分别为 50 mg/L、100 mg/L、150 mg/L、200 mg/L、250 mg/L、300 mg/L、350 mg/L、400 mg/L。将上述溶液同样稀释 400 倍,并加入钼酸铵硫酸溶液及氯化亚锡甘油溶液,静置 5 min 后便成为颜色由浅到深的蔚蓝色标准色液。

③比色法。使用分光光度计或光电比色计,能更迅速、准确地测定蔗汁磷酸值。但须先行测定并绘制磷酸值曲线。由于色值与吸光度成正比,因此该绘图为直线。

磷酸值曲线的绘制:在比色皿中盛入蒸馏水,用红色滤光片(波长约为 660 nm)校正零点。然后用上述已知磷酸值的标准色液分别测出其吸光度。用数理统计的回归法,根据回归方程式作出较准确的磷酸值曲线。

蔗汁磷酸值的测定:在比色皿中盛入未加任何试剂的空白样液,用红色滤光片(波长约为 660 nm)校正零点。然后测定蔗汁配制液的吸光度,从磷酸值曲线中查取对应的磷酸值,或代入回归方程式计算出磷酸值。

④注意事项。绘制磷酸值曲线和测定蔗汁样品磷酸值时,比色皿厚度应选择使仪器透光度读数在 20%～80%之间,并应使用配套的比色皿,配套使用的比色皿在同一光径的透光度之差不大于 0.2%。如改用另一规格的比色皿,则要重新制作曲线。

2. 钙、镁盐含量的测定

(1)目的。糖汁中钙、镁盐的存在,在加热蒸发设备中会形成积累,影响传热效率,增加成品糖灰分。镁盐含量高时,会使赤砂糖含有苦涩味。因此,分析其含量,以便在提净工艺过程中采取相应的措施将其除去,减少其影响。

(2)基本原理。本法基于络合滴定的原理。用络合剂作为标准溶液滴定样品,使之与蔗汁中的钙、镁离子形成络合物。乙二胺四乙酸二钠盐(简称 EDTA)是一种羧络合剂,能在不同 pH 条件下,与钙、镁盐等阳离子形成稳定的络合物,蔗汁中的钙、镁盐大多以 Ca^{2+}、Mg^{2+} 的形式与之形成紫红色的络合物,当全部 Ca^{2+}、Mg^{2+} 都已反应完毕,指示剂变为原有的蓝色,即达到终点。其反应式如下:

$$M^{2+} + HIn^{2-} \longrightarrow MIn^- + H^+$$
(蓝色)　　(紫蓝色)

滴定至终点时:

$$MIn^- + H_2Y^{2-} \longrightarrow MY^{2-} + HIn^{2-} + H^+$$
(紫蓝色)　　　　(蓝色)

上述反应式中,M^{2+} 代表钙、镁离子;HIn^{2-} 代表铬黑 T 指示剂,H_2Y^{2-} 代表 EDTA。

(3)主要仪器、设备及试剂。

1)滴定管:50 mL。

2)锥形瓶:250 mL。

3)0.01 mol/L EDTA 溶液。

4)150 g/L 氢氧化钠溶液。

5)20 g/L 盐酸羟胺溶液。

6)钙指示剂。

7)铬黑 T 指示剂。

8)铵盐缓冲液。

（4）测定方法。

1）钙、镁盐总量的测定。准确吸取蔗汁样品 10 mL，移入 250 mL 锥形瓶中，加入约 100 mL 蒸馏水和 5 mL 铵盐缓冲液，混匀后，再加入 20 g/L 盐酸羟胺 3 滴和铬黑 T 指示剂约 0.1 g，待混合溶解后，溶液呈紫红色。用 0.01 mol/L EDTA 溶液滴定，至溶液由酒红色突然变为蓝色为止。另用 100 mL 蒸馏水按上述方法做空白试验。前后滴定所消耗 EDTA 体积之差，为滴定钙、镁盐总量所耗用的 EDTA 体积。

2）钙盐含量的测定。准确吸取蔗汁样品 10 mL，移入 250 mL 锥形瓶中，加入约 100 mL 蒸馏水及 2 mL 150 g/L 氢氧化钠溶液，混匀后，放置约 3 min，使氢氧化镁沉淀析出，再加入 20 g/L 盐酸羟胺溶液 3 滴及钙指示剂约 0.1 g，待混合溶解后，溶液呈葡萄红色。用 0.01 mol/L EDTA 溶液滴定，至呈现纯蓝色为止。另用 100 mL 蒸馏水按上述方法做空白试验。前后滴定所消耗的 EDTA 体积之差，为滴定钙盐所消耗的 EDTA 体积。

镁盐的含量可用滴定钙、镁盐总量所耗用 EDTA 量与滴定钙盐所耗用 EDTA 量之差来计算。

（5）计算。样品中钙盐（或镁盐）的含量按下列公式计算，以 100 g 蔗汁中所含 CaO（或 MgO）的质量表示：

样品中钙盐（或镁盐）的含量以 100 g 蔗汁中所含 CaO（MgO）的质量表示：

$$CaO = \frac{C \times V_2 \times 0.056\,08 \times 100}{W}$$

$$MgO = \frac{C \times (V_1 - V_2) \times 0.040\,32 \times 100}{W}$$

式中　V_1——滴定钙、镁盐总量所耗用 EDTA 体积（mL）；

　　　V_2——滴定钙盐所耗用 EDTA 体积（mL）；

　　　C——EDTA 标准溶液的浓度（mol/L）；

　　　W——蔗汁样品质量（g）；

　　　0.056 08——1 mmol CaO 的质量（g）；

　　　0.040 32——1 mmol MgO 的质量（g）。

（6）注意事项。

1）测定钙、镁盐含量的蔗汁样品必须在预灰前采集。预灰后的蔗汁中钙盐含量已发生变化，影响分析结果真实性。

2）样品中不能加入汞盐作防腐剂，因为汞离子能消耗 EDTA，会使分析结果偏高。

3）因蔗汁或其他糖汁本身有颜色，故在滴定终点时很难呈蓝色，一般为灰绿色，在判断终点时要仔细观察。同时指示剂用量不能太多，以免颜色过深而影响重点的判断。

4）蔗汁中的某些离子会对分析起干扰作用，其中主要为铁、铝、锰。必要时，可加入氰化钾除去铁，加入三乙醇胺除去铝，加入盐酸羟胺除去锰，以减少它们的干扰。

甘蔗日常分析

水分测定

在制品的色值及混浊度测定

近红外测定中间制品

模块 3
成品糖理化指标分析

任务 3.1　白砂糖蔗糖分测定

● 企业案例

有客户反映最近的产品蔗糖分不达标，因此主任安排小明对该批号白砂糖的蔗糖分进行测定，小明又应该选择什么仪器和使用什么方法来测定呢？

● 任务目标

通过本任务的学习，学生达到以下目标：
(1) 进一步熟悉旋光检糖仪的使用方法。
(2) 掌握一次旋光法测定白砂糖蔗糖分的方法及其计算。

● 素质目标

养成对比分析学习方法的习惯。

● 任务描述

使用旋光仪对白砂糖的蔗糖分进行分析。

● 程序与方法

(1) 原料：白砂糖。
(2) 设备及仪器：旋光仪、200 mm 观测管、分析天平、水银温度计（0～50 ℃，分度为 0.1 ℃）、100 mL 容量瓶、量筒、烧杯。
(3) 试剂：蒸馏水或去离子水，无水乙醇。
(4) 步骤：称取样品白砂糖 [（26.000±0.002）g] 于干燥洁净的小烧杯中，加蒸馏水 40～50 mL，以细玻璃棒搅拌使其完全溶解。倒入 100 mL 的容量瓶中，用少量蒸馏水分 3～5 次冲洗烧杯及玻璃棒，洗水一并倒入容量瓶，每次倒入洗水后，摇匀瓶内溶液，直至加蒸馏水至容量瓶标线下方，至少放置 10 min 使其达到室温，然后加蒸馏水至容量瓶标线下约 1 mm 处，确保容量瓶颈部已洗净，小心勿使溶液夹带气泡，有气泡时，可用乙醇或乙醚消除。垂直拿住容量瓶颈部的上方，使容量瓶的标线与操作者眼睛成水平。对着明亮的背景观察，用长咀的滴管加水至标线，用干净的滤纸吸干容量瓶瓶颈的内壁，将塞子塞紧，充分摇匀。如发现混浊，用滤纸过滤，漏斗上须加盖表面玻璃，将最初的 10 mL 溶液弃去，收集之后的滤液 50～60 mL。

用待测的溶液将观测管至少冲洗 2 次，然后将溶液装满观测管，注意不使观测管内夹带气泡。将观测管置于检糖仪中，测定旋光读数后，立即测定观测管内溶液的温度（准确至 0.1 ℃），记录以上数据。

（5）数据处理及计算。

项目	数据及结果
观测糖度读数 /°Z	
观测 P_1 时糖液温度 /℃	
蔗糖分 /%	

3.1.1 任务相关知识——白砂糖蔗糖分测定

利用蔗糖的旋光性，在规定条件下以国际糖度标尺刻度为 100 °Z 的检糖仪，测定糖样品水溶液的旋光度，从而测定溶液中蔗糖的浓度。白砂糖的纯度已很高，所含杂质很少，对旋光法测定的干扰甚微，故可用一次旋光法测定白砂糖的蔗糖分。但由于糖液的旋光性受温度的影响，所以必须对测定的结果进行温度校正。

1. 仪器设备

（1）检糖仪。检糖仪应是根据国际糖度标尺，按糖度（°Z）刻度的，测量范围为 –30～+120 °Z，并用标准石英管加以核准，可选用三种形式。

1）装有可调整分析器即检偏器的检糖仪（圆盘式旋光计），采用单色光源（波长为 540～590 nm），通常采用绿色的汞光或黄色的钠光。

2）石英楔检糖仪：
①配有单色光源（波长为 540～590 nm）；
②配有白炽灯作为光源，用适当的滤色器分离出有效波长为 587 nm 的光；

3）装有法拉第线圈作为补偿器的检糖仪，采用单色光源（波长为 540～590 nm）。

注：旧糖度 °S 刻度的检糖仪仍然可以使用，但读数 °S 须乘上一个系数 0.999 71 转换为 °Z。

（2）容量瓶。容量：（100.00±0.02）mL，应分别用（20.00±0.1）℃ 的水称量加以校正。容量瓶的容量在（100.00±0.01）mL 范围内，不必更正便可使用；超出此范围应采用与 100.00 mL 相应的校正系数加以更正，方可使用。

（3）旋光观测管。长度：（200.00±0.02）mm，须由法定的计量机构出具合格证明，或用具有该项证明的观测管进行比较检验。

（4）分析天平：感量 0.1 mg。

2. 试剂

蒸馏水：不含旋光物质。

3. 检糖仪的校准

检糖仪要用经法定的计量机构检定合格的标准石英管校准。

（1）石英管旋光度的温度校正。应测定使用检糖仪（没有石英补偿器）读取石英管读数时的温度，并准确至 0.2 ℃，测定旋光度时环境及糖液的温度尽可能接近 20 ℃，应在 15～25℃ 的范围内。如果这个温度与 20 ℃ 相差大于 ±0.2 ℃，则采用以下公式进行标准石英管旋光度的温度校正：

$$\alpha_t = \alpha_{20}[1+1.44\times10^4(t-20)]$$

式中　α_t——t（℃）时标准石英管的旋光值（°Z）；

α_{20}——20 ℃ 时标准石英管的旋光值（°Z）；

t——读数时石英管的温度（℃）。

（2）不同波长下石英管读数（°Z）的换算系数。石英管的糖度读数在不同波长下以绿色汞光（波长 546 nm）为基准，使用其他光源时，应除以表 3-1 中相应的系数进行换算。

表 3-1　不同波长下石英管读数换算系数

光源	波长 /nm	换算系数
经过滤白炽光	587	1.001 809
黄色钠光	589	1.001 898
氦/氖激光	633	1.003 172

4. 溶液的配制

称取样品（白砂糖）26.000 g 于干燥洁净的小烧杯中，加入蒸馏水 40～50 mL，使之完全溶解。移入 100 mL 的容量瓶中，用少量蒸馏水冲洗烧杯及玻璃棒不少于 3 次，并将洗水一并移入容量瓶，摇匀瓶内溶液后，加蒸馏水至容量瓶标线附近。放置至少 10 min 使之与室温平衡，然后加蒸馏水至容量瓶标线下约 1 mm 处。有气泡时，可用乙醚或乙醇消除。加蒸馏水至标线，充分摇匀。

如果发现溶液混浊，用滤纸过滤，漏斗上须加盖表面皿，以减少溶液的蒸发。将最初的 10 mL 滤液弃去，收集之后的滤液 50～60 mL。

5. 旋光度的测定

用待测的溶液将旋光观测管至少冲洗 2 次后将观测管盛满，注意观测管内不能夹带气泡。将旋光观测管置于检糖仪中，目测检糖仪测定 5 次，读数准确至 0.05 °Z；如用自动检糖仪，在测定前，要有足够的时间使仪器达到稳定。

测定旋光读数后，立即测定观测管内溶液的温度，并记录至 0.1 ℃。

6. 计算及结果表示

测定旋光度时环境及糖液的温度尽可能接近 20 ℃，应在 15 ～ 25 ℃ 的范围内。如果这个温度与 20 ℃ 相差大于 ±0.2 ℃，则应进行温度校正。

根据旋光仪的不同，可以分别采用不同的温度校正公式，计算结果以％表示，取一位小数。

采用石英楔检补偿器的检糖仪：

$$P=P_t [1+0.000\,32(t-20)]$$

没有石英楔检补偿器的检糖仪：

$$P=P_t [1+0.000\,19(t-20)]$$

式中　P——蔗糖分（％）；
　　　P_t——观测旋光读数（°Z）；
　　　t——观测时糖液温度（℃）。

7. 允许误差

两次测定值之差不应超过其平均值的 0.05%。

任务 3.2　白砂糖还原糖分测定

● 企业案例

有客户反映某批次白砂糖短期内返潮，主任估计是产品的还原糖分超标，他安排小明对该批次白砂糖进行还原糖的测定，小明又应该选择什么仪器和使用什么方法来测定呢？

● 任务目标

通过本任务的学习，学生达到以下目标：
（1）了解奥夫纳尔法的测定原理。
（2）掌握奥夫纳尔法的测定方法及计算。

● 素质目标

养成对比分析学习的习惯。

● 程序与方法

（1）原料：白砂糖。

（2）设备及仪器：250 mL 碘量瓶、50 mL 滴定管、25 mL 移液管、300 mL 锥形瓶、电炉。

（3）试剂：奥夫纳尔试剂（铜溶液）、0.032 3 mol/L $Na_2S_2O_3$ 溶液、0.016 15 mol/L I_2 溶液、1 mol/L HCl 溶液、10 g/L 淀粉溶液、冰乙酸。

（4）步骤。称取白砂糖 10.00 g，用 50 mL 蒸馏水溶解于 300 mL 锥形瓶中，糖液含转化糖不超过 20 mg，然后加入 50 mL 奥夫纳尔试剂，充分混合，用小烧杯倒置覆盖其上，在电炉上加热，使其 4～5 min 内沸腾，并继续准确地煮沸 5 min（煮沸开始的时间，不是从瓶底发生气泡时算起，而是从液面上冒出大量的气泡算起）。取出，置冷浴中冷却至室温（不要摇动）。再取出，加入冰乙酸 1 mL，在不断摇动下，加入准确计量的碘溶液，视还原的铜量而加入 5～30 mL，其数量以确保碘液过量为准。用量杯沿锥形瓶壁加入 15 mL 浓度为 1 mol/L 的盐酸。立即盖上小烧杯，放置 2 min，不时摇动溶液，然后用 $Na_2S_2O_3$ 溶液滴定过量的碘，滴定至溶液呈浅黄绿色时，加入 2～3 mL 淀粉指示剂（此时呈深蓝色），继续滴定至蓝色刚褪尽为止，记录滴定耗用体积。

（5）数据处理及计算。

项目	数据及结果
滴定用硫代硫酸钠溶液浓度 /mol/L	
滴定用硫代硫酸钠溶液容积 /mL	

续表

项目	数据及结果
滴定用 0.032 3 mol/L 硫代硫酸钠溶液容积 /mL	
加入碘液浓度 /（mol·L^{-1}）	
加入碘液容积 /mL	
加入 0.016 15 mol/L 碘液容积 /mL	
样品还原糖 /%	

● 思考

中间制品和最终产品的还原糖分测定有何异同？请使用对比分析方法进行解答。对比分析方法又体现了什么样的哲学原理？

3.2.1 任务相关知识——白砂糖还原糖分测定

白砂糖还原糖分测定采用奥夫纳尔法。奥夫纳尔法适用于有大量蔗糖存在条件下还原糖的测定。在现行的国家及行业标准中，对各类成品糖中还原糖的测定均指定用奥夫纳尔法。

1. 基本原理

奥夫纳尔法也是一种铜还原法，但其对还原糖的定量方法与兰-艾农法不同。奥夫纳尔法首先利用过量奥氏试剂中的铜离子与样品中的还原糖作用生成氧化亚铜，然后通过采用碘量法去定量生成的氧化亚铜（Cu_2O）含量，从而确定样品中还原糖的含量。该法使用的奥氏试剂由于碱性较弱，可将铜盐与酒石酸钾钠等混合配成溶液，仍可保存较长时间，而酒石酸钾钠也不会将二价铜还原。另外，由于试剂的碱性较弱，还原糖与铜盐的反应过程进行得较慢，蔗糖被氧化的量也大为减少，测定结果比较准确。目前在糖厂中使用该方法进行白砂糖的还原糖分的测定。

测定时，首先取一定量的样品与过量的奥氏试剂共热，样品中的还原糖与铜离子生成 Cu_2O 沉淀，生成的 Cu_2O 沉淀量与样品中还原糖的量成正比。

$$还原糖（定量）+Cu^{2+} \longrightarrow Cu_2O（沉淀）$$

反应完成后，加入乙酸，使过量的 Cu^{2+} 离子生成乙酸铜络合物而失去氧化能力，防止蔗糖被氧化。

用盐酸将 Cu_2O 沉淀溶解：

$$Cu_2O+2HCl \longrightarrow Cu_2Cl_2+H_2O$$

然后准确加入过量的碘标准液将 Cu_2Cl_2 氧化。因加入的碘已准确计量，只要设法求出过量的碘的量，就可以知道与亚铜离子反应的碘的量。

$$Cu_2Cl_2+2KI+I_2 \longrightarrow 2CuI_2+2KCl$$

最后用硫代硫酸钠标准溶液滴定反应中过量的碘，用淀粉溶液作指示剂。

$$I_2(过剩的) + 2Na_2S_2O_3 \longrightarrow Na_2S_4O_6 + 2NaI$$

由消耗的碘就可以知道与还原糖生成的 Cu_2O 量,从而求出还原糖的量。

2. 奥氏试剂

奥氏试剂也称铜溶液,提供反应所需的铜离子。奥氏试剂由硫酸铜、酒石酸钾钠、无水碳酸钠、磷酸氢二钠(或无水氯化钠)组成。奥氏试剂的碱性较弱,所以没有必要将其制备为分别存放的甲液与乙液。因在测定过程中不需要对 Cu^{2+} 离子进行计量,所以奥氏试剂配好后就可以直接使用,不用标定。但要求配制后在沸水浴中加热灭菌 2 h,并储存于棕色的瓶中。

3. 主要仪器设备

(1) 250 mL 碘量瓶。
(2) 50 mL 滴定管,酸式、碱式各一支。
(3) 25 mL 移液管。
(4) 电炉。

4. 主要试剂

(1) 奥氏试剂(铜溶液)。称取结晶硫酸铜($CuSO_4 \cdot 5H_2O$)5 g,酒石酸钾钠($NaKC_4H_4O_6 \cdot 4H_2O$)300 g,碳酸钠(Na_2CO_3)10 g,磷酸氢二钠(Na_2HPO_4)50 g,用 900 mL 冷水溶解后在沸水浴中加热灭菌 2 h,冷却后稀释为 1 000 mL,加入少量活性炭或精制硅藻土过滤,过滤后将滤液储存于有色试剂瓶中。

(2) 0.032 3 mol/L 硫代硫酸钠溶液。准确称取硫代硫酸钠($Na_2S_2O_3$)8.000 0 g,溶于水中,然后移入 1 000 mL 容量瓶,定容后用重铬酸钾标准液标定。

重铬酸钾标准液配制及标定方法。准确称取经 120℃ 干燥至恒重的基准重铬酸钾约 1.58 g(称准至 0.000 2 g),用约 100 mL 水溶解后移入 1 000 mL 容量瓶定容,摇匀后用移液管准确吸取该溶液 25 mL,注入 250 mL 碘量瓶中,加入 2 g 碘化钾及 15 mL 浓度为 2 mol/L 的硫酸,将瓶塞盖紧,轻轻摇匀后置于暗处反应 5 min,加入 100 mL 水,用待标定的硫代硫酸钠滴至淡黄色,加入 5 g/L 淀粉指示剂 2 mL(变为蓝黑色),继续用硫代硫酸钠滴至溶液变为亮绿色。此项标定应进行多次,直至两次的相对误差在 0.2% 以内。同时作空白试验。

硫代硫酸钠标准液的量浓度按下面式子计算:

$$C = \frac{25m}{49.03(V-V_1)}$$

式中 C——硫代硫酸钠的量浓度($mol \cdot L^{-1}$);

m——基准重铬酸钾质量(g);

V——标定时耗用硫代硫酸钠量(mL);

V_1——空白试验时耗用硫代硫酸钠量(mL);

25——换算系数;

49.03——重铬酸钾摩尔质量(1/6 摩尔)(g/mol)。

(3) 0.016 15 mol/L 碘溶液。准确称取化学纯碘 4.100 0 g,碘化钾 20 g,溶于少量水中。然后移入 2 000 mL 容量瓶，加水至标线，摇匀后用标准硫代硫酸钠溶液标定并储存于棕色瓶中。

由于碘几乎不溶于水，但能溶于碘化钾溶液中，所以配制时一定要加入一定量的碘化钾。另外由于碘容易挥发，准确称量有困难，一般是配制浓度较高的储备液，使用前再进行稀释及标定。

（4）10 g/L 淀粉溶液。

（5）冰乙酸。

（6）1 mol/L 盐酸。

5. 测定步骤

称取一定量样品（精制白砂糖、白砂糖、方糖 10.00 g，白冰糖 5.00 g，黄冰糖 1.00 g），用 50 mL 蒸馏水溶解于 300 mL 锥形瓶中。糖液含转化糖不能超过 20 mg，否则要适当减少样品的量。然后加入 50 mg 奥夫纳尔试剂，充分混合，用一烧杯倒置覆盖其上，在电炉上加热，使其在 4～5 min 内沸腾，并继续准确煮沸 5 min（煮沸开始的时间，不是从瓶底发生气泡时算起，而是从液面上冒出大量的气泡算起）。取出，置冷水中冷却至室温（不要摇动，防止生成的 Cu^+ 离子被氧化）。加入冰醋酸 1 mL，在不断摇动下，加入准确计量的碘溶液，视还原的铜量而加入 5～30 mL，其数量以确保碘液过量为准（溶液显黄棕色即为过量）。用量杯沿锥形瓶壁加入 15 mL 浓度为 1 mol/L 的盐酸。塞上瓶盖，放置于暗处反应 2 min，不时摇动溶液。然后用 $Na_2S_2O_3$ 溶液滴定过量的碘。滴至溶液呈浅黄绿色时，加入 2～3 mL 淀粉指示剂（此时呈深蓝色），继续滴定至蓝色刚褪尽为止。记录滴定耗用的 $Na_2S_2O_3$ 溶液体积。

6. 计算方法

（1）所需数据。

1）加入的碘液量 A_1（mL）；

2）碘液的浓度 M_a（mol·L^{-1}）；

3）滴定时消耗的硫代硫酸钠溶液量 B_1（mL）；

4）硫代硫酸钠的浓度 M_b（mol·L^{-1}）。

（2）计算方法。在没有蔗糖存在的情况下，与 Cu_2O 反应的每一毫升浓度为 0.032 3 mol/L 的标准碘溶液相当于样品中含有还原糖 1 mg。若有蔗糖存在，由于蔗糖也会消耗部分碘液，应按标准碘溶液的消耗量，用下面的公式计算（适用于 10 g 蔗糖）或查附表 4 获得一个校正值，从结果中将校正值减去。

$$10 \text{ g 蔗糖校正值 } C = 1.025 + 0.069\,5X - 0.001\,24X^2$$

式中　X——标准碘溶液的消耗量（mL）。

例如，对于白砂糖中还原糖的测定，称取的样品为 10.00 g，测定结果如下：

加入的碘液量为 A_1（mL）；碘液的浓度为 M_a（mol·L^{-1}）；

滴定时消耗的硫代硫酸钠溶液量为 B_1（mL）；硫代硫酸钠的浓度为 M_b（mol·L^{-1}）。

计算样品中还原糖的含量时，首先将碘液与硫代硫酸钠溶液换算为标准浓度对应的体积：

$$A = A_1 M_a / 0.016\ 15 \text{（mL）}$$
$$B = B_1 M_b / 0.032\ 3 \text{（mL）}$$

耗用的碘液量为 （$A-B$）mL

按 $X=$（$A-B$）代入蔗糖校正值计算公式（或查附表4），得 10 g 蔗糖校正值 C。

则样品的还原糖分为：

$$\text{还原糖分（\%）} = \frac{\text{样品中还原糖重}}{\text{样品重}} \times 100\% = \frac{0.001 \times (A-B-C)}{10} \times 100\% = 0.01(A-B-C)\%$$

当蔗糖质量不为 10 g 时，蔗糖校正值可按以下公式计算：

$$\text{蔗糖校正值} = 0.214\ 7\sqrt{X} + 0.113\ 7Y - 0.299\ 8$$

式中 X——耗用的碘液量（$X=A-B$）（mL）；

Y——蔗糖质量（g）。

【例 3-1】称取白砂糖 10 g，用奥夫纳尔法测定还原糖，加入 0.016 3 mol/L 碘液 25.00 mL，滴定耗用 0.037 2 mol/L 硫代硫酸钠溶液 17.25 mL，求白砂糖的还原糖含量。

【解】首先将溶液换算为标准浓度对应的体积。

$A = 25 \times 0.016\ 3\ /\ 0.016\ 15 = 25.23$（mL）

$B = 17.25 \times 0.037\ 2\ /\ 0.032\ 3 = 19.87$（mL）

$X = A - B = 25.23 - 19.87 = 5.36$（mL）

按公式计算 10 g 蔗糖校正值 C：

$10\ \text{g 蔗糖校正值}\ C = 1.025 + 0.069\ 5X - 0.001\ 24X^2$

$\quad\quad\quad\quad\quad\quad\quad = 1.025 + 0.069\ 5 \times 5.36 - 0.001\ 24 \times 5.36^2 = 1.36$

白砂糖还原糖分 $= 0.01 \times (25.23 - 19.87 - 1.36) = 0.040$（%）

7. 注意事项

（1）向锥形瓶内加入碘液前一定要将溶液冷至室温，加入碘液后要加盖，并置于暗处反应 2 min。防止碘在较高温度下挥发和遇强光分解而造成误差。

（2）加入盐酸的量应能将 Cu_2O 沉淀全部溶解，并稍有过量，使溶液呈微酸性。因为在碱性条件下奥夫纳尔试剂中的硫酸铜能与碘液中的碘化钾反应而析出碘。此外，碘与硫代硫酸钠的反应也要求在中性或微酸性条件下进行，否则将会发生副反应使过程复杂，无法进行计算。但溶液也不能过酸，因为在强酸性溶液中，硫代硫酸钠会发生分解。因此，加入酸的量应控制好。

（3）因淀粉呈螺旋结构，碘分子有可能会进入结构中而使其难以参加反应，所以淀粉指示剂不要加入太早，以免产生的蓝色不易消失。

（4）硫代硫酸钠滴定过量后只能重做，不能用碘液回滴，因硫代硫酸钠在强酸性溶液中会发生分解。

（5）硫代硫酸钠溶液的浓度不稳定，容易受空气和蒸馏水中的细菌、CO_2、O_2 的作用而分解，光的照射能加速分解，所以至少每半个月要标定一次。

任务 3.3　白砂糖干燥失重的测定

● 企业案例

主任观察刚搜集到的某批次样品，感觉有点潮湿，认为产品的水分较大，他安排小明对以上样品的干燥失重进行测定，小明应该选择什么仪器和使用什么方法来测定呢？

● 任务目标

通过本任务的学习，学生达到以下目标：
（1）理解常压干燥法测定白砂糖水分的原理。
（2）掌握常压干燥法测定白砂糖水分的方法和计算。

● 素质目标

养成耐心细致的良好习惯。

● 任务描述

使用常压干燥法测定白砂糖水分。

● 程序与方法

（1）原料：白砂糖。
（2）设备及仪器：恒温干燥箱、带温度计玻璃干燥器、玻璃称量瓶（直径为 6～10 cm，深度为 2～3 cm）、分析天平。
（3）试剂：蒸馏水或去离子水（配制所有溶液的用水其电导率应低于 15 μs/cm）。
（4）步骤：将干燥箱预先加热到 130 ℃。将已打开盖的干燥洁净的空称量瓶及其盖子一同放入干燥箱中，干燥 30 min，然后将称量瓶盖上盖子，从干燥箱中取出，放入干燥器中冷却至室温。将称量瓶称量并称取 9.5～10.5 g 样品（应准确至 ±0.1 mg），样品在称量瓶中要摊平，然后将盛有样品的称量瓶及其盖子一同放入预热至 130 ℃ 的干燥箱中，准确干燥 18 min，将称量瓶盖上盖子，从干燥箱中取出，放入干燥器中冷却至室温，称量应准确至 ±0.1 mg。
（5）数据处理及计算。

项目	数据及结果
称量瓶或盛器的质量 /g	
称量瓶或盛器加样品质量 /g	
称量瓶或盛器加样品干燥后质量 /g	
干燥失重 /%	

> **思考**
>
> 为什么要使用常压干燥法对白砂糖进行水分测定？有其他更好的方法吗？

任务相关知识——白砂糖干燥失重的测定

采用恒温干燥箱在较高的温度下，加热蒸发失去水分和挥发性物质而达到干燥目的，由于蒸发失去的不完全是水，故干燥法测得的水分也称"干燥失重"。白砂糖干燥失重是产品质量控制的一个主要指标。

1. 主要仪器、设备

（1）干燥箱：测定过程中，离称量瓶上面（2.5±0.5）cm处的温度要保持在（130±1）℃。

（2）干燥器。

（3）扁形称量瓶：直径为 6～10 cm，深度为 2～3 cm。

（4）分析天平：感量 0.1 mg。

2. 测定步骤

测定方法分为两种，即仲裁法和常规法。

（1）将干燥箱预先加热到 105 ℃（仲裁法）或 130 ℃（常规法）。

（2）将已打开盖的干燥洁净的空称量瓶及其盖子一同放入干燥箱中，干燥 30 min，然后将称量瓶盖上盖子，从干燥箱中取出，放入干燥器中冷却至室温。

（3）称量瓶的质量 W。

（4）用称量瓶尽快称取 20.000 0 g～30.000 0 g（仲裁法）或 9.500 0 g～10.500 0 g（常规法）样品（应准确至 ±0.1 mg），样品和称量瓶的总质量 W_1。

（5）摊平称量瓶中的样品，然后将盛有样品的称量瓶及其盖子放入预热至 105 ℃（仲裁法）或 130 ℃（常规法）的干燥箱中，准确干燥 3 h（仲裁法）或 18 min（常规法）。

不必干燥至恒重。但必须确保在测定的任何阶段，都不能有样品的有形损失，盛皿须用干燥洁净的坩埚夹夹取。

（6）将称量瓶盖上盖子，从干燥箱中取出，放入干燥器中冷却至室温。

（7）称量干燥并冷却至室温的称量瓶（盖上盖子）和样品的质量 W_2，应准确至 ±0.1 mg。

注：现在糖厂中为了加快测定的速度，对于白砂糖，也有采用微波干燥。样品称量后直接置于微波炉中加热 1～2 min，冷却后称重。但这不是国家标准规定的方法，所以只能用于内部指标控制。

白砂糖水分测定已有进口的成套设备，将样品置于设备的盛盘上，推入仪器中，就能自动完成称量及干燥过程。设备使用红外线干燥技术，整个测定过程耗时不超过 2 min，并能自动显示并打印结果。

3. 计算

干燥失重以百分数表示，计算结果取到两位小数。两次测定值之差不得超过其平均值的15%。

$$干燥失重（\%）=100\%（W_1-W_2）/（W_1-W）$$

式中　W——称量瓶或盛器的质量（g）；

　　　W_1——称量瓶或盛器加样品质量（g）；

　　　W_2——称量瓶或盛器加样品干燥后质量（g）。

任务 3.4 白砂糖电导灰分的测定

● 企业案例

厂长在对生产一线进行巡检后,通过生产报表发现中间制品的钙盐含量很高,他判断,有可能会出现一定的设备结垢现象和白砂糖电导灰分过高的情况。假如你是化验室主任,你应该使用什么方法和仪器来测定电导灰分呢?

● 任务目标

通过本任务的学习,学生达到以下目标:
(1) 理解电导法测定白砂糖灰分的原理。
(2) 掌握电导法测定白砂糖灰分的方法和计算。

● 素质目标

养成用联系的哲学观点看问题的习惯。

● 任务描述

使用电导法测定白砂糖电导灰分,并使用合适的方法进行结果计算。

● 程序与方法

(1) 原料:白砂糖。

(2) 设备及仪器:电导仪(DDS-11C 型或 DDS-11A 型)、阿贝折射仪、分析天平、100 mL 容量瓶、烧杯。

(3) 试剂:蒸馏水或去离子水(配制所有溶液的用水,其电导率应低于 15 μS/cm)、0.01 mol/L 氯化钾溶液 [取分析纯等级的氯化钾,加热至 500 ℃(呈暗红炽热)脱水 30 min 后,称取 745.5 mg,溶解于 1 000 mL 容量瓶中,并加水至标线]。

(4) 步骤:称取白糖样品(31.3±0.1)g 于干燥洁净的小烧杯中,加蒸馏水 40～50 mL,以细玻璃棒搅拌,使其完全溶解。然后将糖液小心地注入 100 mL 容量瓶中,以蒸馏水少量多次洗涤烧杯及玻璃棒,洗水并入容量瓶中,加水到标线(此样液为 28 °Bx)。摇匀后,倒入测定电导率专用的干燥洁净的小烧杯内(倒入前需用样液将烧杯冲洗 2～3 次)待测。另取一个干净的小烧杯,注入溶解样品的蒸馏水。用电导仪分别测定样液及蒸馏水的电导率,并记录读数及读数时的温度。测量时的温度范围最好不要超过(20±5)℃。

(5) 数据处理及计算。

项目	数据及结果
样液的电导率 /（μS·cm^{-1}）	
样液的温度 /℃	
蒸馏水的电导率 /（μS·cm^{-1}）	
蒸馏水的温度 /℃	
电导分灰 /%	

● 思考

电导灰分法测定的是什么物质？请同学们学会使用联系的哲学观点对全厂质量控制进行学习。

3.4.1 任务相关知识——白砂糖电导灰分的测定

样品中的灰分有可溶性灰分和非可溶性灰分两部分，构成灰分质量的不溶性物质，可以在测定样品的不溶物时反映出来。而通过电导法测定溶液的电导率，可反映样品中可溶性物质（如盐类、游离酸等离子型物质），能更好地反映样品的品质。电导法操作简单、无污染、重现性好。

1. 测定原理

电导是物质传送电流的能力。在液体中常以电阻的倒数——电导来衡量其导电能力的大小。电导的国际单位是西门子（S）。在实际使用中，一般采用微西（μS），$1S=10^6 \mu S$。由于纯水是几乎不导电的，当水中存在着离子型的电解质时，溶液就有一定的导电能力。溶液中电解质的浓度与种类不同，则溶液电导也不同，即电导能反映出水中存在的电解质的程度。所以可通过测定溶液的导电能力来分析电解质在溶液中的量。这就是电导仪的基本原理。

溶液的导电能力与电极的面积（A）成正比，与电极的距离（L）成反比。当电极面积与距离均为一个单位时，测得的电导称为电导率。常用的电导率单位是微西/厘米（μS/cm）。它表示当电极的面积为1 cm^2、距离为1 cm时溶液的导电能力。

纯蔗糖溶液几乎不导电。当样品中含有离子型非糖分时，溶液就会导电，而这类离子型非糖分就是样品中灰分的主要组成成分。所以可以用样品的电导率来直接反映样品中离子型非糖分的数量，称为电导灰分。

2. 主要设备

（1）电导仪。实验室中测定电导的仪器是电导仪。常用的型号有DDS-11A、DDS-11C等。

测量范围：$0 \sim 10^5$ μS/cm。

测量误差：不应大于满量程的0.5%。

DDS-11A 型电导仪及配用电极如图 3-1 所示。

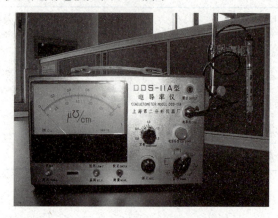

图 3-1　DDS-11A 型电导仪

（2）试剂。

1）蒸馏水或去离子水。精制白砂糖必须用电导率低于 2 μS/cm 的重蒸馏水（蒸馏过两次）或去离子水。对于一般的白砂糖允许用电导率低于 15 μS/cm 的蒸馏水。

2）0.01 mol/L 氯化钾溶液。取分析纯等级的氯化钾，加热至 500 ℃，脱水 30 min，冷却后称取 0.745 5 g，溶解后移入 1 000 mL 容量瓶中，加水至标线。

3）0.002 5 mol/L 氯化钾溶液。用移液管吸取 0.01 mol/L 氯化钾溶液 50 mL 于 200 mL 容量瓶中，加水稀释至标线。此溶液在 20 ℃时的电导率为 328 μS/cm。

3. 测定步骤

（1）样液的制备。用电导法测定样品的灰分，实际上就是测定样液的电导率。样液的电导率受浓度的影响，样液在某一浓度范围内具有最大电导率，在这一浓度范围附近，浓度的变化对电导率的影响很小。为了减小浓度变化对测定结果的影响，用电导法进行灰分测定时，一般将样液的浓度调整至其最大电导率对应的浓度附近。在这一浓度范围附近，浓度对电导率的影响很小，所以可以用粗天平进行称量，方便操作。根据试验数据，对于糖厂的中间制品，浓度应为 5 g/100 mL，对于白砂糖，浓度应为 28 g/100 g（即 28 °Bx）。

1）5 g/100 mL 样液的配制。适用于糖厂的在制品、赤砂糖、红糖、原糖及废蜜的灰分测定。

首先用阿贝折射仪测定样品的折光锤度，各种蔗汁可直接测定折光锤度。浓度较高，含有未溶解蔗糖结晶的样品（如糖膏、糖蜜等），要用已知量的水稀释（通常重量比为 1∶1），然后测定折光锤度。按下式计算出称取样品的质量，加水溶解，移入 100 mL 容量瓶中，加水至刻度，摇匀、待测。

$$应称样品质量（g）= 5 \times 100 / 样品锤度$$

样品原浓度低于 5 g/100 mL 时，可直接测定样品折光锤度后进行灰分测定。

2）31.3 g/100 mL 样液的配制。适用于白砂糖、方糖、冰糖等的电导灰分测定。

称取样品（31.3±0.1）g 于干燥洁净的小烧杯中，加水溶解后，移入 100 mL 容量瓶中，加水至刻度、摇匀后待测（样液的浓度为 28 °Bx）。

（2）测定。按要求选择电导仪的电极常数、温度补偿、测量范围，用上述配好待测的样液，冲洗测定电导率用的电导电极及洁净的小烧杯2次，然后倒入样液，用电导仪测定其电导率。将稀释用的蒸馏水倒入小烧杯，用玻璃棒搅拌，搅拌时间应与溶解样品的时间接近（因搅拌过程会引起空气中的二氧化碳溶入蒸馏水中，使电导率增大），然后测定稀释样品用的蒸馏水的电导率。

（3）计算。

1）数据记录。

①样液温度 T_1；

②蒸馏水温度 T_2（电导仪设有温度补偿时不用记录）；

③样液电导率 C_{1t}（μS/cm）；

④蒸馏水电导率 C_{2t}（μS/cm）。

2）计算。

①样液电导率温度校正：

$$C_1 = C_{1t}[1+0.026(T-20)]$$

注：电导仪设有温度补偿时 $C_1 = C_{1t}$。

②蒸馏水电导率温度校正：

$$C_2 = C_{2t}[1+0.022(T-20)]$$

注：电导仪设有温度补偿时 $C_2 = C_{2t}$。

③白砂糖电导灰分计算（31.3 g/100 mL）：

$$电导灰分（\%）=6\times10^{-4}(C_1-0.35C_2)$$

注：因为测定 C_2 时，单位容积内完全是水，测定结果是表示单位容积内离子型物质的含量，但在测定样液时，由于其中的糖占据部分体积，水在单位容积中的比例下降，所以单位容积中由水带来的离子型物质也相对减少，当蔗糖溶于水后，溶液黏度增大，离子迁移率相应减少，为了补偿这两方面的变化，所以计算公式中蒸馏水的电导率应乘以一校正系数 0.35。

④其他制品电导灰分计算（5 g/100 mL）：

$$电导灰分（对样品质量\%）=18\times10^{-4}(C_1-0.9C_2)$$

【例 3-2】混合汁的观测折光锤为 15.40 °Bx，按公式称出配制浓度为 5 g/100 mL 溶液应需的样品量：5×100/15.4=32.47（g），定容为 100 mL 后，用带温度校正的电导仪测得样液的电导率为 277 μS/cm，蒸馏水的电导率为 0.85 μS/cm，求混合汁的电导灰分。

【解】由于仪器已进行了温度校正，所以可直接进行计算。

电导灰分（对样品质量%）=18×10⁻⁴×（277-0.9×0.85）=0.50

4. 注意事项

（1）新购买的电极要先进行电极常数标定后才能使用，电极应定期进行常数标定。

（2）为确保测量精度，电极使用前应用电导率小于 0.5 μS/cm 的蒸馏水（或去离子水）冲洗两次，然后用被测试样品冲洗三次后方可使用。

（3）电极插头座禁止受潮，以避免造成不必要的误差。

(4) 蒸馏水倒出来后要立刻进行测定，否则会吸收空气中的二氧化碳而增大电导率。

(5) 测定蒸馏水电导率时，应与测定溶糖电导率时的操作一致，也应经搅拌、移入容量瓶、摇匀等操作，并且每一步的时间应基本相同。否则会产生较大的误差。白砂糖溶解需搅拌较长时间，而蒸馏水经长时间搅拌后，电导率会增大许多。

5. 电导仪的使用

(1) 仪器的使用。为保证测量准确及仪表安全，须按以下要求使用。

1) 通电前，检查表针是否指零，如不指零，可调整表头中的调整螺钉，使表针指零。

2) 当电源线的插头被插入仪器的电源孔（在仪器的背面）后，开启电源开关，灯即亮。预热后即可工作。

3) 将范围选择器扳到所需的测量范围，如不知被测量对象的大小，应先调至最大量程位置，以免过载打弯表针，以后逐挡改变到所需量程。

4) 把电极插头插入插座，使插头的凹槽对准插座的凸槽，然后用食指按一下插头顶部，即可插入，拔出时捏住插头的下部，往上一拔即可。

5) 被测定为低电导（5 μS/cm 以下）时，溶液中的离子较少，为了减少电极对离子的吸附造成电导率下降，应选用表面积较小的光亮型电极；被测液电导率在 5 μS/cm 以上时，通过的电流较大，电解作用明显，电极的极化作用显著，应选用表面积较大的铂黑型电极，以减小电流密度，避免极化作用。

6) 将电极常数旋钮调至该电极常数的位置。

7) 将温度补偿旋钮调至被测溶液的温度，如果不需要温度补偿，则把温度补偿旋钮调至 20 ℃ 位置即可。

8) 将校正测量换挡开关扳向"校正"，调整校正调节器，使指针作满刻度指示。

9) 将开关扳向"测量"，将指示电表中的读数乘以范围选择器上的倍率，即得被测溶液的电导率。读数时，若量程开关指向红点，则按红色的标尺读数；指向黑点，则按黑色的标尺读数。

10) 当被测样液的电导率大于 10 000 μS/cm 时，可选用 DJS-10 铂黑型电极，此时电极常数旋钮调至该电极常数的 1/10 位置。例如：若电极的常数为 9.8，则应使电极常数旋钮调至 0.98 的位置，还应将测得的读数乘以 10，得到被测溶液的电导率。

11) 在测量中要经常检查"校正"是否改变，即将开关扳向"校正"时，指针是否仍作满刻度指示。

(2) 电极常数的标定。电极常数是一个重要的数据，直接影响测定的结果，所以新购买的电极在使用前应进行电极常数的标定。方法如下：

1) 将分析纯以上等级的氯酸钾在 200 ℃ 下干燥 2 h，然后在 500 ℃ 下脱水 30 min。

2) 取经脱水处理的氯酸钾 715.5 mg，用电导率低于 2 μS/cm 的蒸馏水溶解于 1 000 mL 容量瓶中，加水至标线，此溶液浓度为 0.01 mol/L。

3) 吸取 0.01 mol/L 氯酸钾溶液 250 mL，稀释至 1 000 mL，此溶液浓度为 0.002 5 mol/L，测量溶液的温度，用此标准溶液对电极进行标定。

4) 将电导仪的电极常数旋钮调至 1.0 处（若有温度补偿旋钮，将温度补偿旋钮调至相应于溶液的温度处）。

5）用待标定电极测定标准氯酸钾溶液的电导率 S。

6）查标准氯酸钾溶液电导率表得氯酸钾的电导率 K。

7）用以下公式计算电极的常数 Q：

$$Q=K/S$$

式中　Q——电极常数；

　　　S——用待标定电极测得的氯酸钾溶液的电导率（μS/cm）；

　　　K——由表 3-2 查得的氯酸钾电导率（μS/cm）。

表 3-2　标准氯酸钾溶液电导率表（μS/cm）

温度 /℃	浓度 /（0.1 mol·L^{-1}）	浓度 /（0.01 mol·L^{-1}）	浓度 /（0.002 5 mol·L^{-1}）
15	10 480	1 147	194
16	10 720	1 173	300
17	10 950	1 198	307
18	11 190	1 225	314
19	11 430	1 251	321
20	11 670	1 278	328
21	11 910	1 305	335
22	12 150	1 332	342
23	12 390	1 359	348
24	12 640	1 386	355
25	12 880	13 413	362

任务 3.5　白砂糖色值及混浊度的测定

● 企业案例

最近的白砂糖样品色值比较高，部分客户不满意，主任让小明检测一下相应产品的色值及混浊度，请你帮助小明选择合适的仪器和方法。

● 任务目标

通过本任务的学习，学生达到以下目标：
（1）理解在制品及白砂糖国际糖色值、混浊度的测定原理。
（2）掌握色值、混浊度的测定方法及计算。

● 素质目标

养成一丝不苟的工作习惯。

● 任务描述

测定某白砂糖样品的色值和混浊度。

● 程序与方法

（1）原料：白砂糖。

（2）设备及仪器：721型分光光度计（723OG型分光光度计）、比色皿、阿贝折光计、酸度计、薄膜过滤器、微孔膜（孔径0.45 μm）、抽滤瓶，真空泵。

（3）试剂：0.5 mol/L HCl、0.5 mol/L NaOH、0.1 mol/L NaOH、三乙醇胺－盐酸缓冲溶液。

1）三乙醇胺－盐酸缓冲溶液制备。称取三乙醇胺[（$HOCH_2CH_2$）$_3$N] 14.920 g，用蒸馏水溶解并定容于1 000 mL容量瓶中，然后移入2 000 mL烧杯内，加入0.1 mol/L盐酸溶液约800 mL，搅拌均匀并继续用0.1 mol/L盐酸调到pH7.0（用酸度计的电极浸于此溶液中测量pH值）。储存于棕色玻璃瓶中。

2）0.1 mol/L盐酸溶液制备。用吸液管吸取9.0 mL浓盐酸（比重为1.19）于预先放有适量蒸馏水的1 000 mL容量瓶中，然后稀释至刻度。

（4）步骤：称取白砂糖样品100.0 g于200 mL烧杯中，加入三乙醇胺－盐酸缓冲溶液135 mL，搅拌至完全溶解，测定其折光锤度及温度，留下一部分（作测定过滤前糖液吸光度用），其余的倒入已预先铺好0.45 μm孔径的微孔膜过滤器中，在真空下抽滤，弃去最初的50 mL滤液，收集滤液应不少于50 mL，用折光计及温度计分别测定滤液的折光锤度及温度，将过滤前后的糖液分别盛于比色皿中，在分光光度计上用420 nm波长测定其吸光度，并用经过过滤的三乙醇胺－盐酸缓冲溶液作为零点色值的参比标准。

（5）数据处理及计算。

名称	项目	数据及结果
白砂糖	过滤前样液的观测折光锤度 /°Bx	
	过滤前样液的温度 /℃	
	过滤前样液的折光锤度 /°Bx	
	过滤后样液的观测折光锤度 /°Bx	
	过滤后样液的温度 /℃	
	过滤后样液的折光锤度 /°Bx	
	过滤前样液的吸光度	
	过滤后样液的吸光度	
	色值 /IU	
	混浊度 /MAU	

● 思考

测定色值要使用多大空隙的膜进行样品过滤？

任务相关知识——白砂糖色值与混浊度的测定

色值与混浊度是衡量制糖生产过程工艺效果、特别是有色物质的除去率及胶体、悬浮粒子除去率的重要指标，也是目前糖厂中对产品的等级影响最大的指标之一。在糖厂中，对产品色值与混浊度的测定通常采用分光光度法。

1. 分光光度法基本原理

当平行的单色光通过均匀、非散射溶液时，光的一部分被吸收，一部分透过溶液，光的强度就会因能量被吸收而减弱，光的强度减弱的程度与溶质的浓度有一定的比例关系，利用分光光度计，就可以测定光强度减弱的程度，从而确定某一组分的含量。这一方法称为分光光度法。分光光度法的测定误差通常为1%～5%，在测定含量较高的组分时，其准确度要低于容量分析和质量分析，但对于微量组分的测定，这样的准确度已能满足要求了。

光强度的变化，可用透光率 T 或吸光度 A（也称消光度 E）来表示。设入射光的强度为 I_0，透过光的强度为 I，则将透光率 T 定义为

$$T = \frac{I}{I_0}$$

在仪器的刻度上，通常用 T（%）来表示，即

$$T(\%) = \frac{I}{I_0} \times 100\%$$

溶液的浓度 C 与透光率的负对数 A 成正比，即

$$A=-\lg T=-\lg \frac{I}{I_0}=kcl$$

式中　A——吸光度；

　　　k——吸光系数，它是某一物质在一定波长下的特征常数；

　　　c——溶液的浓度；

　　　l——光线通过液层的厚度。

一般的分光光度计均可同时用透光率 T（%）及吸光度 A 表示测定结果。在指针式仪表上，透光率的刻度是等分的，而吸光度的刻度的间隔是不均匀的。

糖厂在制品及成品白砂糖的溶液都不是非散射性溶液，含有引起光散射的悬浮粒子，对这些溶液进行测定时，其测定结果为吸收和散射两者引起光衰减的总和，要进行单纯的色值测定，必须对溶液进行有效的过滤，以消除溶液中悬浮粒子对测定的影响。

2. 白砂糖色值及混浊度的测定

（1）主要设备与试剂。

1）分光光度计。

①测量范围：透过率 0～100%。

②波长范围：360～800 nm。

③波长误差：在 420 nm 处波长误差不大于 ±1 nm。

2）阿贝折射仪。蔗糖质量分数锤度（°Bx）0～95，最小分度值为 0.2。

3）微孔膜过滤器。滤膜应厚薄均匀，膜面上分布着对称、均匀、穿透性强的微孔，孔径为 0.45 μm，孔隙度达 80%，孔道呈线性状而互不干扰，滤膜与直径 150 mm 的糖品过滤器配套使用。

4）三乙醇胺 – 盐酸缓冲液。称取三乙醇胺 [（$HOCH_2CH_2$）$_3$N]14.92 g，用蒸馏水溶解并定容于 1 000 mL 容量瓶中，然后移入 2 000 mL 烧杯中，加入 0.1 mol/L 盐酸约 800 mL，置于搅拌器上，放入酸度计电极，在不断搅拌情况下用 0.1 mol/L 盐酸调节到 pH7.00±0.02。储存于棕色玻璃瓶中。

（2）测定步骤。

1）样液制备。称取白砂糖样品 100.0 g 于 200 mL 烧杯中，加入三乙醇胺 – 盐酸缓冲溶液 135 mL，搅拌至完全溶解。

2）过滤前吸光度测定。取部分制备好的糖液用折光计测定滤液的折光锤度，盛于比色皿中，在分光光度计上用 420 nm 波长测定其吸光度，用未经过滤的三乙醇胺 – 盐酸缓冲溶液作为零点的参比标准。

3）过滤后吸光度测定。其余样液倒入已预先铺好 0.45 μm 孔径的微孔膜过滤器中，在真空下抽滤，弃去最初的 50 mL 滤液，收集滤液应不少于 50 mL，用折光计测定滤液的折光锤度，将过滤后的糖液盛于比色皿中，在分光光度计上用 420 nm 波长测定其吸光度，并用经过滤的三乙醇胺 – 盐酸缓冲溶液作为零点的参比标准。

（3）计算方法。

1）数据记录。

①过滤前样液观测折光锤度 B_1。

②过滤后样液观测折光锤度 B_2。

③比色皿厚度。

④过滤前样液吸光度 A_1。

⑤过滤后样液吸光度 A_2。

2）计算方法。

①色值。

$$C = \frac{A_2}{bc} \times 1\,000$$

式中　C——色值，单位为国际糖色值单位（IU）；

　　　A_2——过滤后样液吸光度；

　　　b——比色皿厚度（cm）；

　　　c——样液浓度（g/mL）（由改正到20℃的折光锤度乘以校正系数0.986 2，然后查表3-3得）。

②混浊度。按国家标准，混浊度也采用国际色值单位 MAU（衰减单位）。计算结果取到个位。

$$MAU = \left(\frac{A_1}{b_1 c_1} - \frac{A_2}{b_2 c_2}\right) \times 1\,000$$

式中　A_1——过滤前样液吸光度；

　　　b_1——测定 A_1 时比色皿厚度（cm）；

　　　c_1——测定 A_1 时样液浓度（g/mL）；

　　　A_2——过滤后样液吸光度；

　　　b_2——测定 A_2 时比色皿厚度（cm）；

　　　c_2——测定 A_2 时样液浓度（g/mL）。

$$c = 折光锤度 \times 相应视密度 / 100$$

注：当使用三乙醇胺作为 pH 缓冲液时，由于缓冲液会带入干固物，所以测得的折光锤度在用于计算及查表前应乘以校正系数 0.986 2，校正三乙醇胺对锤度的影响。

对于白砂糖，样液浓度 c（g/mL）可以直接查蔗糖溶液折光锤度与每毫升含蔗糖质量（在空气中）查表3-3。

也可以用以下公式计算：

$$样液浓度\ c\ (g/mL) = 0.014\,1B - 0.094\,16$$

式中　B——经校正后样液的折光锤度（°Bx）。

表 3-3 蔗糖溶液折光锤度与每毫升含蔗糖克数（在空气中）对照表

折光锤度 /°Bx	g 蔗糖/mL（在空气中）	折光锤度 /°Bx	g 蔗糖/mL（在空气中）	折光锤度 /°Bx	g 蔗糖/mL（在空气中）
40.0	0.470 2	41.7	0.493 8	43.4	0.517 8
40.1	0.471 5	41.8	0.495 2	43.5	0.519 2
40.2	0.472 9	41.9	0.496 6	43.6	0.520 6
40.3	0.474 3	42.0	0.498 0	43.7	0.522 1
40.4	0.475 7	42.1	0.499 4	43.8	0.523 5
40.5	0.477 1	42.2	0.500 8	43.9	0.524 9
40.6	0.478 5	42.3	0.502 2	44.0	0.526 3
40.7	0.479 9	42.4	0.503 6	44.1	0.527 8
40.8	0.481 2	42.5	0.505 1	44.2	0.529 2
40.9	0.482 6	42.6	0.506 5	44.3	0.530 6
41.0	0.484 0	42.7	0.507 9	44.4	0.532 1
41.1	0.485 4	42.8	0.509 3	44.5	0.533 5
41.2	0.486 6	42.9	0.510 7	44.6	0.534 9
41.3	0.488 2	43.0	0.512 1	44.7	0.536 4
41.4	0.489 6	43.1	0.513 5	44.8	0.537 8
41.5	0.491 0	43.2	0.515 0	44.9	0.539 2
41.6	0.492 4	43.3	0.516 4		

【例 3-3】称取白砂糖 100.0 g，按色值规定方法进行测定，测得的数据如下。

（1）过滤前样液折光锤度 B_1：43.48 °Bx。

（2）过滤后样液折光锤度 B_2：43.29 °Bx。

（3）比色皿厚度：3 cm。

（4）过滤前样液吸光度 A_1：0.225。

（5）过滤后样液吸光度 A_2：0.184。

求样品的色值与混浊度。

【解】

过滤前样液浓度校正：0.986 2×43.48=42.88（g/mL）

查表 3-3 得过滤前样液的浓度为 0.509 8（g/mL）。

过滤后样液浓度校正：0.986 2×43.29=42.69（g/mL）

查表 3-3 得过滤后样液的浓度：0.507 8（g/mL）。

过滤前样液的色值为：

$$IU_1 = \frac{0.225}{3 \times 0.509\,8} \times 1\,000 = 147$$

过滤后样液的色值为：

$$IU_2 = \frac{0.184}{3 \times 0.509\,8} \times 1\,000 = 121$$

混浊度：147−121=26（MAU）

则白砂糖样品的色值为 121 IU，混浊度为 26 MAU。

（4）注意事项。

1）微孔滤膜切忌折叠，应平整存放，使用时正面（平滑有光亮）向上，平放于滤板上。使用前应用中性蒸馏水，在室温下浸渍约 2 h，使膜体充分润湿。由于膜体含有水分，所以最初部分的滤液应弃去，以减少对浓度的影响。

2）由于糖液中的色素对 pH 非常敏感，pH 每增加 0.1 个单位，色值会增大 25%，因此 pH 应严格控制在规定的范围内。

3）抽滤糖液时，真空度应控制在 50～55 kPa 为宜，真空度太大容易损坏滤板。

4）测定糖液吸光度时，比色皿应放在紧靠光敏元件一端，以降低光的散射作用对测定的影响。

5）同一样品的色值会随着波长的增大而减小，而波长受仪器的准确度影响，要经常对仪器的波长进行校正。另外仪器的光谱宽度（色纯度）也影响结果的准确性，通常低挡仪器的光谱宽度较宽，测定结果会偏小，在仪器升级换代时对这一问题要加以注意。

任务 3.6　白砂糖二氧化硫含量的测定

● 企业案例

同地区的兄弟企业最近产品质量下降得很快，主要原因是白砂糖中的二氧化硫含量过高，假如你是化验室分析员，你应当使用什么方法来检测二氧化硫的含量？

● 任务目标

通过本任务的学习，学生达到以下目标：
（1）了解白砂糖中二氧化硫的测定原理。
（2）掌握白砂糖中二氧化硫的测定方法及计算。

● 素质目标

养成提升食品安全的意识。

● 任务描述

测定某批次样品的二氧化硫含量。

● 程序与方法

（1）原料：白砂糖。
（2）设备及仪器：分光光度计、25 mL 带塞比色管 6～12 支。
（3）试剂：四氯汞钠吸收液、显色剂（盐酸副玫瑰苯胺溶液）、2 g/L 氨基磺酸铵溶液、0.2% 甲醛溶液、10 g/L 淀粉指示剂、0.05 mol/L 碘溶液、0.1 mol/L 硫代硫酸钠标准溶液、0.25 mol/L 硫酸溶液、0.5 mol/L 氢氧化钠溶液、二氧化硫标准液（每毫升约含二氧化硫 1.5 mg，须标定求得其准确数字）、二氧化硫标准应用溶液（每毫升接近于 5 mg，不一定是整数，但要知道准确数字，最好是准确 5 mg）。
（4）实验步骤：
1）标准曲线的绘制。用微量吸管准确吸取二氧化硫标准应用溶液 2.0 mL、4.0 mL、6.0 mL、8.0 mL、10.0 mL、12.0 mL，分别置于 25 mL 带塞比色管中，依次加入四氯汞钠吸收液 18.0 mL、16.0 mL、14.0 mL、12.0 mL、10.0 mL、8.0 mL（总体积为 20.0 mL，相当于每 10 mL 中二氧化硫的含量为 5 mg、10 mg、15 mg、20 mg、25 mg、30 mg），在各管中加入 12 g/L 氨基磺酸铵溶液 2 mL，0.2% 甲醛溶液 2 mL，显色剂 2 mL，盖严摇匀，于 15～25 ℃室温下（或用水溶液控制温度），放置 20 min，即在分光光度计上用 550 nm 波长，并以 1 cm 装盛溶液的比色皿来测定其吸光度，将蒸馏水调零，测出结果与标准应用液每 10 mL 含二氧化硫质量相同，绘制标准曲线，或用回归法求得回归方程。

2）样品配制及测定。准确称取白砂糖样品 10.0 g（可视样品含二氧化硫的高低而减增），用约 25 mL 蒸馏水溶解，并移入 50 mL 容量瓶中，准确加入 0.5 mol/L 氢氧化钠溶液 2.0 mL，盖严容量瓶塞摇匀，5 min 后，再准确加入 0.25 mol/L 硫酸溶液 2.0 mL，盖严后摇匀，立即加入四氯汞钠吸收液 10 mL，加水至标线，摇匀备用。

准确吸取配制好的样液 10 mL，置于带塞比色管中，加四氯汞钠吸收液 10 mL，加入 12 g/L 氨基磺酸铵溶液 2 mL，加 0.2% 甲醛溶液 2 mL，显色剂 2 mL，盖塞摇匀，在 15～25 ℃温度下，放置 20 min，立即在分光光度计上用 550 nm 波长，用 1 cm 比色皿盛装样液，并以糖液作空白测定其吸光度。从标准曲线查得相应的二氧化硫含量（或代入回归方程计算得出二氧化硫的含量），然后进行计算。

3）数据处理、计算及作曲线图。

项 目		编 号					
		1	2	3	4	5	6
标准曲线的绘制	二氧化硫标准应用溶液毫升数	2.0	4.0	6.0	8.0	10.0	12.0
	四氯汞钠吸收液毫升数	18.0	16.0	14.0	12.0	10.0	8.0
	吸光度						
测定	吸光度						
	从曲线（或计算）得到相当 SO_2 的微克数						
	测定时所吸取样液中所含样品的克数						
	二氧化硫（mg/kg）						

● 思考

二氧化硫含量高对人体有何影响？为了保障人民群众的身体健康，我们要注意食品质量安全，生产出合乎健康标准的白砂糖，此时我们应该如何做？

任务相关知识——白砂糖二氧化硫含量的测定

SO_2 含量是白砂糖的一项重要的质量指标。糖厂测定白砂糖 SO_2 含量的方法主要有盐酸副玫瑰苯胺比色法和碘量法。

1. 盐酸副玫瑰苯胺比色法

（1）基本原理。二氧化硫被四氯汞钠吸收后，生成一种稳定的二氯亚硫酸盐，再与甲醛及盐酸副玫瑰苯胺作用，生成紫红色，可与标准比色而定量，其反应过程如下：

$$(HgCl_4)^{2-} + SO_2 + H_2O \longrightarrow (HgCl_2SO_3)^{2-} + 2Cl^- + 2H^+$$

$$(HgCl_2SO_3)^{2-} + HCHO + 2H^+ \longrightarrow HOCH_2SO_3H + HgCl_2$$

$$(H_2NC_6H_4)_2C \cdot C_6H_4 \cdot NH \cdot HCl + 3HCl \longrightarrow$$

（有色）

$$(ClH_3N \cdot C_6H_4)_2C \cdot Cl \cdot C_6H_4 \cdot NH_2 \cdot HCl + 2HOCH_2SO_3H \longrightarrow$$
（无色）

$$(HSO_3 \cdot CH_2 \cdot HN \cdot C_6H4)_2C \cdot C_6H_4 \cdot NH \cdot HCl + 3HCl + 2H_2O$$
（紫红色）

碱性品红与盐酸加成后变成非醌型的无色化合物，后与二氯亚硫酸盐及甲醛作用，再生成醌型结构的有色化合物，呈紫红色。

由于白砂糖中含有部分结合的无机或有机盐类，其中亚硫酸钙等溶解度较低。二氧化硫与葡萄糖的加成产物均须借助氢氧化钠与之反应，使二氧化硫释出，才能测得白砂糖中二氧化硫的全含量。

（2）计算方法。

$$二氧化硫（mg/kg）= A / W$$

式中　A——从曲线（或计算）得到相当 SO_2 的微克数；
　　　W——测定时所吸取样液中所含样品的克数。

2. 碘量法

（1）测定方法及步骤。

1）在 250 mL 锥形瓶中分别加入 150 mL 蒸馏水、5 g/L 淀粉溶液 5 mL 和 3 mol/L 盐酸溶液 10 mL，摇匀。

2）用 0.002 5 mol/L 碘标准溶液滴定至浅蓝色出现以补偿蒸馏水和试剂所消耗的碘液。

3）称取 50.0 g 白砂糖样品放入以上锥形瓶中，摇动至溶解。

4）用 0.002 5 mol/L 碘标准溶液滴定至浅蓝色出现，并记录体积 V。

（2）计算方法。

$$二氧化硫含量（mg/kg）= V \times N \times 64.06 \times 1\,000 / m$$

式中　V——耗用碘液量（mL）；
　　　N——标准碘液浓度（0.002 5 mol/L）；
　　　m——称取样本质量（g）。

拓展学习——近红外测定成品糖

近红外光谱分析技术具有前处理简单、无损、快速等优点。光照射在蔗糖制品表面发生漫反射，光的反射情况反应样品的颗粒信息，再从反射光中分离出近红外光谱，并检测其强度变化，从而实现检测蔗糖样品各项指标的目的。

1. 主要设备

本节以聚光科技 SupNIR-4000 型长波近红外分析仪为例，包括近红外主机、漫反射探头、厚度探测器、光纤附件、CM-2000 化学计量学软件。近红外主机内采用的性能参数：波段范围为 1 000~1 800 nm，采样间隔为 1 m，光谱仪分辨率小于 7 nm，波长准确性小于 ±0.2 nm，杂散光小于 0.1%（1 692 nm），全谱扫描时间小于 0.2 s。CM-2000 化学计量学软件用于建模，其遵循 ASTM 定量和定性分析方法规范。

2. 主要试剂

分析各中间制品所需要的试剂，具体见本模块各样品分析方法所需主要试剂。

3. 测定步骤

（1）建模。

1）光谱预处理。先选全波段光谱，用偏最小二乘法建模，考察 415 种预处理方法组合的优劣。根据结果和经验，选择"标准化 +Savitzky--Golay 一阶求导 + 正交信号校正（OSC）"联合处理为预处理方法，以消除温度波动、波长漂移和基线漂移产生的影响，更好的提取颗粒信息。处理后近红外光谱如图 3-2 所示。

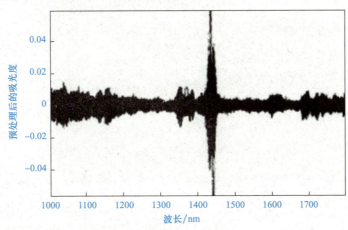

图 3-2 "标准化 +Savitzky-Golay 一阶求导 +OSC" 处理后的图谱

2）波段选择。选择全波段建模。

3）模型。利用 PLS 算法作为数据分解方式，从复杂的图谱数据中提取主成分，作为神经网络的输入变量，以减少输入变量并提高网络预测精度。以 RMSEC 最小为原则，为各个参数选择最佳值，利用不同参数共建立 58 个模型进行优化比较。最佳参数为：输入层节点数 4，隐含层的节点数 5，隐含层的转化函数 tansig，输出层的转化函数 logsig，初始学习速率 0.9，动量项 0.9，迭代最大次数 200。最佳 ANN 模型的 RMSEC 为 2.971 8。

4）应用 ANN 模型在线测定。将 ANN 算法建立的白砂糖粒度模型嵌入到近红外主机，对样品进行分析测定，连续运行 168 h。每隔 1 h 对样品进行采样分析，对比国标法测定值和近红外预测值之间的偏差，要求粒度预测误差控制在 ±5 之内。

不溶于水杂质的测定

赤砂糖还原糖分的测定

模块 4
化验室标准溶液的配制

模块 4 化验室
标准溶液的配制

Module 1
Basic knowledge of laboratory

Task 1.1   Instrument identification and titration operation

Enterprise case

Xiaoming is a new analyst. One day, the laboratory director gave him a bottle of HCl solution and a bottle of NaOH solution and asked him to select a suitable dropper and indicator to neutralize and titrate the above solutions. If you were Xiaoming, what would you do?

Mission objectives

Through the study of this task, students can achieve the following goals:

(1) Familiar with common glassware in the laboratory and basic knowledge of the laboratory;

(2) Master the titration operation technique of the burette;

(3) Learn to observe and judge the end point of titration.

Quality objectives

Develop the habit of being meticulous and careful.

Task description

Read and understand the basic laboratory knowledge necessary for this module. Titrate HCl and NaOH solutions with 50 mL acid and base burettes.

Implementation conditions

(1) Instruments: one 50 mL acid and one basic burette, three 250 mL conical flasks, one 250 mL and one 400 mL beaker, one 10 mL and one 100 mL measuring cylinder.

(2) Reagents: 0.1 mol/L HCl solution, 0.1 mol/L NaOH solution, phenolphthalein indicator, and methyl orange indicator.

Procedures and methods

(1) Carefully read the basic knowledge of the laboratory attached to this module, and complete the relevant exercises before class.

(2) Titration exercise.

Clean the prepared acid burette, coat the cock with vaseline, check for leakage, rinse with 0.1 mol/L HCl solution three times (5~10 mL each time), fill the HCl solution to above the "0" scale, remove the bubbles at the lower end of the burette, and adjust the liquid level to "0.00" mL.

Clean the prepared alkaline burette, check for leakage, rinse it with 0.1 mol/L NaOH solution three times (5~10 mL each time), fill it with NaOH solution to above the "0" scale, remove the bubbles at the lower end of the burette, and adjust the liquid level to "0.00" mL.

1) Discharge the solution from the burette: Accurately discharge 20 mL of HCl solution from the burette into a 250 mL conical flask.

2) Titration: Add 2 drops of phenolphthalein indicator to the above conical flask containing HCL solution and titrate with NaOH solution from an alkaline buret. Observe the change in colour around the dropping point.

3) Judgement of titration endpoint: At the beginning of titration, there is no obvious colour change around the dropping point, and the titration rate can be slightly accelerated. When a temporary colour change (light pink) occurs around the drop, NaOH solution should be added drop by drop. As the loss of colour slows down, the solution should be added slowly. By the time the endpoint is approached, the colour spreads throughout the solution and shakes 1 or 2 times before disappearing, at which point a drop should be added and shaken a few times. Finally, add half a drop of the solution and rinse the walls of the conical flask with distilled water. Keep dropping until the solution suddenly changes from colourless to light pink and does not disappear within 30 s that is the endpoint, and note down the reading.

In order to practice correctly judging the endpoint of the titration, continue to accurately add a small amount of HCl solution into the conical flask to make the colour fade, and then titrate with NaOH solution to the endpoint according to the above method. Repeat this many times until the end point of titration can be judged skillfully and the difference between the endpoint reading and the amount of NaOH solution is not more than 0.02 mL.

Accurately add 20 mL NaOH solution into a 250 mL conical flask according to the above method, add 2 drops of methyl orange indicator, and titrate with 0.1 mol/L HCl solution until the solution changes from yellow to orange. Practice over and over again.

Raw data records

1. Phenolphthalein as indicator

Order	1	2	3
Consumption of 0.1 mol/L NaOH solution / mL			

2. Methyl orange as indicator

Order	1	2	3
Consumption of 0.1 mol/L HCl solution / mL			

Reflection

How to complete the neutralization titration quickly and well?

Task 1.2   Analytical balance use exercise

Enterprise case

After Xiaoming completed the acid-base neutralization titration task, the director of the laboratory gave him an analytical balance and an electronic balance and asked him to weigh some samples.

Mission objectives

Through the study of this task, students can achieve the following goals:

(1) Be familiar with the structure of analytical balance;

(2) Master the weighing operation of electronic balance, and be able to weigh the sample correctly.

Quality objectives

Develop a good habit of patience.

Task description

Weigh with an analytical balance and record the data.

Implementation conditions

(1) Instruments: analytical balance, tray balance, medicine spoon, small beaker, conical flask.

(2) Specimens: powder or granular.

Procedures and methods

(1) Be familiar with the structure of analytical balance;

(2) Complete the sample weighing task assigned by the teacher;

(3) Record data accurately.

Reflection

Analytical balance is gradually eliminated by electronic balance, What laws of human development have been confirmed? At this stage, what purpose can be achieved by learning the use of analytical balance?

1.2.1 Task-related knowledge—basic knowledge of laboratory safety

1.2.1.1 Safety knowledge of chemical laboratory

(1) The laboratory technician shall have a serious working attitude, be familiar with the business, and strictly observe the operation procedures.

(2) The laboratory shall be equipped with various fire prevention equipment. Such as sandbags, sandboxes, fire extinguishers, etc., the laboratory technician must be familiar with the use of various fire prevention equipment and fire extinguishing objects.

(3) Inflammable, explosive and toxic substances shall be kept by special personnel.

(4) All electrical equipment in the laboratory shall be well insulated, and the instruments shall be properly grounded.

(5) All experiments that can produce toxic or irritating gases shall be carried out in the fume hood, and the head shall not be put into the fume hood, gas masks shall be provided.

(6) The reagent bottles, samples and solutions containing drugs shall be labelled. Never fill the container with items that do not match the label.

(7) Handle flammable and highly toxic substances carefully. Use such substances in a well-ventilated area away from sources of ignition. Metallic mercury is volatile. If the thermometer is broken carelessly, collect the mercury beads as soon as possible and cover the liquid mercury with sulfur powder to convert the metallic mercury into non-volatile mercury sulfide.

(8) Do not leave the post during heating. Before heating the test tube, the water drops on the outer wall should be wiped off; when heating the test tube, do not face others or yourself, and do not look down at the liquid being heated.

(9) When opening the stopper of hydrochloric acid, nitric acid, ammonia, hydrogen peroxide and other reagent bottles, be careful that the gas suddenly rushes out. When smelling, do not put your nose directly close to the mouth of the bottle, but use a hand fan to smell. When using concentrated acid, alkali and lotion, avoid contacting with skin or splashing on clothes, and pay more attention to eye protection.

(10) Dilution of sulfuric acid must be carried out in a hard, heat-resistant beaker or conical flask. Keep in mind that the sulfuric acid can only be poured into the water slowly and stirred while pouring. When the temperature is too high, it should be cooled or cooled before continuing. It is strictly prohibited to pour water into the sulfuric acid. When sulfuric acid is diluted, a large amount of heat will be generated, and the density of concentrated sulfuric acid is greater than

that of water. When water is poured into sulfuric acid, the water will inevitably float on the top of sulfuric acid. The heat generated when mixed with sulfuric acid may boil the solution, and the sulfuric acid will splash out and hurt people.

(11) When using various electrical appliances, pay attention to the matching of voltage, current and power, and do not touch the power plug with wet hands. Be familiar with the installation of water, electricity and gas in the laboratory, the location of the main gate and the storage location of fire extinguishing equipment for emergency use.

(12) It is forbidden to use laboratory utensils to hold food, teacups and tableware to hold medicines, and beakers or measuring cups should not be used as tea sets.

1.2.1.2 Handling of laboratory accidents

(1) Cut treatment: After smearing iodine on the wound, apply a band-aid, and send the severe case to the hospital in time.

(2) Scald treatment: Smear scald medicine on the wound or wet the wound with 10% $KMnO_4$ solution until the skin turns brown, or smear the wound with 5% picric acid solution.

(3) Acid and alkali corrosion: When acid and alkali are splashed on clothes or skin, dry cloth or absorbent paper should be used to absorb them, and wash them with plenty of water immediately. In case of acid burn, the part shall be washed with water and then treated with saturated sodium bicarbonate, dilute ammonia solution or soapy water; in case of alkali burn, the part shall be washed with water and then treated with 2%~5% acetic acid or 3% boric acid solution. If acid splashes into the eyes, flush first with copious amounts of water, then with 1%~3% sodium bicarbonate solution, and finally with copious amounts of water. In serious cases, they should be sent to the hospital immediately after the above treatment. If alkali splashes into the eyes, flush with plenty of water, then flush with 3% boric acid solution, and then send to the hospital immediately.

(4) When toxic gases such as bromine, chlorine, and hydrogen chloride are inhaled, a small amount of steam mixed with alcohol and ether can be inhaled to detoxify, and fresh air should be breathed outdoors. In case of inhalation of hydrogen sulfide and carbon monoxide, please go outside to breathe fresh air immediately.

(5) When the poison enters the mouth, a cup of dilute copper sulfate solution can be taken orally, and then the finger can be inserted into the throat to induce vomiting, and then immediately sent to the hospital.

(6) In case of electric shock, first cut off the power supply, and use insulating materials such as dry wooden sticks or bamboo poles as soon as possible to separate the electric shock victim from the power supply. If necessary, artificial respiration should be carried out and immediately sent to the hospital for rescue.

1.2.1.3 General knowledge of fire prevention and extinguishing in the laboratory

1. Main causes of fire in the laboratory

(1) The flammable substance is too close to the fire source.

(2) Wire ageing, poor plug contact or electrical failure, etc.

2. Mixture or contact of several substances liable to cause fire

(1) Mixing activated carbon with ammonium nitrate;

(2) Clothes contaminated with strong oxidants (such as potassium chlorate);

(3) Rags and concentrated sulfuric acid;

(4) Combustible substances (wood or fibre, etc.) and concentrated nitric acid;

(5) Organics and liquid oxygen;

(6) Aluminum and organic chloride;

(7) Contacting phosphine, silane, alkyl metal, white phosphorus, etc. with air.

3. Extinguishing method

In case of fire or fire in the laboratory, do not panic. Calmly and decisively take extinguishing measures (Table 1-1) and give an alarm in time.

Table 1-1 Description of Extinguishing Methods for Combustible Materials

Comburent	Extinguishing method	Explain
Paper, textile or wood	Sand, water, fire extinguisher	Need to cool down and isolate the air
Oil, benzene and other organic solvents	CO_2, dry powder fire extinguisher, asbestos cloth, dry sand, etc.	It is suitable for fire extinguishing on valuable instruments
Alcohol, ether, etc.	Water	It needs to be diluted, cooled and isolated from the air
Electricity meter and instrument combustion	CCl_4, CO_2 and other fire extinguishers	Do not use water and foam extinguishers to extinguish the fire because the fire extinguishing materials are not conductive
Contact of active metals (such as potassium, sodium, etc.) and phosphides with water	Dry sand and dry powder fire extinguisher	Never use water or foam, CO_2 extinguishers
Clothes on the body	Roll on the spot, press the fire extinguishing stuffing or take off clothes, and cover the fire with special fireproof cloth	Do not run, otherwise it will intensify the combustion

1.2.2 Use of glassware

1.2.2.1 Type, specification and washing method of glassware

1. Type of glassware

Because glass has a series of valuable properties, it has high chemical stability, thermal stability, good transparency, certain mechanical strength and good insulation properties. Therefore, a large number of glass instruments are used in various laboratories.

The chemical composition of the glass is mainly SiO_2, CaO, Na_2O and K_2O. With the introduction of B_2O_3, Al_2O_3, ZnO, BaO and so on, various glasses with different compositions have different properties and uses.

The extremely hard glass and hard glass contain more SiO_2 and B_2O_3, which belong to the second class of high borosilicate glass. They have high thermal stability, good acid and water resistance in chemical stability, and slightly poor alkali resistance. Generally, instrument glass and container glass are soft glass with poor thermal stability and corrosion resistance.

The chemical stability of glass is good, it is not free from erosion, just the amount of corrosion meets certain standards. Due to the corrosion of glass, trace ions enter the solution and the glass surface absorbs the ions to be analyzed in the solution, which is a problem that should be paid attention to in microanalysis. Hydrofluoric acid has a strong corrosive effect on glass, so glass instruments can not be used for experiments containing hydrofluoric acid. Lye, especially concentrated or hot lye, has an obvious corrosive effect on glass. Glass instruments storing lye, if with the grinding mouth, will also cause the grinding mouth to stick together and can not be opened. Therefore, glass containers should not be used to store lye for long periods of time.

Quartz glass belongs to special instrument glass. Its physical and chemical properties are similar to those of glass. It has excellent chemical stability and thermal stability, but its price is relatively expensive.

2. Commonly used glassware

There are many kinds of glassware used in laboratories. According to their uses, glassware can be roughly divided into three categories: containers, measuring instruments and other commonly used utensils. Some special glass instruments are also used in laboratories of different specialities.

Here, we mainly introduce the knowledge of general glass instruments and some ground glass instruments. See Table 1-2 for the names, specifications and uses of common glassware.

Table 1-2 List of names, specifications and uses of common glass instruments

Instrument	Specifications	General use	Precautions for use
Test tube	Such as 25 mm×150 mm, 10 mm×75 mm	Reaction vessel. Convenient operation, observation and medication. The amount is small	1) The test tube can be directly heated by fire, but can not be quenched. 2) When heating, use a test tube clamp to hold the tube. The tube mouth should not face people, and the test tube should be moved continuously to make it heated evenly. The liquid contained should not exceed 1/3 of the volume of the test tube. 3) The small test tube is generally heated by a water bath
Centrifuge tube	Such as 25 mL, 15 mL, 10 mL	A small amount of sediment Identification and separation	Do not heat directly with fire
Beaker	Such as 1,000 mL, 250 mL, 100 mL, 50 mL, 25 mL	Reaction vessel. Use when there are many reactants	1) Can be heated to high temperatures. Care should be taken not to cause the temperature to change too drastically during use. 2) When heating, the bottom is padded with asbestos net to make it heated evenly
Flask	Such as 500 mL, 250 mL, 100 mL, 50 mL	Reaction vessel. It is used when the reactants are more and need to be heated for a long time	1) The flask can be heated to high temperatures. Care should be taken not to cause the temperature to change too drastically during use. 2) When heating, the bottom is padded with an asbestos net to make it heated evenly
Conical flask	Such as 500 mL, 250 mL, 100 mL	Reaction vessel. It is convenient to shake and suitable for titration	1) The conical flask can be heated to high temperatures. Care should be taken not to cause the temperature to change too drastically during use. 2) When heating, the bottom is padded with an asbestos net to make it heated evenly

Instrument	Specifications	General use	Precautions for use
Iodine flask	Such as 250 mL, 100 mL	Used in iodometry	1) Pay attention not to scratch the frosted part of the stopper and the edge of the bottle mouth to avoid leakage. 2) Open the stopper at the time of dropping, and wash the bottle mouth and the iodine solution on the stopper into the bottle with distilled water
Volumetric flask	Such as 1 000 mL, 500 mL, 250 mL, 100 mL, 50 mL, 25 mL	Used for preparation of solution with accurate concentration	1) Do not heat. 2) Solids cannot be dissolved in it
Funnel	Such as 6 cm long necked funnel, 4 cm short necked funnel	For filtering or pouring liquid	Do not heat directly with fire
Dropper	Material: pointed glass tube and rubber nipple	1) Suck or add a small amount (a few drops or 1 to 2 cm^3) of liquid. 2) Pipetting the supernatant of the precipitate to separate the precipitate	1) When dropping, keep vertical, avoid tilting, especially avoid standing upside down. 2) The pipe tip shall not contact with other objects to avoid contamination
Measuring cylinder, measuring cup	Measuring cylinder: such as 100 mL, 50 mL, 25 mL, 10 mL Measuring cup: Such as 100mL	For volumetric metering of liquids	It cannot be heated

continued

Instrument	Specifications	General use	Precautions for use
Pipet, pipette	Pipet: such as 10 mL, 5 mL, 1 mL Pipette: Such as 50 mL, 10 mL, 5 mL, 1 mL	Used to accurately measure a certain volume of liquid	It cannot be heated
Buret, buret Rack	The burette is divided into basic form (a) and acid form (b), Colorless and brown. Such as 50 mL, 25 mL	1) The burette is used for titration or to accurately measure a certain volume of solution. 2) The burette holder is used to hold the burette	1) Alkaline burette holds the alkaline solution, the acid burette holds acid solution, and the two can not be mixed. 2) The basic burette shall not contain oxidant. 3) Brown burette should be used for the titrant which is easy to decompose when exposed to light. 4) The piston of acid burette shall be fixed with a rubber band
Wash the bottle	Material: Plastic. Such as 500 mL	Wash the precipitate and container with distilled or deionized water	
Titer plate	Material: White porcelain plate Specification: 12 points, 9 points, 6 points, etc. According to the number of concave points	Precipitation reactions for spot reactions which generally do not require separation, in particular it's a color reaction	1) No heating. 2) It cannot be used for the reaction with hydrofluoric acid and concentrated alkali solution
Drier	18 cm, 15 cm, 10 cm	1) During quantitative analysis, placing a burnt crucible in the crucible for cooling. 2) It is used to store the sample to prevent the sample from absorbing moisture	1) The temperature of the burned object should not be too high before it is put into the dryer. 2) Check whether the desiccant in the dryer is invalid before use

3. Washing method for glass instrument

In the work of laboratory analysis, cleaning glass instruments is not only necessary preparatory work before the experiment but also technical work. Whether the instrument washing meets the requirements has an impact on the accuracy and precision of the analysis. Different analytical work (such as industrial analysis, general chemical analysis, microanalysis, etc.) have different instrument cleaning requirements; how mainly introduce the washing methods of glass instruments based on general quantitative chemical analysis.

(1) General steps for washing the instrument.

1) Brush with water. Prepare brushes for washing instruments of various shapes, such as test tube brushes, beaker brushes, bottle brushes, burette brushes, etc., brush the instrument with a brush dipped in water and wash the soluble substances and brush the dust adhered to the surface.

2) Scrub with detergent powder, soap or synthetic detergent. Decontamination powder is a mixture of sodium carbonate, clay, fine sand, etc. It's alkalinity can remove oil stains, and the friction of solid substances can remove a variety of dirt. The disadvantage is that the degreasing effect is not well, which will damage the glass, so it is strictly prohibited to wipe the burette and other instruments and the transparent surface of optical glass such as cuvette with decontamination powder.

In recent years, synthetic detergents have been used to clean glass instruments due to their strong ability to remove oil stains. When the cleaned instrument is inverted, there is no water drop on the wall of the instrument after the water flows out. A small amount of pure water can be used to brush the instrument three times to remove the impurities brought by tap water, and then it can be used.

(2) Use of various washing liquids. According to the nature of instrument contamination, different washing liquids can be used to clean the instrument effectively in Table 1-3. It should be noted that when using various detergents with different properties, it is necessary to remove the previous detergent before using another one, to avoid interaction, and the resulting products may be more difficult to clean.

Table 1-3　Several common detergents

Chromic acid solution: Dissolve 20 g of ground potassium dichromate in 40 mL of water and slowly add 360 mL of concentrated sulfuric acid	It is used to remove the residual oil stain on the wall. Use a small amount of lotion to brush or soak overnight. The lotion can be reused
Industrial hydrochloric acid (concentrated or 1∶1)	It is used to wash away alkaline substances and most inorganic residues. Do not heat the alkali-ethanol lotion
Alkaline lotion: 10% aqueous solution of sodium hydroxide or ethanol solution	The aqueous solution is heated (can be boiled) for use, and the oil removal effect is particularly good. Note that too long a boiling time will corrode the glass

continued

Alkaline potassium permanganate lotion: Dissolve 4 g of potassium permanganate in water, add 10 g of sodium oxide, and dilute with water to 100 mL	Clean greasy dirt or other organic matter, and remove brown manganese dioxide separated from that contaminated part of the container with concentrate hydrochloric acid or oxalic acid lotion, ferrous sulfate, sodium sulfite and other reducing agents
Oxalic acid lotion: Dissolve 5–10 g of oxalic acid in 100 mL of water and add a small amount of concentrated hydrochloric acid	Washing manganese dioxide generated after washing with potassium permanganate washing solution, and heating for use if necessary
Iodine–potassium iodide solution: Dissolve 1 g of iodine and 2 g of potassium iodide in water and dilute to 100 mL with water	The black–brown stain left after washing the used silver nitrate titrant can also be used to scrub the white porcelain sink stained with silver nitrate
Organic solvent: Benzene, ether, acetone, dichloroethane, etc.	1) Oil stains or organic substances soluble in the solvent can be washed away, and attention should be paid to their toxicity and flammability when using them. 2) The scale of indicator solution prepared with ethanol can be washed with hydrochloric acid–ethanol (1 ∶ 2) washing solution
Ethanol, concentrated nitric acid (do not mix in advance)	This method can be used for organic matters that are difficult to cleane by ordinary methods: add not less than 2 mL of ethanol into a container, add 10 mL of concentrated nitric acid, let it stand for a while, and then a violent reaction will occur immediately, releasing a large amount of heat and nitrogen dioxide. After the reaction stops, wash it with water. The operation should be carried out in a fume hood, and the container should not be blocked for protection

In recent years, the dosage of chromic acid pickling solution has been gradually reduced due to its high toxicity. Synthetic detergents and organic solvents are mostly used to remove oil stains. Chromic acid pickling solution is not used as far as possible, but sometimes it is still used, so it is also included in the table.

These washing methods use the physical (inter solubility) and chemical properties of substances to achieve the purpose of cleaning instruments. Waste acid and recovered organic solvents can be separately collected and utilized in the laboratory.

(3) Washing of sand core glass filter.

1) Before use, the new filter should be cleaned with hot hydrochloric acid or chromic acid solution while pumping and filtering and then washed with distilled water. It can be repeatedly pumped and washed with water in the positive or reverse position.

2) For different precipitates, use appropriate detergent to dissolve the precipitate first, or repeatedly wash the precipitate with water, then rinse it with distilled water, dry it at 110 ℃, and

then store it in a dust-free cabinet or a covered container. Otherwise, it is difficult to clean the filter holes because of the accumulated dust and sediment.

(4) Washing methods for special requirements. It is effective to wash with steam after washing with the general method. Some experimental instruments require steam washing. The method is to install a steam pipe with a flask, turn the container to be washed upside down and blow it with water vapour.

Some trace analysis has high requirements for instrument washing. It is required to wash away impurity ions at the ppb level (ppb is a way of expressing concentration, which is expressed by the ratio of solute mass to total solution mass, also known as the concentration of parts per billion). The cleaned instrument should be soaked in 1 : 1 HCl or 1 : 1 HNO_3 for several hours (up to 24 h), so as not to adsorb inorganic ions. Then rinse it with pure water. Some instruments need to be burned at a temperature of several hundred degrees Celsius to meet the requirements of trace analysis.

1.2.2.2 Preparation and use of glassware

1. Drying and storage of glassware

(1) Drying of glassware. Glassware should be kept in a dry and clean place, and the instruments often used in experiments should be cleaned and dried after each experiment. Instruments used in different experiments have different requirements for drying. Generally, instruments such as beakers and conical flasks used in quantitative analysis can be used after cleaning, while many instruments used in organic chemistry experiments or organic analysis require drying, some require no watermarks, and some require no water. The instrument shall be dried according to different requirements.

1) Dry. The instrument that is not in a hurry to use and requires general drying. After brushing with pure water, put it in a dust-free place to control the moisture, and then dry it naturally.

2) Drying. The cleaned instrument to remove water, and drying the instrument in an electric oven at the temperature of 105 to 110 ℃ for about 1 hour. It can also be dried in an infrared lamp-drying oven. This method is suitable for general instruments. When heating, the temperature should be raised gradually to avoid sudden cooling and heating. The weighing bottle used for weighing shall be cooled and stored in the dryer after drying. When drying instruments with solid glass plugs and thick walls, pay attention to slowly warming up and the temperature should not be too high to avoid cracking. Note that the measuring instrument should not be heated, should not be used to store concentrated acid or alkali, and should not be dried in an oven.

The hard test tube can be heated and dried with an alcohol lamp. It should be baked from the bottom, and the mouth of the test tube should be downward, to avoid the explosion of the test tube caused by the backflow of water droplets. After baking until there are no water droplets, the mouth of the tube should be upward to remove the water vapour.

3) Drying by hot (cold) air. For instruments that are in a hurry to dry or larger instruments

that are not suitable to be placed in an oven, the method of drying can be used. Usually, a small amount of ethanol and acetone (or ether at last) is poured into the instrument that has been controlled to remove moisture and shaken (the solvent should be recovered), and then blown by an electric blower for 1-2 min with cold air. When most of the solvent is volatilized, hot air is blown to completely dry the solvent. Then the residual steam is blown away by cold air so as not to be condensed in the container. This method requires good ventilation to prevent poisoning, and no contact with open fire to prevent the explosion of organic solvent vapor.

4) Do not pour hot solution or hot water into thick-walled instruments.

(2) Custody of glassware. Glassware should be stored in separate categories in the storeroom for easy access. Frequently used glass instruments should be placed in the experimental cabinet, and the high and large ones should be placed inside. Here are some ways to keep the instrument.

1) Pipette. After the pipette is cleaned, wrap the two ends with clean filter paper. If the pipette is used for experiments with higher requirements, wrap it with filter paper to prevent contamination.

2) Burette. Brush it with pure water, fill it with pure water, cover it with a short glass test tube or plastic sleeve, or place it upside down on the burette clamp.

3) Cuvette. Place filter paper under a small porcelain plate or a plastic plate, after clean, invert and dry the cuvette, and put the cuvette into a cuvette box or a clean utensil.

4) Instrument with grinding plug. Instrument volumetric flasks or colour comparison tubes with ground stoppers should be tied with small strings or rubber bands before cleaning, so as not to break the stoppers or confuse each other. A piece of paper should be placed between the plug and the mouth of the grinding instrument that needs to be stored for a long time, so as not to stick for a long time. Burettes that have not been used for a long time should be padded with paper after removing Vaseline, and the piston should be tied with a rubber band for storage. If there are sand grains between the grinding plugs, do not rotate them with force, so as not to damage their accuracy. In the same way, do not use detergent powder to scrub the grinding area.

5) Complete sets of instruments such as Soxhlet extractor and gas analyzer should be cleaned immediately after use and stored in special cartons.

In a word, in the spirit of being responsible for the work, all the glass instruments should be cleaned and kept according to the requirements, and good working habits should be developed. Do not leave grease, acid, corrosive substances (including concentrated alkali) or toxic drugs in the containers, to avoid future trouble.

(3) Methods for opening the ground plus. When the grinding piston can not be opened, it is easy to be broken by screwed hard, and corresponding measures should be taken to open it according to different situations.

1) The piston is stuck by an oily substance such as Vaseline, it can be heated slowly with a hair dryer or a slow fire to reduce the viscosity of the oil, it can be opened by tapping the plug with a wooden stick after melting.

2) Pistons are stucked by dust for a long time idle, can be soaked in water and opened after a few hours.

3) The piston stuck by the alkaline substance can be heated to boiling water, and then the stopper can be tapped with a wooden stick.

4) The stopper of the reagent bottle with reagent can not be opened, it is necessary to accumulate some experience and take appropriate methods to open the stopper of the grinding bottle.

5) There is a corrosive reagent in the bottle, such as concentrated sulfuric acid, etc., a plastic drum should be placed outside the bottle to prevent the bottle from breaking. The operator should wear a plexiglass mask. When operating, do not make the face too close to the bottle mouth. Open the bottle mouth of toxic steam (such as liquid bromine) in a fume hood, tap the bottle cap with a wooden stick or wash the bottle mouth. Blow out a little distilled water from the washing bottle to moisten the grinding mouth, and then tap the bottle cap.

6) Corks stuck by crystallization or alkali metal salt deposition and strong alkali, the bottle mouth can be soaked in water or dilute hydrochloric acid for a period of time before trying to open.

2. Preparation and use of titration analysis instrument

Titration analysis is based on the volume and concentration of the standard solution consumed during titration to calculate the analysis results. Therefore, in addition to accurately determining the concentration of the standard solution, its volume must also be accurately measured. The error of solution volume measurement is the main source of error in titration analysis. If the volume measurement is not accurate (for example, the error is greater than 0.2%), even if the other operation steps are done accurately, it is futile. Therefore, it is necessary to measure the volume of the solution accurately to make the analysis results meet the required accuracy. On the one hand, it depends on whether the volume of the volumetric instruments used is accurate; on the other hand, it also depends on whether these instruments can be used correctly.

Volumetric instruments used to measure the exact volume of a solution in titrimetric analysis include the burette, pipet, pipette and volumetric flask. Burette, pipet and pipette are "measuring out" type measuring instruments, which are marked with "A", but in China, "Ex" is used to measure the volume of liquid discharged from the measuring instrument. Generally, the volumetric flask is a "measuring in" type measuring instrument, which is marked with "E", but in our country, the word "in" is used to indicate "measuring in". It is used to determine the volume of liquid injected into the measuring device. The other is the "measuring out" type volumetric flask, which is marked with "A" or "Ex", indicating that the volume of the liquid is the same as the volume marked on the bottle when it is poured out according to a certain method after the liquid is filled to the scale of the marked line at the marked temperature.

The burette is used to accurately measure the volume of the solution released during titration. It is a slender glass tube with a scale. According to its capacity and scale value, the burette can be divided into three types: constant burette, semi-micro burette and micro burette. According to different requirements, there are a "blue ribbon" burette and a brown burette (used to hold potassium permanganate, silver nitrate, iodine and other standard solutions). According to the different structures, it can be divided into ordinary burette and automatic burette. According to different uses, it can be divided into acid burette and alkali burette.

An acid burette, with a ground glass stopcock to control the flow of droplets, is used to hold acids or oxidizing solutions. However, the alkaline solution can not be filled, because the ground cock will be corroded by the alkali and will not be able to rotate. A latex tube with glass beads control the liquid drop, and an alkali burette (alkali tube for short) is connected to a sharp glass tube at the lower end. It is used to hold alkaline solution and non-oxidizing solutions, and cannot hold $KMnO_4$, $AgNO_3$, I_2 and other solutions, to prevent the tube from being oxidized and deteriorating.

The buret should be clamped vertically to the buret holder for titration. Use of acid tube: bend the ring finger and little finger of the left hand to the palm, gently stick to the outlet tube, and use the other three fingers to control the rotation of the piston, but be careful not to pull the cock outward, so as not to push out the cock and cause leakage, and do not buckle inward too much, so as not to make it difficult for the cock to rotate and operate freely. Use of alkali tube: clamp the outlet tube with the ring finger and little finger of the left hand, squeeze the latex tube aside with the thumb and forefinger at the position where the glass bead is located, move the glass bead to the palm side, make the solution flow out from the gap beside the glass bead, do not pinch the glass bead hard, and do not make the glass beads move up and down; do not pinch the latex tube at the lower part of the glass bead, so as to prevent air from entering and forming bubbles, which will affect the reading. When stopping the titration, loosen the thumb and index finger first, and then loosen the ring finger and little finger.

To control the titration speed, three dropping methods must be mastered: continuous dropping drop by drop, that is, the general titration speed, the method of "see the drop into a line"; adding only one drop, to achieve the skilled operation of adding only one drop; making the drop suspended without falling, that is, adding only half a drop, or even less than half a drop. Close the cock immediately when titrating to the endpoint, and be careful not to let the solution in the burette flow out slightly, otherwise, the final reading will be affected. The reading shall be read to the second place after the decimal point, that is, it is estimated to be 0.0 mL. If the reading is 25.33 mL, recorded the data immediately.

At the end of the titration, the remaining solution in the burette should be discarded and should not be poured back into the original reagent bottle, so as not to contaminate the whole bottle of operating solution. Immediately wash the burette and place it upside down on the burette stand.

1.2.3 Record of analysis data

The results of measurement are approximate data with errors, that is, the true value is impossible to measure. Therefore, when recording and calculating, the number of bits and the position of the data should be determined based on the possible accuracy of measurement. If the number of digits of the data involved in the calculation is less, the accuracy of the measurement results will be impaired; if the number of digits is too large, it is easy to be mistaken for high measurement accuracy, and unnecessary calculation workload is increased.

1.2.3.1 Operation rules of significant figures

1. Significant digits

Significant figures are those that can be measured in experimental work. There is always a limit to the accuracy of the scale by the instrument. For example, in a 50 mL measuring cylinder, the minimum scale is 1 mL, and one more digit can be estimated between the two scales, so the actual measurement can be read to 0.1 mL, the minimum scale is 0.1 mL, and one more digit is estimated, which can be read to 0.01 mL. Since the last digit is estimated, this digit is called a suspicious digit. For example, when reading the liquid level position of the burette, A may read 21.32, B may read 21.33, and C may read 21.31. It can be seen that 21.3 is displayed on the burette, that is, the minimum graduation of the burette is 0.1, and the last digit is estimated by the experimenter. Different estimates may be obtained by different experimenters, but this estimated number is objective, so it is a significant number. That is to say, the significant figure is the exact number measured plus one suspicious digit.

From the above, we can see that significant figures have different meanings from mathematical numbers. Mathematical numbers only represent magnitude, while significant figures not only represent the magnitude of the quantity but also reflect the accuracy of the instrument used (the minimum scale of the instrument). For example, "Take 6.5 g NaCl", which not only indicates that the mass of NaCl is 6.5 g, but also indicates that it can be weighed with a platform scale with a sensitivity of 0.1 g. If "Take 6.500 0 g NaCl", it indicates that it must be weighed on the analytical balance.

The number of significant digits of the instrument reading is determined by the performance of the instrument. For example, the analytical balance can be weighed to 0.000 2 g, and the burette can be read to 0.01 mL. When recording measured values, one and only one digit of the estimate must be retained. For digital instruments, the displayed digits are significant and need not be estimated.

2. Number of significant digits

The greater the number of significant digits, the smaller the relative error of the determination. Therefore, when recording measurement data, do not write casually, otherwise, it will exaggerate or reduce the accuracy.

The number of significant digits of a number from the first non-zero digit on the left to the rightmost digit.

For example, 0.023 has two significant digits and 230.40 has five significant digits.

Note that all zeros before the first non-zero digit on the left are not significant digits, and are merely used to indicate the position of the decimal point. However, all zeros after the last non-zero digit are significant digits. "0" plays a different role in numbers. Sometimes it is a significant number and sometimes it is not, depending on the position of "0" in the number.

(1) "0" precedes a number and only plays a positioning role. "0" itself is not a significant digit. For example, in 0.027 5, the two zeros before the number 2 are not significant digits. The number has only 3 significant digits.

(2) "0" is a significant digit in numbers. For example, both zeros in 2.006 5 are significant digits, and 2.006 5 has 5 significant digits.

(3) "0" is also a significant digit after the decimal digit. For example, the three zeros in 6.500 0 are all significant digits. The three zeros before the number 3 in 0.003 0 are not significant digits, and the zeros after the number 3 are significant digits. So, 6.500 0 is 5 significant digits. 0.003 0 is 2 significant digit.

(4) A positive integer ending in "0" with an indefinite number of significant digits. Such as 54 000, may be 2, 3, or 4 or even 5 significant digits. This number should be rewritten in exponential form depending on the number of significant digits. Write 5.4×10^4 for 2 bits, 5.40×10^4 for 3 bits, and so on.

In short, in order to correctly identify and write significant figures, the data recorded in the determination process should be significant figures. Some figures are listed below, and significant digits is indicated:

7.400 0	54 609	5 significant digits
33.15	0.070 20%	4 significant digits
0.027 6	2.56×10^{-1}	3 significant digits
49	0.000 40	2 significant digits
0.003	4×10^{-5}	1 significant digit
63 000	200	Indefinite number of significant digits

3. Operation of significant figures

In number arithmetic, generally speaking, two numbers should be added or subtracted so that they have the same degree of accuracy, and two numbers should be multiplied or divided so that they have the same degree of accuracy, that is, each number retains the same number of significant digits, and so does the result of the calculation.

Attention should be paid to the following points in the approximate operation.

(1) Added or subtracted, the number of significant digits reserved for sum or difference shall be based on the number with the least number of digits after the decimal.

For example, 0.031 2+23.34+2.503 81. On the basis of 23.34, the other figures are rounded to the second decimal place and then added.

$$
\begin{array}{r}
0.03 \\
23.34 \\
+)\ 2.50 \\
\hline
25.87
\end{array}
$$

(2) When multiplying and dividing, the number of significant digits depends on the number with the largest relative error or the number with the least number of significant digits.

For example

$$\frac{0.023\ 4 \times 4.033 \times 71.07}{127.5} = 0.052\ 604\ 2$$

The result of the calculation shall be taken as 0.052 6, i.e. the same number of digits as 0.023 4 (the number with the least number of significant digits).

However, in the process of operation, the result of each operation can be reserved for one more digit than the number with the least number of significant digits.

For example, $0.023\ 4 \times 4.303 = 0.100\ 690\ 2$

0.100 7 can be taken to continue the operation (one more significant digit than 0.023 4).

$0.100\ 7 \times 71.07 = 7.156\ 749$

7.157 proceed to the next operation:

$7.157 \div 127.5 = 0.056\ 133\cdots\cdots$

The result shall be taken as 0.056 1.

(3) In analysis and calculation, some fractions are often encountered, for example, the equivalent of a substance is equal to 1/2 of its component, or 25 mL of 250 mL of test solution is absorbed, that is, 1/10 is absorbed. Here, "2" and "10" can be regarded as sufficiently effective, but, they can not be regarded as having one or two significant digits, thus serving as the basis for judging the number of significant digits of the calculation result. This class of numbers is called exact numbers. An exact number can have a "0" added to the right of its decimal part to increase number of significant digits.

(4) If the first significant digit of a number is greater than or equal to 8, the number of significant digits can be counted as one more digit, such as 8.37 may be considered to be four significant digits.

(5) Number rounding (rounding) rule. The error caused by the rounding of numbers is called "rounding error" or "rounding error". In order to avoid the influence of the rapid accumulation of errors on the accuracy of the measurement results, the rounding rule of "rounding six, rounding five and leaving two" is usually used in the calculation:

1) If the first digit to be rounded off is a number from 0 to 4, the last digit retained remains unchanged;

2) If the first digit to be rounded off is a number from 6 to 9, the last digit to be retained is increased by 1;

3) If the first digit to be rounded off is 5 and the digits to the right of it are all 0, the last digit to be retained is added by 1 when it is an odd number and remains unchanged when it is an even number.

In the analysis work of chemical management, whether the analysis results are accurate and reliable is a very important issue. The wrong analysis results often have serious consequences. Therefore, the original data should be recorded realistically during the measurement, and the measured data should be processed after the measurement. If it is found that the conclusion of the analysis is inconsistent with the actual situation, it should be carefully checked based on these original data to find out the cause of the error, and it is not allowed to achieve the so-called "consistency" by changing the data.

1.2.3.2　Precision and accuracy of analysis

The main purpose of analysis is to measure the content of a certain component in a sample, and hope that the same result can be obtained by multiple measurements of the same sample. However, in the actual measurement, although the true value is objective, in the limited measurement, due to the influence of instruments and other factors, the true value of the measured object can not be determined, and there will be differences between the results of multiple measurements.The closer the measured value is to the true value and the smaller the difference between the results of repeated measurements, the higher the quality of the analysis results, which is the accuracy and precision of the analysis work.

1. Accuracy

The degree of agreement between the measured value and the true value. It indicates the correctness of the measured value.

2. Precision

Precision refers to the degree of closeness between the results of a group of parallel measurements under the same conditions. It describes the reproducibility of the measured data.

3. Relationship between accuracy and precision

(1) Precision is the prerequisite to ensure accuracy. If the precision does not meet the requirements, it means that the measured results are unreliable and the premise to measure accuracy is lost.

(2) High precision can not guarantee high accuracy.

In other words, accurate experiments must be precise, and precise experiments are not necessarily accurate.

The concepts of accuracy and precision can be illustrated by the distribution of impact points in target practice. Fig. 1-1 shows that the accuracy of B is higher, and Fig. 1-2 shows that the precision of A is higher.

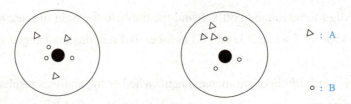

Fig.1-1　The accuracy of B is higher　　Fig.2-1　The precision of A is higher

1.2.3.3　Error of analysis process

1. Types and sources of errors

(1) System error. The systematic error is caused by some regular reasons in the determination process. Its influence on the analysis results is relatively fixed, and the error is often reproducible, which makes the determination results often higher or lower. The systematic error can be eliminated through correction after finding out the cause of the error. The main sources of systematic errors are as follows:

1) The analysis method is not perfect. For example, the precipitation in gravimetric analysis has a certain solubility, and the equivalent point and the endpoint in titration are not very consistent; or the test method itself is not perfect, for example, the reaction is incomplete or there are side reactions, dissolution of the precipitation, coprecipitation and dissociation of the complex; the empirical formula used does not fully conform to the real situation and is only approximate, or when the theoretical formula is used, the experimental conditions do not fully conform to the conditions required for the establishment of the theoretical formula, and the background or blank value correction is incorrect, which will cause errors.

2) The parts of the instrument are not precisely manufactured, the accuracy is reduced due to wear caused by long-term use, the instrument is not adjusted to the best state and the utensils are not corrected, the defects of the instrument itself or the use of uncorrected instruments. For example, the two arms of the balance are not equal in length, and the actual value of the burette is not exactly equal to the market value.

3) The reagent or distilled water used contains impurities.

4) Physiological characteristics of the operator. It is often associated with the operator, one is the operator's personal physiological function defects, such as poor eye discrimination, not correctly reading the scale value and distinguishing the color tone and shade, for example, in the titration analysis the colour change of the indicator is insensitive. Then there is the operator's subjective prejudice and inherent habits, such as the latter reading is affected by the former reading, unconsciously hoping to obtain the same or similar results for the two determinations; the advance or lag of the change of the observation endpoint. And the unreasonable operation, such as poor representativeness of sampling, inappropriate selection of ignition precipitation temperature and improper control of reaction conditions, etc.

5) The error caused by the environment not completely conforming to the conditions

required for the measurement. For example, the change of environmental temperature causes the change of the precision of measuring instruments and utensils; the serious influence of atmospheric pollution on the blank value of trace component determination; the influence of humidity and vibration on the accurate weighing; and the influence of parallax caused by the change of illumination on the reading, etc.

(2) Accidental error. This is an error due to many variable causes. Although the operator operates carefully and controls the important external conditions as consistently as possible, the measured series of data are inevitably different, and the size and positive and negative of the measured data error are not certain, which is an accidental error. The occurrence of this kind of error is irregular, so it can not be eliminated by the correction method, but can only be reduced to a minimum through standardized operation. The causes of accidental errors are as follows:

1) Inconsistent readings differences in personal discrimination. For example, when reading an eye dropper, the value of the second place after the decimal point is estimated to be inaccurate, resulting in reading errors.

2) There is a slight change in the temperature, humidity, pressure, etc. of the measurement environment.

3) Loss, contamination, misreading of weights, recording and calculation errors due to negligence in operation.

For example, when the concentration of lye is determined by acid-base titration during volumetric analysis, the volume of lye used, V_{alkali}, and the concentration of the standard acid solution, N_{acid}, are both constant values. When the analysis principle is inappropriate (such as incorrect selection of indicators) or the operation is wrong, the error is ultimately reflected in the high or low volume of V_{acid} consumed during titration. Several situations in which errors may occur are shown in the Table 1-4.

Table 1-4 Error when titrating alkali solution with standard acid solution

Cause of error	V_{acid} consumed for titration	N_{alkali} result bias ($N_{alkali} = N_{acid} V_{acid}/V_{alkali}$)
Bubble not discharged at the end of acid burette	On the high side	On the high side
Acid burette not rinsed with standard acid	On the high side (due to the low concentration of acid actually used)	On the high side
Erlenmeyer flask for titration was rinsed with alkali solution by mistake	On the high side (due to the high volume of alkali actually used)	On the high side
Water drops in the pipette, not rinsed with unknown lye	On the low side (due to the low concentration of lye in actual application)	Low

For beginners, often due to carelessness, do not comply with the operating procedures and other reasons, and the set of many operational errors, if the formation of bad habits, will have a serious impact on the analysis results. Therefore, it is the minimum work discipline that every analyst must observe to strictly follow the operating rules and operate meticulously and conscientiously.

In the actual measurement process, many factors play a role at the same time, and there are interaction effects among the factors, so the situation is more complex. The possible causes of errors listed above are only to point out the possible general direction for finding the source of errors, which can not replace to find the real causes of errors for specific occasions. The task of analysts is to find ways to reduce errors and constantly improve the precision and accuracy of the determination results.

2. Methods to reduce errors

(1) Correct use of instruments and equipment. Use the instrument correctly or calibrate the instrument according to the regulations. For example, when weighing the sample and sediment in gravimetric analysis and weighing the reference substance and sample in volumetric analysis, the same balance and the same box of weights should be used as much as possible (do not use three weights of 2 g, 2 g and 1 g instead of 5 g when only one 2 g weight is used without an asterisk) the error between the balance and the weight can be offset for the most part. If the same interval of the same burette is used as far as possible when the standard solution is calibrated and the measured solution is titrated, the error caused by the inaccurate calibration of the burette can also be offset.

(2) Carry out the analysis operation correctly according to the procedure. The error can be reduced by strictly abiding by the operating procedures, carefully making the original records, and carefully reviewing the calculation. The uniformity and representativeness of the sample composition, the operating environment, temperature and humidity of the test have a considerable impact on the accuracy of the analysis results, which must also be paid enough attention to. For example, if the same indicator is used to calibrate the standard solution and titrate the measured solution, the error that the discolouration point of the indicator can be eliminated. If possible, the calibration and determination should be carried out by the same analyst. The error caused by the difference in individual knowledge of the endpoint can also be eliminated.

(3) Carry out a blank test. The blank test is to carry out the analysis test according to the same operation procedures and conditions as the sample analysis under the condition of not adding the sample or replacing the sample with distilled water. The result obtained in this way is called the blank value. By deducting the blank value from the analysis result of the sample, the error caused by the impurity of the reagent or water can be corrected.

(4) A controlled trial was conducted. The standard sample with a similar composition to the sample is selected and operated under the same conditions, and the analysis result of the standard sample is compared with true value, so that the error caused by inaccurate method and impure

reagent can be measured, and the error can be corrected in the analysis result of the sample.

(5) Increase the number of parallel measurements. On the premise of eliminating the systematic error, the more times of measurement, the closer the average value of the results is to the true value, the accuracy and precision of the results can be improved.

Module 2
Daily analysis of raw materials and Semi-product

Task 2.1 Measuring Brix of WIP by density method and Refractometry

Enterprise case

The director takes back part of the intermediate cane juice from the production line and then arranges for Xiaoming to measure the concentration of the solid solution of the above samples. What instrument should Xiaoming choose and what method should he use to measure it?

Mission objectives

Through the study of this task, students can achieve the following goals:

(1) Understand and master the concepts of soluble solids, dry substance, apparent solids, Brix and refractometer Brix.

(2) Master the determination methods of Brix: density method and Refractometry.

Quality objectives

Develop a good habit of comparative learning.

Task description

The Brix of the sample was determined using a Brix hydrometer and a refractometer respectively.

Implementation conditions

(1) Sample: Syrup

(2) Equipment and instruments: Brix hydrometer, Abbe refractometer, thermometer (0 ~ 100 °C), volumetric cylinder, beaker.

Procedures and methods

1. Brix was determined by density method

(1) Sample processing.

(2) Determination of Brix.

2. Determination of refractive Brix by refractometry

(1) Sample processing.

(2) Determination of Brix.

For details, please refer to Task 2.1 of this module.

3. Data processing and calculation

(1) Calculation.

1) Density method. The Brix spindle takes 20 °C as the standard. If the reading temperature is not 20 °C, it is necessary to check the Brix temperature correction table for correction. When the temperature is lower than 20 °C, the Brix is observed minus the value of the lookup table; and when the temperature is higher than 20 °C, the Brix is observed plus the value of the lookup table, which is the corrected Brix.

For the diluted sample, the Brix of the diluted sample multiplied by the dilution factor is the Brix of the original sample.

2) Refractometry. Abbe refractometer takes 20 °C as the standard. If the reading temperature is not 20 °C, check the sugar liquid refraction Brix temperature correction table for correction. When the temperature is lower than 20 °C, the measured value is subtracted from the lookup table value; and when the temperature is higher than 20 °C, the measured value is added to the lookup table value, which is the corrected refractive index value.

For the diluted sample, the refractive Brix of the diluted sample multiplied by the dilution factor is the refractive Brix of the original sample.

(2) Data and result processing.

1) Density method.

Project	Data and results
Brix of sample solution /°Bx	
Temperature of sample solution /°C	
Temperature correction number of Brix	
Brix of sample solution /°Bx	
Dilution factor	
Brix of original sample /°Bx	

2) Refractive method.

Project	Data and results
The refractive index of the sample	
Brix of observed refraction of sample /°Bx	
Sample temperature /°C	
Temperature correction number	
Brix /°Bx	

Precautions

(1) When using the brix spindle, it must be taken vertically to prevent breaking. When putting in the measuring cylinder, it is necessary to wait for the Brix spindle to float before letting go, so as to avoid the damage caused by the sudden fall of the Brix spindle and the collision with the bottom of the measuring cylinder. When observing the reading, the Brix should not stick to the edge of the container, and the care should be taken to keep it clean so that there is a good meniscus in contact with the liquid surface.

(2) The refractometer prism is made of soft glass. Do not touch it with hard materials to prevent damage.

(3) Specimens with very dark colors, such as molasses, can be measured by refractometer using reflected light by adjusting the reflector so that no light enters from the entrance prism and at the same time opening the side cover of the refracting prism so that light enters from the side hole of the refracting prism.

Reflection

Refractometer requires more sample than gravity brix spindle?

2.1.1 Task-related knowledge—sample collection and handling

2.1.1.1 Terms related to chemical management of sugarcane sugar industry

1. Sugar

(1) A general term for sugars.

(2) A general term for products produced by sugar mills that consist essentially of sucrose. It is customarily called sugar.

2. Granulated sugar

Granulated sugar is sucrose in the form of sand grains. Due to the different particle size, there are young sand, medium sand and coarse sand. Because of the different quality and color, there are white granulated sugar and brown granulated sugar.

3. White granulated sugar

Sugar cane juice, beet juice or crude sugar solution is purified by sulfitation or carbonation, and then concentrated, crystallized, purged and dried to obtain clean white granulated sugar.

4. Brown granulated sugar

Brownish red or yellowish brown sugar with molasses.

5. Chopped or (and) shredded cane

The crushed material of sugarcane after being crushed by a sugarcane cutter (or a ripper).

6. Cane juice

Juice extracted from sugarcane.

7. Juice

A general term for all kinds of thin sugar solutions that have not been evaporated in sugar mills. Such as mixed juice, neutralized juice, and clarified juice, etc.

8. First expressed juice

Cane juice extracted by the first two rollers of the press unit.

9. Last expressed juice

Cane juice extracted by the last two rollers of the press unit.

10. Mixed juice

Cane juice extracted by pressing or exudation and sent to cleaning treatment.

11. Bagasse

The residue of cane material after pressing or exudation.

12. Preliming juice

Sugarcane juice after preliming treatment.

13. Limed juice

Sugar juice treated with lime (plus lime).

14. Neutralized juice

Sugar juice obtained by sulfurous fumigation and neutralization with lime.

15. 1st carbonatation juice

Sugarcane juice after the first carbonation treatment.

16. 1st carbonatation clear juice

The clear sugarcane juice is obtained by solid-liquid separation of carbonation juice.

17. 2nd carbonatation juice

Sugarcane juice after the second carbonation treatment.

18. 2nd carbonatation clear juice

The clear sugarcane juice is obtained by filtering the 2nd carbonatation juice.

19. Sulfitated juice

Sugarcane juice that has been treated with sulfur.

20. Clarified juice

Clear sugarcane juice was obtained after clarification treatment.

21. Filtered juice

Clear sugarcane juice filtered from a filter.

22. Mud juice

Sugarcane juice with thick suspension discharged from the clarifier (or thickening and filtering equipment).

23. Filter cake, Mud

Filter residue discharged from the filter.

24. Syrup

A highly concentrated sugar solution is obtained by concentrating a sugarcane juice.

25. Purified syrup

Syrup after purification.

26. Condensate (Steam condensate)

Water condensed by steam or juice after heat exchange.

27. Crystallization of sucrose (sucrose)

The process in which crystals are precipitated and grown by maintaining a certain supersaturation coefficient during the concentration or cooling of syrup or molasses.

28. Solubility coefficient

The ratio of the sucrose solubility of an impure sugar solution to the solubility of a pure sucrose solution at the same temperature.

29. Coefficient of supersaturation

The ratio of the number of parts of sucrose dissolved in water per part of supersaturated sugar solution to the number of parts of sucrose dissolved in water per part of saturated sugar solution at the same temperature and purity.

30. Nucleus

A tiny crystal grain in a supersaturated sugar solution that is the nucleus of crystal growth.

31. Seed

Boiled in a seed crystal pot or mixed with low-purity granulated sugar and molasses, containing a certain amount of crystal grains to support the material used for crystallization. It is customary to call seeds.

32. Quincunx Conglomerate

A crystal group consisting of two or more crystals facing each other at irregular angles.

33. Aggregate, Mounted grain

A group of crystals formed by the aggregation of many crystals.

34. Twin crystals, Married grain

A crystal formed by twinning two crystals symmetrically about a common plane.

35. False grain

In the process of sucrose crystallization, new and unnecessary tiny grains are produced.

36. Magma

A pasty mixture obtained by mechanically mixing granulated sugar with a sugar solution. There are two kinds of crystal seed magma and molasses-washed magma.

37. Massecuite, Strike

A mixture of crystals and mother liquor obtained by boiling sugar. According to the order of boiling sugar, it can be divided into A, B and C massecuite and so on.

38. Molasses

A generic term for the mother liquor separated from a massecuite or molasses-washed magma. There are green molasses, white molasses, final molasses and so on.

39. Green molasses

The mother liquor separated directly from the massecuite during the separation of molasses.

40. White molasses

Thin molasses obtained by steam washing or water washing in the process of separating molasses.

41. Final molasses, Blackstrap molasses

The mother liquor separated from the last massecuite.

42. Semi-product, Stock in process

In the production process of sugar mills, except for shredded cane, finished sugar, final molasses, bagasse and filter cake, the materials (or products) being processed or to be processed in each process are collectively referred to as products in process. For example, various sugarcane juices, syrups, massecuite, molasses, etc.

43. Sugar products

Materials containing more sucrose in the process of sugar production. It usually refers to sugar, semi-product and products.

44. Normal weight

According to the international uniform regulations, prepare a certain weight of pure sucrose into 100.000 cm^3 aqueous solution at 20.00 °C, and measure its optical rotation with a 200.000 mm polarimeter tube and a polarimeter with an international sugar scale. The reading is 100 °S (international sugar scale), and this weight is called the normal weight.

A weight of 26.000 g (26.016 0 g under vacuum) is weighed in air with a brass weight as a normal weight.

45. Soluble solids

Solid matter (including sucrose and non-sucrose substances) dissolved in sugar juice.

46. Solids by drying (Dry substance, Total solids)

The residue obtained after sugar has been dried.

47. Apparent solids

Soluble solids in the sugar solution are measured with a brix spindle or a sugar refractometer (it is only an approximate solid solution).

48. Brix (°Bx)

Read with a Brix spindle at 20 °C. For pure sucrose solution, it is the weight percent of sucrose; for impure sucrose solution, it is the weight percent of apparent solid solution in the solution.

49. Refractometer Brix

Refractometer reading at 20 °C with sugar. For pure sucrose solution, it is the weight percent of sucrose; for impure sucrose solution, it is the weight percent of apparent solid solution in the solution.

50. Pol (Polarisation)

The approximate value of the weight percentage of sucrose in the sugar product is measured by the single rotation method.

51. Purity

The percentage of sucrose in the solid solution of a sugar product.

52. True purity (or T. P.)

The percentage of sucrose in the dry substance of a sugar product.

$$\text{True purity (\%)} = \frac{\text{Sucrose}}{\text{Solids by drying}} \times 100\%$$

53. Gravity purity (or G. P.)

The percentage of sucrose in the apparent solid solution of a sugar (measured with a Brix spindle).

$$\text{Gravity purity (\%)} = \frac{\text{Sucrose}}{\text{Brix}} \times 100\%$$

54. Apparent purity (or A. P.)

The percentage of pol in the apparent solid solution of a sugar product.

$$\text{Apparent purity (\%)} = \frac{\text{Pol}}{\text{Brix}} \times 100\%$$

Note: Brix in the formula can be measured in the following ways.

(1) Measured by Brix spindle in cane sugar factory.

(2) It is measured by sugar refractometer in the cane sugar factory, and the calculation result is called refractometer purity.

55. Conductivity ash%

The percent by weight of ash in the sugar was measured by the conductometric method.

56. Colour (chroma) Colour

The value indicating the color of the sugar measured according to the specified method (expressed in ICUMSA colour).

57. ICUMSA colour

The internationally unified unit of colour of sugar products. In a certain range of pH value, the absorbance coefficient of the solution at the specified wavelength (420 nm for white granulated sugar and light colored sugar, and 560 nm for dark colored sugar) with appropriate liquid layer thickness and concentration is multiplied by 1000, which is the international sugar colour. The symbol is IU_x (X is the wavelength used, when 420 nm is used, X may not be marked).

$$IU_x = \frac{-\log T}{bc} \times 1\,000$$

Where $-\log T$——absorbance at the specified wavelength;

b——cuvette thickness (cm);

c——concentration of sugar solution (g/cm³).

58. Turbidity

A measure of the degree of light scattering in a sugar juice caused by the presence of colloidal particles.

59. Lime content, Calcium salt content.

Weight percent of calcium salt in sugar. Expressed in grams of calcium oxide per 100g of sample.

60. Fiber cane

Weight percent of water insoluble matter in sugarcane tissue to sugarcane.

61. Sucrose

Determine the weight percentage of sucrose in the sugar product by the specified method.

62. Reducing sugar

Determine the weight percentage of reducing substances (expressed as invert sugar) in the sugar product by the specified method.

63. Total sugar

The total amount of sucrose and reducing sugar in the sugar is measured by the specified method.

64. Cane sugar cell fragmentation (degree of fragmentation)

In the process of sugarcane treatment (cutting, crushing, tearing, pressing, etc.), most of the cells containing sucrose in sugarcane tissue are broken due to mechanical action. The percentage of the number of broken cells to the total number of unbroken cells before treatment. The number of cells was represented by the amount of sucrose (or pol) contained in the cells. That is

$$\text{Degree of fragmentation (\%)} = \frac{\text{Sucrose fraction contained in cells ruptured after treatment (Pol)}}{\text{Sucrose fraction contained in all unbroken cells before treatment (Pol)}} \times 100\%$$

65. Purification effect

The percentage of non-sucrose substances removed in the purification process relative to the non-sucrose substances originally contained in the sugar juice.

$$\text{Purification effect (\%)} = \frac{100(P_2 - P_1)}{P_2(100 - P_1)} \times 100\%$$

Where P_1 —— purity of sugar juice before purification (%);

P_2 —— purity of sugar juice after purification (%).

66. Absorption rate of sulfur dioxide

The percentage of the amount of sulfur dioxide absorbed by the sugar juice during the fumigation process to the total amount of sulfur dioxide entering the fumigator.

67. Intensity of sulfitation

The amount of SO_2 absorbed by the juice. Expressed as the volume of 1/64 mol/L iodine solution consumed to titrate 10 mL of sulfur smoke juice.

68. Phosphate in juice

The amount of soluble phosphate contained in the juice itself. It is usually expressed in weight of phosphorus pentoxide per liter.

69. Alkalinity

The amount of alkali present in a sugar juice when it is in an alkaline reaction. Expressed in weight of calcium oxide in 100 mL of sugar juice.

70. Natural alkalinity

The amount of alkali metal carbonate remaining in the sugar juice after the alkali metal hydroxide contained in the sugar juice itself is saturated with two carbons to precipitate the calcium salt is expressed in equivalent weight of calcium oxide.

71. Total calcium (Total calcium oxide)

The amount of unreacted calcium oxide and calcium oxide that has reacted to form calcium carbonate and a portion of the alkali metal contained in 100 cm^3 of the main ash juice or one carbon juice is equivalent to weight of calcium oxide.

72. Purity difference

The difference in purity between two materials before and after treatment in the sugar process.

73. Decolorization rate (Percent of decolorization, Decolorization ratio)

The percentage reduction in the colour of sugar juice before and after decolorization.

74. Juice extraction rate (Draft)

The percentage of the weight of juice extracted from the percolator or press to the weight of the treated sugar.

75. Sucrose (or Pol) extraction

The percentage of sucrose (or Brix) extracted from the sugar stock during the juice extraction process.

$$\text{Sucrose (or Pol) extraction (\%)} = \frac{\text{Weight of sucrose in extracted sugarcane juice}}{\text{Weight of sucrose in raw materials}} \times 100\%$$

76. Reduced sucrose (or pol) extraction

The actual extraction rate was converted into the extraction rate based on the standard sugarcane fiber content (12.5%).

$$E_{12.5}=100-\frac{(100-E)(100-F)}{7F}$$

Where $E_{12.5}$— comparison extraction rate (%);

E— actual extraction rate (%);

F— sugarcane fiber content (%).

77. Boiling house recovery

Weight percentage of sucrose (or Pol) of finished white granulated sugar (including brown granulated sugar, etc., and white granulated sugar prepared from products in process) to sucrose (or Pol) in the mixed juice. It represents the percentage of sucrose (or Brix) weight actually recovered by the refining process from the sucrose (or Brix) of the juice mix (or juice bleed).

$$\text{Boiling house recovery (\%)}=\frac{\text{Weight of sucrose (or Pol) in finalised and unfinalised white granulated sugar}}{\text{Weight of sucrose (or Pol) in mixed juice (or exudate juice)}}\times 100\%$$

78. Reduced boiling house recovery

There are many factors affecting the recovery rate of boiling, which is objectively affected by the purity of sugar. Internationally, the process technology is compared by comparing the recovery rate of boiling, that is, the theoretical purity of final molasses is calculated from the actual purity of mixed juice (or percolating juice) and the recovery rate of boiling, and then converted to the recovery rate when the purity of mixed juice (or percolating juice) is 85. It is not affected by the purity of mixed juice (or exudated juice) and can be used as an index for direct comparison of the recovery rate of each factory.

In 1965, the International Society of Sugar Cane Technologists (ISSCT) formally adopted Deerr's calculation formula:

$$M_v=\frac{J(100-R)}{10\,000-JR}\times 100\%$$

$$R_{85}=\frac{100(85-M_v)}{85(100-M_v)}\times 100\%$$

In 1971, ISSCT proposed to adopt Gundu-Rao's formula instead of Deerr's formula.

The formula used by Gundu-Rao to calculate the comparative scouring recovery is:

$$R_{85}=R+\frac{M}{1-M}\times\frac{0.85-J}{0.85J}$$

Where M_v— theoretical final purity of final molasses (%);

M—final purity of final molasses;

J—purity of mixed juice (or exuded juice) (Deerr formula as percentage and Gundu-Rao formula as ratio);

R—boiling recovery rate, same as above;

R_{85}—compared with the recovery rate of boiling, as above.

79. Total recovery (sucrose recovery)   Overall recovery, Total recovery

The weight percent of sucrose (or degree of sugar) in the finished sugar and in the product that can be made into sugar to the sucrose (or degree of sugar) in the sugar material.

$$\text{Total recovery (\%)} = \frac{\text{Weight of sucrose (or Pol) in finished and in products that can be made into sugar}}{\text{Weight of sucrose (or Pol) in the raw material}} \times 100\%$$

80. Reduced overall recovery

The product of the comparative extraction rate and the comparative scouring recovery rate is divided by 100%.

$$\text{Reduced overall recovery (\%)} = \frac{E_{12.5} \times R_{85}}{100} \times 100\%$$

Where $E_{12.5}$— comparison extraction rate (%);

R_{85}—compared with the recovery rate of boiling (%).

81. Crystallization percent in massecuite

Weight percent of crystalline sucrose in massecuite to massecuite solid solution.

$$\text{Crystallization percent in massecuite (\%)} = \frac{\text{Purity of massecuite} - \text{purity of molasses}}{100 - \text{Purity of molasses}} \times 100\%$$

82. Sugar yield (equal yield of white granulated sugar)

Weight percentage of finished and unfinished white granulated sugar to sugar material in this period.

$$\text{Sugar yield (\%)} = \frac{\text{Weight of finished and unfinished white granulated sugar for the period}}{\text{Weight of actual sugar material handled}} \times 100\%$$

83. Undetermined losses

The unmeasured loss is equal to the total sucrose (or Pol) loss in the sucrose (or Pol) balance minus the measured loss, including the loss of sampling, weighing, analytical error, leakage sugar, chemical conversion and decomposition. Cane sugar mills are expressed as a percentage of the weight of sucrose (or Pol) in the cane.

2.1.1.2 Task-related knowledge—sampling purpose and sampling period

1. Purpose of sampling

In the chemical management of sugar factories, the objects to be analyzed are various, some of which are large in number and uneven in composition. For example, sugarcane, sugar, bagasse, limestone, coal and so on, while laboratory analysis can not analyze all the materials, only a small number of samples can be processed. In this case, to provide reliable and valid analytical data for production, the sample must be representative of the average composition of a large number of materials. Sampling is the first step of analysis. Whether a small amount of sample can represent the average composition of the material determines whether the analysis results are representative. Under normal circumstances, the error caused by sampling is often greater than the error caused by the analysis process. Therefore, strengthening the management of sample collection and using scientific methods to sample are important means to improve the accuracy of analysis.

There are many kinds of materials in the sugar industry, there are mainly three types of materials: liquid, solid and gas, among which liquid materials account for the majority.

2. Sample cutting period

See the Table 2-1 for the sampling method and sampling period of various materials in the sugar factory.

Table 2-1　Summary of Sampling Methods and Sampling Period

Sample name	Sampling method	Intercept the sample cycle	Preservatives used	Remarks
Sugarcane	Once every 2 h	2 h or 4 h		(1) Formaldehyde 2/1 000 sample weight; (2) Mercuric chloride solution 1/1 000 of sample amount; (3) Mixed juice for determination of silt and suspended matter. Formaldehyde was used as a preservative to sample once every 1 h. Calcium samples were determined without preservatives; (4) The clarified juice is used to determine the colour, and the samples of acid value and calcium salt are not added with preservatives. Filter juice determination of colour, acid value of the sample without preservatives
First expressed juice	Continuous	2 h or 4 h	Formaldehyde	
Bagasse	Once every 15 min	2 h or 4 h	Formaldehyde	
First expressed juice	Continuous	2 h or 4 h	Formaldehyde	
Mixed juice	Continuous	2 h or 4 h	Mercuric chloride solution	
Last expressed juice	Continuously or once every 15 min	2 h or 4 h	Formaldehyde	
Preliming juice	Continuous	1 h or 2 h		
1st carbonatation juice / 1st carbonatation clear juice / 2nd carbonatation clear juice	Continuous	2 h		
Clarified juice	Continuous	2 h or 4 h	Mercuric chloride solution	
Filtered juice	Continuous	2 h or 4 h	Formaldehyde	
Purified syrup	Continuous	1 h or 2 h		
Filter cake	Once every 15 min	2 h or 4 h		
Massecuite		Every time it's cooked into a can		
Green molasses and white molasses	Once every 15 min	When each number of massecuite is finished		
Final molasses	Continuous	When each number of massecuite is finished		
White granulated sugar	Continuous or once every 15 min	When each number of massecuite is finished		
Brown granulated sugar	Once every 15 min	When each number of massecuite is finished		

2.1.1.3 Sampling method

1. Collection of liquid and gas samples

The composition of liquid and gas materials is relatively uniform and the fluidity is good. Generally, the continuous sampling method can be used for sampling. Sugar juice, syrup and other liquid materials with low concentration and little suspended solids in sugar mills can use this sampling method. For materials with high concentration (such as final molasses and massecuite) or materials with low concentration but more suspended solids (such as lime milk, 1st carbonatation juice and 2nd carbonatation juice), indirect sampling method can be used, that is, manual (or automatic) timing to collect quantitative samples. After the samples collected within the specified time are fully mixed, a part of them is taken for analysis.

As for the gas, it is necessary to introduce into the gas analyzer from the storage tank or pipeline by using a small pipe.

2. Collection of solid samples

Because the composition of solid materials is not as uniform as that of liquids or gases, obtaining representative solid samples has always been a difficult problem to solve satisfactorily. This is particularly the case with the collection of raw sugarcane. Bagasse and filter cake are sampled at five or seven points. Bagasse is sampled at five points equally divided by the length of the press roll. Filter cake can be sampled at five points in the corner and center of the filter cake hopper. If a vacuum filter is used, it can be sampled at five or seven points according to the length of the drum. As for the finished sugar, a quantitative sample is taken from each package when it is packed and weighed. The number of samples collected (except sugarcane) was reduced by the "quartering method", and an appropriate amount of samples was taken for analysis.

3. Sample collection method for each process

(1) Pressing process.

1) Sugarcane. Before the sugarcane is unloaded into the sugarcane conveying belt, two representative whole−stem sugarcanes shall be collected every two hours and placed in a cool place where they are not exposed to the sun and rain, and the samples shall be cut at the specified time for the determination of sugarcane inclusions or other items.

Note: sugarcane samples must be handled with care to prevent inclusions from falling off.

2) Bagasse. At the bagasse outlet of the last press, take samples once every 15 min (in case of roller blocking or suspension of infiltration water injection, take samples on time). When sampling, the length of the press roller is divided into five points (the midpoint of the roller and the midpoint to both ends are equally divided into three sections and equally divided into five points) as sampling points. Take the whole layer of bagasse sample with both hands (one up and one down) or with a clamp and put it into a bagasse sample storage barrel with preservatives (10 mL formaldehyde moistened with degreased cotton, put it into an iron box with small holes, and put it at the bottom of the barrel), and cover it tightly, and collect it to the specified time for sample cutting. When cutting the sample, the sample shall be mixed evenly in the storage barrel,

and then a part (not less than 2 kg) shall be put into the sample barrel for the determination of pol and moisture.

3) First expressed juice. A chute is installed at the place where the cane juice flows out of the front roller of the first press. The chute is covered with a semicircular steel wire mesh to remove the cane crumbs. A juice guide groove is installed at the notch. The cane juice is led out by a copper wire. The flow rate is adjusted to make it drip into the sample barrel evenly and continuously. Excess cane juice is refluxed to the original cane juice tank. The sample bucket shall be filled with 4mL of formaldehyde as preservative. The sample volume shall be 2 000~3 000 mL during sample cutting.

4) Mixed juice. Connect a small tube at the appropriate position of the pipeline behind the mixed juice pump, and connect two copper wires at the tube orifice, so that the mixed juice can flow into two glass bottles evenly, one of which is added with 3 mL of mercuric chloride alcohol saturated solution as preservative, and the other is not added with preservative (for calcium salt determination). A glass funnel is installed on the bottle mouth to reduce the evaporation of water in the cane juice, the sample collected in each bottle should not be less than 2 000 mL per hour (note: for every 100 mL of sample, the amount of preservative is 1 mL). Cut the sample once every 2 hours, shake up the cane juice in the bottle, filter it with a strainer to remove the cane crumbs, and take the sample according to the following points ① and ②. In addition, the samples for the determination of suspended solids in the mud and sand shall be collected according to the point 3 below.

① Samples for determination of Brix, pol, sucrose content, reducing pol and phosphate value: the same amount of samples with preservatives added shall be taken in glass bottles every 2 hours, and the samples shall be cut off at the specified time, and the samples shall not be less than 2 000 mL.

② Sample for determination of calcium salt: take the same amount of sample without preservative in a glass bottle every 2 hours, and cut the sample at the specified time. The sample shall not be less than 1 000 mL.

③ Samples for determination of suspended solids in silt and sand: Take samples with a spoon every 1 hour before preliming, about 200 mL, and put them into a sample bucket with 10 mL formaldehyde as preservative, and then add 5–10 mL formaldehyde every 3 h to strengthen anticorrosion, and collect them to the specified time for sample cutting.

5) Last expressed juice. Install a chute at the bottom of the rear roller of the last press. The cane juice extracted from the top and rear rolls was sampled continuously or intermittently and stored in a sample bucket containing 10 mL of formaldehyde. In case of intermittent sampling, the bagasse samples shall be sampled every 15 min immediately after the bagasse samples are collected, and the samples shall be collected at the specified time, and the sample volume shall be 2 000~3 000 mL.

6) Preliming juice. After the preliming treatment of the mixed juice and before the

neutralization, continuous sampling shall be carried out at the appropriate position, and the sample shall be cut off at the specified time, and the sample shall not be less than 1 000 mL.

(2) Purification and evaporation process.

1) Sulfitated juice. Intermittent or continuous sampling method shall be used at the appropriate position of the sulfur fumigation or sulfur fumigation pipeline, and the sample shall be collected to the specified time, and the sample shall not be less than 200 mL.

2) 1st-carbonatation juice. 200 mL of samples were taken at irregular intervals at the time of analysis on the 1st-carbonatation juice outlet pipe.

3) 1st carbonatation clear juice. Conduct continuous sampling at the proper position of the pipeline before the first carbon meter enters the second carbon saturation, and cut the sample after accumulation to the specified time. The sample shall not be less than 1 000 mL.

4) 2nd carbonatation juice. Install a small cock on the outlet pipe of the second carbonatation juice, and the operator of the production post shall take samples and check them irregularly as required.

5) 2nd carbonatation juice. Conduct continuous sampling at the proper position of the two-carbon clear juice pipeline, and cut the sample after accumulating to the specified time. The sample shall not be less than 1000mL.

6) Clarified or sulfitated clear juice. Samples shall be taken continuously at the proper position of the pipeline before the clarified juice or sulfur bleached clear juice enters the clear juice storage tank, and the sample shall be intercepted once every 2 h. Not less than 2 000 mL, stir the sample evenly, and reserve the sample according to the following methods.

① Samples for determination of colour, acid value and calcium salt: take the same amount of samples in a glass bottle every 2 h, and cut the samples at the specified time. The samples shall not be less than 1 000 mL.

② Samples for determination of Brix, pol, sucrose content and reducing pol: take the same amount of samples in glass bottles every 2 h. Add 2 mL of mercuric chloride alcohol saturated solution into the bottle in advance as a preservative, and cut off the sample at the specified time. The sample shall not be less than 2 000 mL.

7) Filtered juice. Samples were continuously taken from the filtered juice pipeline, and formaldehyde was used as a preservative for the determination of Brix and pol. Preservatives are not added to the samples for determination of colour and acid value.

8) Raw syrup. Conduct continuous or indirect sampling at the appropriate position of the raw syrup pipeline without sulfur fumigation, and intercept the sample once every 2 hours. The sample shall not be less than 2 000 mL. After the sample is stirred evenly, keep the sample according to the following methods.

① Samples for determination of Brix, colour, acid value and calcium salt: Take the same amount of samples in a glass bottle every 2 h, and cut the samples at the specified time. The samples shall not be less than 1 000 mL.

② Samples for determination of Brix, Pol, sucrose content and reducing pol: take the same amount of samples in a glass bottle every 2 h, and cut the samples at the specified time. The samples shall not be less than 2 000 mL.

9) Purified syrup. After sulfur bleaching, the raw syrup shall be sampled continuously at the appropriate position of the clean syrup pipeline, and the sample shall be cut off at the specified time, and the sample shall not be less than 1 000 mL.

10) Filter cake.

① Sampling method on the mud filter hopper: A bamboo or metal semi-cylinder with a diameter of about 3 cm and a length of about 60 cm was used as the sampling tool. After the filter cake is discharged from the hopper, the tool is deeply inserted into the mud at the four corners and in the middle of the hopper to sample (five points). Scrape the collected sample into the sample bucket. Take the same amount of samples from the filter cake of each hopper truck according to the above method, and cut the samples at the specified time.

② Sampling method in the spiral conveying trough: If the unloaded filter cake is transported outside the factory through the spiral conveying trough, the same amount of sample can be taken with a long-handled wooden shovel at the outlet of the spiral trough every 15 min and put into the sample bucket. Accumulate to the specified time and cut the sample.

③ During the filtration operation, the vacuum filter can be divided into three points (both sides and the middle) according to the length of the drum and sampled once every 15 min. The samples are put into the sample bucket and accumulated to the specified time for sample cutting.

(3) Crystallization process.

1) Massecuite.

① Boiling into massecuite: After each can of massecuite is boiled, take about 1 kg of sample with a spoon in the center of the launder and put it into the sample cup during the sugar unloading process.

② Boiling massecuite: When the materials in the tank are fully convected, take about 1 kg of sample from the sampler of the boiling sugar tank and put it into the sample cup.

2) Molasses.

① White molasses: Intermittent sampling shall be carried out at the notch or appropriate position where the molasses flows out under the molasses separator. The same amount of sample shall be taken every 15 min from 5 min after the start of molasses separation to the end of molasses separation, and then put into the sample bucket.

② Green molasses: The sampling method is the same as that of white molasses.

3) Final molasses. From the beginning to the end of each number of massecuite, take the same amount of samples in the final molasses box once every full box, and take samples in proportion if less than one box. Or at the final molasses outlet of the final molasses scale, take the same amount of samples for each weighing, and the sampling times shall be determined according to the weight of each weighing. Generally, take the same amount of samples once with

1~2 d, but the samples collected for each number shall not be less than 6 times. The concentration of the sample taken must be consistent with the concentration of the final molasses at the time of measurement or weighing.

(4) Finished sugar and raw sugar.

1) White granulated sugar. Each separated can of massecuite is a number. When weighing and packaging, about 5 kg of samples are collected continuously and placed in a container with a lid. After mixing, it is a number sample. In addition to the sample for number analysis, another 0.5 kg is taken and placed in a container with a lid. After 24 h of accumulation, it is a daily collection sample.

2) Refined sugar. Same as white granulated sugar.

3) Brown (yellow) granulated sugar.

When the brown (yellow) granulated sugar is weighed and packed, the same amount of samples shall be collected continuously and put into the sample bucket. The samples shall be intercepted once per shift and fully mixed. About 1 kg of samples shall be taken for analysis. In addition, take the same amount of samples from the samples of the three shifts and mix them as the daily collection samples.

4) Brown sugar. Same as brown granulated sugar.

5) Crystal sugar. The number of products packed in each shift is one. Randomly take 0.5 kg samples (including basin ice, basin bottom and column ice) from the packaging department every 2 hours, slightly crush and mix them, put them into double-layer food bags, and accumulate them for 8 hours as shift samples.

6) Sugar cube. The product produced and packaged in each shift (or each can) is a number. At the end of the conveyor belt of the sugar cube machine, 0.5 kg of samples shall be taken at random every 2 hours and put into a container with a lid. After 8 hours of accumulation, it shall be used as a shift sample.

7) Liquid sugar. The product produced and packaged in each shift (or each can) is a number. About 0.5 kg of samples shall be randomly collected at the appropriate position of the packaging machine or conveying pipeline every 2 hours and put into a container with a lid for 8 hours as the shift sample.

8) Raw sugar. After the raw sugar arrives at the sugar factory, when unloading and weighing, collect samples according to the provisions of the national standard of *Raw Sugar* (GB/T 15108—2017), put them in a container with a lid, mix the samples and take out about 1 kg of samples for analysis by quartering.

(5) Auxiliary materials.

1) Limestone. When each batch of limestone enters the plant, sampling shall be conducted on the ship (or vehicle). First observe the quality and color of limestone, and then select representative stones at each point on the ship (or vehicle) and break them into small pieces with a diameter of about 5 cm one by one. In each pile of small stones, take the same amount of

samples, and the total amount shall not be less than 3 kg. The obtained sample shall be reduced into small sample (not less than 0.5 kg) by quartering method, and crushed for analysis.

2) Lime. After each batch of lime enters the factory, first observe the quality of lime, the color of debris and lump lime, and then take about 3 kg of the most representative lime samples from various places. For the carbonation plant, when unloading the lime in the lime pit, observe the quality and debris of the lime in the unloading lime pile, and take 3 kg of the largest representative lime sample at five points.

Break the collected lime into even small pieces, repeatedly treat them by quartering until about 60 g is left, break them into powder, and pass them through a sieve with an aperture of 0.15 mm; note that the fine particles finally sieved are often the most difficult to break, but they must not be discarded, so as not to affect the representativeness of the sample.

3) Sulfur (aperture 0.28 mm). When each batch of sulfur enters the plant, select 5% bags (or boxes) from the total number of bags (or boxes) of sulfur to sample at 2/3 of the depth; if the sulfur is in bulk, pay attention to selecting representative samples at different positions, mix them and divide them into about 2 kg by quartering. It is then ground and passed through a 10-mesh sieve (aperture 2 mm) and divided into two parts by quartering: One part is used for the determination of moisture; the other part is grinded and divided into two parts, which pass through a 30-mesh sieve (aperture 0.63 mm) and a 60-mesh sieve (aperture 0.28 mm) respectively, and are divided into two parts by quartering respectively, one part is used for determination, and the other part is stored for reexamination.

4) Phosphoric acid. When each batch of phosphoric acid enters the factory, take out the sample from 10% of each batch with a clean and dry glass tube. When sampling, the glass tube shall be vertically inserted into the container for uniform sampling. The total amount of each batch of samples shall not be less than 500 mL. The samples shall be collected in a clean and dry glass bottle with a ground stopper after being uniformly mixed, labeled and sent to the laboratory for analysis.

5) Flocculant. When each batch of flocculant enters the factory, 5 barrels are randomly selected, and the same amount of samples are collected from each barrel and mixed evenly for analysis.

(6) Other.

1) Coal. When each batch of coal is unloaded from the truck or ship into the plant (or at the coal pile), a representative sample of about 5 kg (large and small blocks shall be taken into account) shall be collected from the top center and bottom of the coal pile and placed in a sealed sample bucket. The large coal sample shall be slightly crushed to a particle size of about 13 mm by hammering, and then quickly reduced to about 1 kg by quartering for analysis.

2) Lubricating oil. When each batch of lubricating oil enters the factory, 5 barrels shall be randomly selected, and a clean and dry glass tube shall be inserted into the oil layer for about 10 cm to collect the same amount of samples and mix them for analysis.

3) Boiler water. Take the water in the boiler water drum passing through the cooling device at the sampling place beside the boiler water drum. The water accumulated in the pipe must be drained before sampling. Rinse the container twice and collect a small cup.

4) Boiler feed water. Take the sample at the sampling place of the hot water drum in the boiler. Before sampling, drain the water accumulated inside, wash the container twice, and then collect a small cup.

5) Condensation water. Take at the steam separator of the evaporation tank or sugar boiling tank and heater. Before sampling, the water accumulated inside should be drained, and the container should be rinsed twice, and then collected a small cup.

6) Condensed water. Install two small tubes for continuous sampling at a height of about 1 m above the ground on the condenser drain pipe of the evaporation tank and the sugar boiling tank. Make sure that the pressure in the sampling bottle is equal to that in the drain pipe before the water is drawn out. At the same time, pay attention to adjusting the flow of water so that the samples can be evenly during the whole sampling time. Accumulate to the specified time and cut the sample.

7) Kiln gas. In the upper part of the kiln gas storage tank, a small tube is used to introduce the kiln gas into the laboratory. It passes through a large glass bottle to separate water, then through a U-shaped tube, in which glass fibers are used to filter out dust. Then it is connected to the gas analyzer and equipped with an exhaust pipe. Before analysis, the last residual kiln gas in the tube should be removed.

2.1.1.4 Sample pretreatment

1. Pretreatment of materials

(1) Sample with relatively uniform composition. For example, finished sugar, filter cake, sugar solution, etc. These samples are relatively uniform in composition, and a part of them can be used as a representative analysis sample by slightly mixing the samples.

(2) Sample with very heterogeneous composition. It is a complex operation to select a representative uniform sample for some samples with uneven particle size, uneven composition and uneven composition. In order to make the sample representative, a certain number of particles of different sizes must be taken from different parts of the material according to a certain procedure. The greater the number of samples taken, the closer the composition of the sample is to the average composition of the material being analyzed. According to experience, the average sample selection amount is related to the uniformity, particle size and breakability of the sample, which can be estimated according to the following formula:

$$Q=Kd^a$$

Where Q—the minimum weight of the average sample (kg);

d—diameter of the largest particle in the sample (mm);

K and $α$ are empirical constants, which can be determined by the uniformity and fragmentation of the material. The K value is between 0.02 and 0.15, and the $α$ value is usually between 1.8 and 2.5.

Take for example $K = 0.1$ and $α = 2.1$

If $d = 1.2$ mm

Then: $Q = 0.1 \times 1.2^{2.1} = 0.15$ (kg)

If $d = 0.8$ mm

Then: $Q = 0.1 \times 0.8^{2.1} = 0.063$ (kg)

Depending on the items that need to be analyzed for each sample, they are treated differently. However, the general steps can be divided into four steps: crushing, sieving, mixing and splitting.

1) Crushing. The sample shall be gradually crushed by mechanical or manual methods until the required particles are obtained. Screening shall be carried out after each crushing, and the coarse particles screened shall be crushed again and shall not be thrown away to ensure the representativeness of the sample.

2) Splitting. After each crushing of the sample, take out a part of the representative sample mechanically or manually and continue to crush it. In this way, the amount of the sample is gradually reduced for easy handling.

The commonly used manual division method is the "quartering method". First, mix the broken sample thoroughly, pile it into a cone, press it into a cake shape, cut it into four equal parts through the center according to the "cross" shape, and discard the two parts of any diagonal. Because the particles of different sizes and densities in the sample are generally evenly distributed in the mixed material, the amount of the remaining sample is half of the original sample, but it can still represent the composition of the original sample. Repeat until the amount of sample is appropriate.

2. Dilution of samples

(1) Multiple dilution. It is usually used for dilution of high concentration or solid samples, such as syrup, massecuite, molasses, etc. The multiple referred to here is the weight multiple, that is, the ratio of the weight of the diluted sample solution to the weight of the original sample:

Dilution factor = weight of diluted sample solution / weight of original sample

When multiple dilution is used, the amount of water added is usually measured with a measuring cylinder. At this time, the density of water is regarded as 1, and each milliliter of water weighs 1 g.

(2) A normal weight of dilution. The specified dilution method is mainly used for polarimetric determination. 26.000 g is referred to as a normal weight in the sugar mill. The normal weight of dilution is to prepare 26.000 g of sample into 100 mL in a volumetric flask, or to expand or reduce it in the same proportion. For example, preparing 52.000 g of sample into 200 mL is also called the normal weight of dilution. The concentration of the diluted sample

solution is called a normal weight. 13.000 g of sample can also be prepared into 100 mL, which is called semi-specified quantitative dilution, and the concentration of the diluted sample solution is called half of the normal weight. Similarly, 1/3 of the normal weight can also be weighed for dilution, and the concentration of the diluted sample solution is called 1/3 of the normal weight.

When the normal weight is used for dilution, it shall be ensured that the volume of the diluted sample solution must reach the specified value. If 26.000 g of sample is weighed to prepare the sample solution with a specified concentration, the total volume of the sample solution must be exactly 100 mL. This is guaranteed for fully soluble samples. However, for some samples that can not be completely dissolved, due to the incomplete dissolution of the sample, the solid part will inevitably occupy a certain volume in the volumetric flask, resulting in the real volume of the sample solution less than the nominal volume of the volumetric flask. That is to say, the amount of water added during dilution is less than the theoretical calculation value, and the concentration of the solution higher than the theoretical calculation value. In actual operation, in order to make up for this error, the weight of the sample can be reduced appropriately when the sample can not be completely dissolved is diluted by a normal weight. For example, in a sugar mill, the normal weight of filter cake is 25.0 g.

The dilution factor of the sample can also be calculated when the normal weight of dilution is used. The calculation method is as follows:

Dilution factor = (volume of diluted sample solution × density of diluted sample solution) / weight of original sample

The density of the diluted sample solution can be calculated by looking up the table according to the observed Brix of the diluted sample solution or by the following formula:

The density of the diluted sample solution = $0.998\,2 + 0.003\,7B + 0.000\,018\,16B^2$

Where B—observed Brix of sample solution (°Bx)

(3) Arbitrary dilution. In production, it is sometimes desirable to quickly provide data on the purity of a sample. For example, boiled sugar ingredients. Arbitrary dilution is mainly used for the determination of purity. Since the purity of the sample is not affected by dilution, when rapid purity determination is required, an appropriate amount of sample can be taken and water can be added to the appropriate concentration for determination. The purity of the diluted sample solution is the purity of the original sample.

3. Preservation and clarification of samples

(1) Preservative. Samples from sugar mills are rich in sugar, which is a good nutrient for microorganisms. When the temperature and pH are suitable, microorganisms are easy to reproduce. Under the action of microorganisms, the proportion of the original components in the sample will change, and new components will be produced, which can not represent the original situation. Therefore, in sugar factories, for samples that are easy to reproduce microorganisms, if can not be tested immediately, preservatives are generally needed to add the collected samples.

Preservatives commonly used in sugar mills are as follows:

1) Mercuric chloride ($HgCl_2$) solution. Mercuric chloride is a heavy metal salt that coagulates proteins, is highly toxic, and has a strong bactericidal effect. It is usually used for the preservation of liquid samples in sugar factories, and the dosage is 0.5 mL of mercuric chloride alcohol solution per liter of liquid samples. Because mercuric chloride is a heavy metal salt, the solution after hydrolysis is acidic, so the sample for measuring ash and pH value can not be used.

2) Formaldehyde (industrial formaldehyde solution contains 30%~40% formaldehyde). Formaldehyde has strong permeability and is easy to penetrate into bacteria to denature proteins, so it can restrain microbial reproduction and achieve the purpose of preservation. Generally, industrial formaldehyde (containing 30%~40% formaldehyde) is used. The dosage is 0.5~1 mL per liter of sample solution. Often used as a preservative for bagasse. Formaldehyde solution is acidic and reducing, so it is not suitable for the sample to be tested for pH value and reducing sugar.

3) Ammonia-chloroform solution. Used as a preservative for bagasse samples. It is usually adsorbed by cotton and placed at the bottom of the sample barrel.

Amount of preservative:

1) Mercuric chloride ($HgCl_2$) solution: add 0.5 mL per liter of sugar juice sample.

2) Formaldehyde: add 0.5~1 mL per liter of sugar juice sample.

(2) Type of clarifier. Most of the samples to be analyzed in the sugar factory are colored. Often in a cloudy state. But also contain varying degrees impurities that affect the determination. In order for the analysis of the sample solution to proceed smoothly, it is necessary to clarify it. For example, sucrose (or polarimetry) is determined by polarimetry for most samples, and the sample solution is required to be light and transparent for polarimetry observation, so clarification is required before polarimetry observation.

Many types of clarifiers have been described in the literature. But the most commonly used is lead salt. Because lead ions can combine with many anions (such as Cl^-, PO_4^{3-}, SO_4^{2-}, CO_3^{2-}, tartrate, etc.) , insoluble precipitates are formed, and some impurities are removed by adsorption.

1) Alkaline lead acetate (also known as low lead acetate or basic lead acetate). It is a mixture of 4 parts $3Pb(CH_3COO)_2 \cdot PbO$ and 3 parts $Pb(CH_3COO)_2 \cdot PbO$ and has the approximate formula $3Pb(CH_3COO)_2 \cdot 2PbO$. It's clarification ability is strong, the formed precipitate can adsorb and remove part of the decomposition products of reducing sugar, coking products of sucrose, reducing sugar, etc. It can make some colloidal substances condense under alkaline conditions, so it is often used as a clarifier when determining the sucrose content of the sample solution. However, because it will precipitate part of the reducing sugar, it can not be used as a clarifier in the sample solution for the determination of reducing sugar.

2) Neutral lead acetate [$Pb(CH_3COO)_2 \cdot 3H_2O$]. Compared with alkaline lead acetate, its clarification and decolorization ability is poor. Because the clarification ability of lead acetate clarifier is related to the content of PbO in the molecule. It can not completely clarify the dark

sugar solution, but it will not or rarely precipitate the reducing sugar in the sugar solution during clarification, so it is suitable for the clarification of the reducing sugar sample solution to be tested.

3) Basic lead nitrate [Pb(NO$_3$)$_2$ · Pb(OH)]. It has the strongest clarification ability and is suitable for the clarification of final molasses and stubborn sugarcane juice. When in use, the same volume of the Pb(NO$_3$)$_2$ solution and NaOH solution is added to the sample to be clarified in turn.

The disadvantage of this clarifier is that it brings a considerable amount of NaNO$_3$ into the sample solution, which affects the optical rotation of sucrose. At the same time, it can also precipitate some reducing sugar, so it is generally not used for clarification when it is not necessary.

Lead salt clarifier has been widely used in sugar analysis in sugar mills for a long time because of its good effect. However, it is poisonous and has adverse effects on human health, so special attention should be paid to its use.

(3) Dosage of clarifier. The dosage of clarifier must be appropriate. If the dosage is too small, the purpose of clarification will not be achieved. If the dosage is too large, lead errors will occur due to the formation of lead sugar compounds. Usually, about 1 g of solid clarifier is added per 100 mL of sample solution. The clarifier used in the sugar mill samples and the amount used are shown in the Table 2-2 and Table 2-3.

Table 2-2 Clarifying agent used in cane sugar factory and its dosage

Sample	Analyze the project	Clarifying agent	Dosage
First expressed juice	Pol	Basic lead acetate powder	Use about 1 g per 100 mL
Mixed juice	Pol, sucrose content	Basic lead acetate powder	Use about 1 g per 100 mL
	Reducing sugar	54 °Bx neutral lead acetate	Use about 2 mL per 100 mL
Last expressed juice	Pol	Basic lead acetate powder	About 0.2~0.3 g per 100 mL of bagasse leachate
Bagasse	Pol	Basic lead acetate powder	Use about 1 g per 100 mL
Clarified juice	Pol	Basic lead acetate powder	Use about 1 g per 100 mL
	Reducing sugar	54 °Bx neutral lead acetate	Use about 2 mL per 100 mL
Filtered juice	Pol	Basic lead acetate powder	Use about 1 g per 100 mL
Filter cake	Pol	54 °Bx alkaline lead acetate	2~7 mL for 25 g sample

continued

Sample	Analyze the project	Clarifying agent	Dosage
Raw syrup	Pol	Basic lead acetate powder	Use about 1 g per 100 mL
	Sucrose content	Basic lead acetate powder	0.5 g per 100 mL for 4-fold dilution
	Reducing sugar	54 °Bx neutral lead acetate	Use 2 mL per 100 mL for 4-fold dilution
Remelt syrup	Pol	Basic lead acetate powder	Use 1 g per 100 mL for 4-fold dilution
Massecuite, molasses	Pol	Basic lead acetate powder	Use 1 g per 100 mL for 6-fold dilution
Blackstrap molasses	Pol	Basic lead acetate powder	Use 1 g per 100 mL for 6-fold dilution
	Sucrose content	Basic lead acetate powder	1/3 of the normal weight of diluent 3.5 g per 200 mL
	Reducing sugar	54 °Bx neutral lead acetate	1/3 of the normal weight of diluent 1 mL per 50 mL
White granulated sugar	Sucrose content	Disuse	
Brown granulated sugar	Sucrose content	Basic lead acetate powder	Prescribed amount of dilution 1 g per 100 mL
	Reducing sugar	54 °Bx neutral lead acetate	Dilute 50 mL of the normal weight with 1 mL

Table 2-3　Clarifying agent and its dosage for samples from beet sugar factory

Sample	Analyze the project	Clarifying agent	Dosage
Beet	Sucrose content	Alkaline lead acetate (d=1.235~1.240)	26 grams of beet paste plus 7 mL
Beet shreds	Pol	Alkaline lead acetate (d=1.235~1.240)	26 grams of beet paste plus 4 to 7 mL
Juice	Pol	Basic lead acetate powder	The right amount
Exudate	Pol	Basic lead acetate powder	The right amount
Waste dregs	Pol	Basic lead acetate powder	The right amount
Waste dregs	Pol	Basic lead acetate powder	The right amount
2nd carbonatation clear juice	Pol	Basic lead acetate powder	The right amount

continued

Sample	Analyze the project	Clarifying agent	Dosage
Filter cake	Sucrose content	Alkaline lead acetate (d=1.235~1.240 g/mL)	Sample 50 g plus 4 mL
Massecuite, molasses	Pol	Basic lead acetate powder	Add an appropriate amount of the normal weight of diluent
Blackstrap molasses	Pol	Basic lead acetate powder	The right amount
Blackstrap molasses	Reducing sugar	Neutral lead acetate (d=1.235~1.240 g/mL)	Add 10 mL of 1/2 of the normal weight of diluent per 100 mL
No.2 and No.3 granulated sugar and rescreened sugar	Pol	Basic lead acetate powder	Use 0.5g for the normal weight

4. Removal of lead and calcium from samples

The residual lead salt in the sample solution clarified by a proper amount of lead acetate clarifier has little effect on the polarimetric determination, so it is not necessary to remove the lead. However, in the determination of reducing sugar, the residual lead salt in the clarified sample solution and the calcium salt brought by the sample itself have an impact on the determination. Lead salts may combine with reducing sugars (especially fructose) to form lead sugar compounds, while calcium salts can form complexes with glucose and fructose, the reducing Pol of the measured sample liquid may be falsely lowered. Therefore, lead and calcium must be removed from the sample solution when measuring reducing sugar.

In the chemical management of sugar mills, disodium hydrogen phosphate ($Na_2HPO_4 \cdot 12H_2O$) and potassium oxalate ($K_2C_2O_4 \cdot H_2O$) are usually used as lead removers, which are added in the form of solutions when used. 1 L of the lead remover solution contains 70 g of disodium hydrogen phosphate and 30 g of potassium oxalate. When measuring reducing sugar, add 1.5~3 mL of lead remover to 50 mL of sample solution. If the reducing sugar is determined by the Aufner method, only disodium hydrogen phosphate is used as a lead remover because the sample solution is acidic and oxalate is reductive at this time. EDTA disodium can also be used instead of potassium oxalate to remove calcium when the reducing sugar is determined by the Lan-Eynon method or the Lan-Eynon constant solution method. The specific method and dosage will be introduced in the following chapters.

2.1.2 Task-related knowledge—daily analysis of WIP

The total amount of sucrose and non-sucrose substances in the sugar solution is called dry substance (solid solution). When the water in the sugar solution evaporates to dryness, the resulting residue is called a true solid solution. The true solid solute in pure sucrose solution is sucrose; the true solid solute in impure sucrose solution includes sucrose and all non-sucrose impurities dissolved in water.

In the chemical management of sugar factories, there are three analytical methods for determining solid solution: Drying method, density method and Refractometry.

1. Determination of dry substance by drying method

The drying method is to evaporate the water in the sample to be measured by heating until the weight is constant, and then calculate the true solid content. Because the determination process of this analysis method is too long and the procedure is more troublesome, therefore this method is not used to determine samples with more moisture in sugar mills. Instead, the density method and Refractometry, which are fast, simple and convenient, are used to determine the solid solution. The solid solute contained in the sample determined by the density method and the Refractometry is the myopic value of the true solid solute content, which is called the apparent dry solid content.

Brix determination is one of the important basic data in chemical management of sugar mills, which can be measured by density sugar Brix spindle and Abbe refractometer.

Brix refers to the value measured with a Brix spindle at a temperature of 20 °C. For pure sucrose solutions, it is the weight percent of sucrose, and for impure sucrose solutions, it is the approximate weight percent of soluble solids in the solution.

There are two common methods to determine the Brix of sucrose-based solution: Density method and Refractometry. This method is suitable for the determination of soluble solid matter content in the solution mainly containing sucrose. Products in process (such as cane juice, syrup, massecuite and various massecuites, molasses, etc.) Are applicable. Samples containing crystalline sugar must be dissolved and diluted before determination.

2. Determination of dry substance by density method

(1) Principle of determination of dry substance by density method.

1) Density. Density is the mass per unit volume of a substance (used to weigh). In sugar mills, the weight of a cubic centimeter of sugar solution in the air at 20 °C is called the apparent density of the sugar solution, which is usually called the density.

2) Relationship between density and dry substance content. The density of sucrose solution increases with the increase of concentration, and the concentration of the solution can be determined by measuring the density of the solution.

(2) Structure and calibration method of brix spindle.

1) Hammermeter structure. The hammer gauge is a glass tube with a thick lower part and a thin upper part and a scale. According to the principle that "the weight lost by an object in a liquid is equal to the weight of the liquid it displaces in the liquid", when a Brix spindle with a certain weight floats in a solution, the greater the density of the liquid, the smaller the volume immersed by the Brix spindle; Conversely, The brix spindle sinks, and the reading of the scale on the upper part of the brix spindle becomes smaller. On the contrary, it becomes larger. Because the upper part of the brix spindle is very thin, the volume per unit length is very small, so it can

clearly reflect the small changes in density.

2) Calibration method of Brix spindle. The calibration of the Brix spindle is based on the weight percentage concentration of pure sucrose solution at 20 °C. For example, if the Brix spindle is placed in a pure sucrose solution with a weight percentage concentration of 5% at 20 °C, the position of the liquid level is set as 5 °Bx, while if the same Brix spindle is placed in pure water at 20 °C, the position of the liquid level is set as 0 °Bx, and different scales can be divided equally between them. For a pure sucrose solution, the Brix is the weight percent of the sucrose content of the solution. But this relationship is only true for pure sucrose solutions. When the solution contains other impurities, because the density of impurities in aqueous solution is not necessarily the same as that of sucrose, there must be a difference between its density and that of pure sucrose solution, and the determination of Brix with the same Brix spindle will produce errors. The more impurities, the greater the error. For example, when a 10% salt solution by weight is measured with a Brix spindle, the result displayed is not 10 °Bx, but 18 °Bx. This example fully illustrates the influence of impurities on Brix determination. Because sugar mills also contain non-sugars in their products, the result of the determination of Brix should be the approximate weight percent of the solid solution, expressed in Brix.The higher the pol of the solution, the closer the Brix to the content of solid solution.

(3) Use of brix spindle.

1) Specification. It is uniformly stipulated that the brix spindle with 20 °C as the standard shall be used. As shown in Fig. 2-1. The general requirements of the brix spindle are as follows: the total length is preferably not more than 30 cm, the minimum division value is 0.05 °Bx, the thermometer is attached, the temperature range is 0~40 °C, and the division value is 1 °C. The commonly used measurement ranges of the brix spindle are as follows: 0~6 °Bx; 5~11 °Bx; 10~16 °Bx; 15~21 °Bx; 20~26 °Bx.

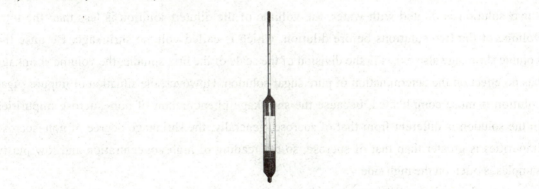

Fig. 2-1　Brix spindle

2) Usage and precautions.

① During the production period, the Brix spindle must be placed in a container with clear water and a soft rubber gasket at the bottom.

② When using, it must be taken out vertically, and the upper section can not be taken obliquely to prevent breaking.

③ Before measuring the sample solution, wipe off the attached water with a dry and clean soft cloth, or rinse with the sample solution, and then gently and vertically put it into the sample solution until the brix spindle floats. Do not let the brix spindle fall suddenly, so that the brix spindle will be damaged by collision with the measuring cylinder.

④ When observing the reading, the Brix spindle shall not adhere to the edge of the measuring cylinder, and must be kept clean and free of oil stains, so that there is a good meniscus in contact with the liquid surface, otherwise the measurement is not accurate.

⑤ If a domestic Brix spindle is used for reading, read according to the instructions attached to the meter (the domestic Brix spindle is generally read according to the upper edge of the liquid meniscus), and if a import one is used, read at the horizontal sight line of the liquid level.

⑥ After use, gently take it out vertically, rinse it with water and put it back into the original container. Be careful not to collide with other brix spindles in the container.

(4) Factors affecting the determination of Brix.

1) Temperature. The scale of the brix spindle must be 20 °C as the standard, otherwise the temperature correction shall be conducted. When the temperature is higher than 20 °C, the density will decrease due to the increase of the volume of the sugar solution; on the contrary, when the temperature is lower than 20 °C, the density will increase accordingly. Therefore, the temperature of the sugar solution should be as close as possible to 20 °C.

2) Impurities. Because the density of non-sucrose substances is different from that of sucrose, the Brix spindle for sucrose is based on pure sucrose solution, so the less the impurity content, the more accurate the determination result. When the content of non-sucrose impurities (especially inorganic non-sugar substances) is high, the reading is generally high.

3) Volume contraction phenomenon. When sucrose is dissolved in water or concentrated sugar solution is diluted with water, the volume of the diluted solution is less than the total volume of the two solutions before dilution, which is called volume shrinkage. Because the volume shrinkage also exists in the division of the scale of the brix spindle, the volume shrinkage has no effect on the determination of pure sugar solution. However, the situation of impure sugar solution is more complicated, because the shrinkage phenomenon of non-sucrose impurities in the solution is different from that of sucrose, generally, the shrinkage degree of non-sucrose impurities is greater than that of sucrose, so the reading of high concentration and low purity samples is often on the high side.

4) Surface tension. When the sample contains more non-sucrose impurities, the downward vertical component force between the sugar solution and the Brix spindle is reduced due to the impurities of some surface active substances, so that the reading is higher.

5) Dilution. Dilution will enlarge the observation error, and the hydration of ions in impure sugar solution can affect the determination results, so the determination should be diluted according to the uniform dilution ratio, so that the determination results can be compared with

each other.

(5) Determination of Brix of mix juice.

1) Instruments and equipment used.

① Sugar brix spindle. The standard temperature is 20 °C, and the scale must meet the following requirements.

Scale range (°Bx) : 0~6 °Bx, 5~11 °Bx, 10~16 °B$_X$, 15~21 °Bx, 20~26 °Bx。

Scale scale 0.1 °Bx.

② Thermometer 0~100 °C, scale 1 °C.

2) Sample treatment and determination. Mixed juice samples collected from the workshop should be filtered with a screen to remove the sugar cane chaff, otherwise the sugar cane chaff will float on the liquid surface, making it difficult to read.

Wash the inner wall of the Brix measuring cylinder with a small amount of the sample solution after mixing. Then fill the cylinder with the sample solution and let it stand until all the bubbles in the sample solution float to the liquid surface and are removed. Slowly insert the Brix spindle washed with the sample liquid into the cylinder (if there is no thermometer in the Brix spindle, insert another thermometer) . When the thermometer correctly indicates the temperature of the sample liquid, read the Brix according to the reading method specified by the Brix spindle, and record the temperature of the sample solution during the measurement.

Note: There are two reading methods specified by the brix spindle: one is to read the true liquid level height of the sample liquid with the horizontal line of sight, and the other is to read the upper edge of the meniscus formed by the contact between the liquid level and the brix spindle. Therefore, read the instructions of the brix spindle carefully before use, and read the readings according to the correct method.

3) Temperature correction. Since the volume of the solution increases and the Brix decreases as the temperature rises, there should be a corresponding temperature for the determination of the Brix. The current Brix spindle takes 20 °C as the standard. If the reading temperature is not 20 °C, it is necessary to check the Brix Temperature Correction Table (attached list 1) add or distvact, the value of the table is added to the observed Brix, which is the corrected Brix. In the absence of any indication, Brix means corrected Brix.

[Example 2-1] The observed Brix of the mixed juice at 19 ℃ is 20.00 °Bx. What is the Brix of the mixed juice?

[Solution] The temperature correction at 19 °C obtained from attached list 1 is 0.06 °Bx.

Then the correction is: 20.00−0.06 = 19. 94 (°Bx).

[Example 2-2] The observed Brix of the mixed juice at 22 ℃ is 19.50 °Bx. What is the Brix of the mixed juice?

[Solution] Check the temperature correction at 22 °C in attached list 1 = 0.12 °Bx,

Then the corrected Brix is: 19. 50+0.12=19.62 (°Bx).

Calculators and computers are widely used, the corrected Brix can also be calculated by the following formula:

$$\text{The corrected Brix} = B + 0.001T^2 + 0.015T - 0.0029\sqrt{B} - 0.17$$

Where T — sample solution temperature (°C);

B — observed Brix (°Bx).

For the above example 2–1, the corrected Brix is calculated as follows:

$$\text{The corrected Brix} = 19.50 + 0.01 \times 22^2 + 0.015 \times 22 - 0.0029\sqrt{19.50} - 0.17 = 19.80$$

(6) Determination of Brix of syrup and massecuite.

1) Instruments and equipment used. Same as the juice mix.

2) Sample treatment and determination. Syrup can be diluted four times, massecuite, molasses, massecuite can be diluted six times, and must completely dissolve its crystals, after adding water to fully stir evenly.

The rest of the steps are the same as the mixed juice.

3) Brix calculation. Same as the juice mix. However, after the Brix of the sample solution to be measured is calculated, it should be multiplied by the dilution factor to obtain the Brix of the original sample.

(7) Use of thermometer. Thermometer is one of the most commonly used instruments in the laboratory.

1) Specification.

① Thermometer for sugar solution conversion: An ordinary mercury thermometer with a scale of 0~100 °C and 1 °C can be used, with a length of about 270 mm and a diameter of less than 10 mm. It is required that the distance between the 60 °C scale and the mercury ball should not be less than 170 mm, so that the scale can be inserted into a 100 mL volumetric flask to expose the bottle mouth for easy observation.

② Temperature when measuring the optical rotation reading: A mercury thermometer with a 0~40 °C (or 0~60 °C) graduation of 0.1 °C can be used, and the mercury bulb should be shorter (to facilitate the measurement of the temperature of the sugar solution in the side funnel-shaped polarimeter tube).

③ Thermometer for oven: a mercury thermometer with a scale of 0~150 °C, a graduation of 1 °C and a longer distance from the mercury ball can be used.

2) Usage and precautions.

① When measuring the temperature of the solution, insert a dry and clean thermometer carefully, and read it after the temperature is constant.

② The mercury column in the capillary tube of the thermometer is easy to be separated into two or several sections, so that the measured temperature is not accurate enough, and special attention should be paid.

③ Do not place the thermometer in a temperature area higher than the measuring range of the thermometer to avoid breakage of the thermometer.

3) Verification method.

There may be some errors in the temperature scale of some thermometers when they leave the factory, so the new thermometers should be uniformly verified before use.

① Verification of temperature below room temperature: Tie the standard thermometer and the thermometer to be verified together with cotton thread, so that the mercury bulb at the lower end of the thermometer is flush, the upper end is fixedly hung on an iron ring attached to an iron frame, and the lower end is placed in a 250 mL beaker filled with water, and add ice or a proper amount of coolant (such as 100 g of water at 10~15 °C, 30 g of ammonium chloride is added to reduce the temperature to −5.1 °C, or 60 g of ammonium nitrate is added to reduce the temperature to −13.6 °C), so that the temperature can be reduced to the temperature required for detection, and the water is stirred to ensure that the water temperature is uniform. When the standard thermometer reaches the verified temperature (must be an integer), immediately read out the temperature of the verified thermometer, and the difference between the two is the correction of the verified thermometer. Then let the water temperature rise, and check one by one according to the above method (or check according to the required degree).

② Verification of the temperature higher than the room temperature: Heat the above incinerator to make the temperature rise gradually, and verify it degree by degree. If the temperature in 100~200 °C is verified, glycerin can be used instead of water, and the verification method is the same as above.

3. Determination of dry substance by refractometry

(1) Principle of determining dry substance by refractometry.

1) Reflection phenomenon and reflection law of light. When a beam of light passes through the interface of two media, it will change its direction of propagation, but still propagate in the original medium. This phenomenon is called reflection of light.

For example, when a beam of light AO in the darkroom strikes the plane mirror MM (as shown in Fig. 2-2), a beam of light OB is reflected from the point O of the mirror surface; AO is called the incident line, point O is called the incident point, and OB is called the reflected line. A straight line OL perpendicular to the mirror surface drawn from the point of incidence is called the normal, the angle α between the incident ray and the normal is called the angle of incidence, and the angle β between the reflected ray and the normal is called the angle of reflection.

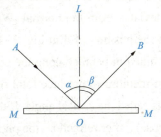

Fig. 2-2 **Reflection of light**

If the mirror is rotated to change the angle of incidence, the angle of reflection will also change. It has been proved by experiments that the reflection of light conforms to the following laws:

① The incident ray, the reflected ray and the normal line are always in the same plane, and the incident ray and the reflected ray are on both sides of the normal line;

② The angle of incidence is equal to the angle of reflection.

2) Refraction phenomena and laws of light. When light is reflected from one medium (such as air) to another medium (such as water), part of the light is reflected back to the first medium, and the other part enters the second medium and changes its direction of propagation. This phenomenon is called light refraction.

As shown in Fig. 2-3, the OD is called the refracted ray, and the angle γ between the refracted ray and the normal is called the angle of refraction. The angle of refraction changes as the angle of incidence changes. Experiments have shown that light is refracted according to the following law.

① The incident ray, the normal and the refracted ray are in the same plane, and the incident ray and the refracted ray are on both sides of the normal;

② No matter how the incident angle changes, the ratio of the incident angle to the sine of the refraction angle is always equal to the ratio of the propagation velocities of light in the two media.

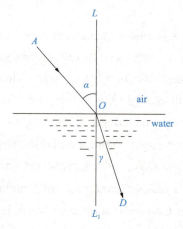

Fig. 2-3   Refraction of light

3) Total reflection. When a ray of light passes from an optically dense medium into an optically sparse medium, the refracted ray deviates from the normal. When the incident angle α_2 gradually increases to a certain angle, such as the position of 4 in Fig. 2-4, the refracted ray 4' coincides with OM, and the refracted ray will be emitted in parallel along the contact surface OM of the two media, and will not enter the optically thin medium to produce total reflection. The angle of incidence at which the angle of refraction is 90° is called the critical angle. If the light is reversed from the range of 1' to 4' (from light sparse to light dense), after refraction, the phenomenon of OU straight line, bright on the left and dark on the right is produced (see Fig. 2-4). The critical angle α_2 can be measured

by experiment. The refractive index of the sugar solution can be calculated when the total reflection occurs when the light enters the sugar solution from the prism.

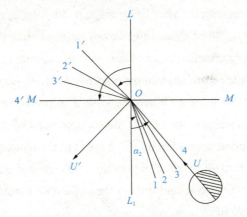

Fig. 2-4 Total reflection of light

The refractive index of pure sucrose solution increases with the increase of concentration. The comparison table of refractive index and concentration of pure sucrose solution has been worked out through experiments, and the refractive index of a sugar solution has been measured. According to this, the weight percentage of sucrose in sugar solution or the percentage of apparent dry substance can be found out.

(2) Types and structures of commonly used refractometers.

1) Type of refractometer hammer. The specifications for refractometers used in sugar mills are as follows:

① The standard temperature: 20 °C;

② Measuring range (°Bx) : 0~30, 0~85 or 0~95;

③ Minimum division value of scale scale (°Bx) : 0.1, 0.25 or 0.5, and (the error not exceed half a division).

The refractometers used in the laboratory are mainly of the following types:

① Binocular Abbe refractometer;

② Monocular Abbe refractometer;

③ Automatic Abbe refractometer;

④ Digital full-automatic Abbe refractometer.

2) Refractometer hammer structure. The structure of the optical part of each type of refractometer is basically the same, and the main difference lies in the reading and mode. In principle, it is based on the refraction of light at the interface of sugar solution and prism, and the total reflection under certain conditions, to measure the refractive index of sugar solution, and at the same time to reflect the reading on the dial.

The binocular Abbe refractometer is taken as example for illustration.

The optical system of the Abbe refractometer consists of an observation system and a reading system. The light is reflected from the reflector and refracted by the incident prism, the refracted prism and the thin layer of sugar solution between them. The dispersion produced by the refracted prism and the sugar solution is eliminated by the Amisi prism. The light and shade boundary produced by the objective lens is imaged on the reticle. After being magnified by the eyepiece, the light is imaged in the line of sight of the observer. When measuring, we can judge whether the boundary line of light and shade just passes through the intersection point of the cross line according to the visual field. The knob is used to adjust the achromatic prism between the objective lens and the refractive prism to make the boundary between light and shade clear. When the knob is adjusted, the prism swings to make the boundary between light and shade pass through the intersection point of the cross line. At this time, the refractive index or sugar solution concentration can be read in the reading cylinder. In a dense optical system, light is reflected by a small reflector, passes through a frosted glass to a dial, passes through a steering prism and an objective lens, and images the scale on a reticle, which is magnified by an eyepiece and imaged in the eye of an observer.

The structure of the Abbe refractometer is shown in Fig. 2-5.

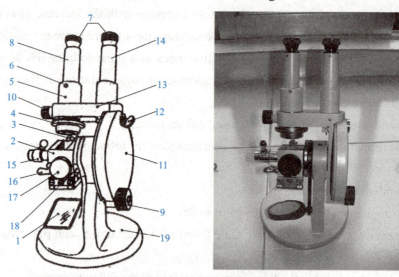

Fig. 2-5 Abbe refractometer

1—Reflector; 2—Prism set (including light inlet prism and reflector); 3—Prism backhand; 4—Dispersion value scale (Amisi prism set); 5—Objective lens; 6—Light and shade dividing line adjustment knob; 7—Eyepiece; 8—Observation lens tube; 9—Light and shade dividing line adjustment knob; 10—Dispersion adjustment knob; 11—Disc set (with scale plate inside); 12—Small reflector (for reading light transmission); 13—Bracket; 14—Reading lens tube; 15—Thermometer socket; 16—Thermostatic water bath connector; 17—Protective cover; 18—Shaft; 19—Base

When the knob is turned to swing the prism and the boundary line of light and shade in the field of vision passes through the intersection point of the cross line, it indicates that the incident angle of light from the prism into the sugar solution has reached the critical angle. Because the measured sugar solution concentration is different, the refractive index is different, so the value of the critical angle is also different. Because the refractive index of the sugar solution is proportional to the sine of the critical angle, the refractive index or the Brix of the sugar solution is directly engraved on the dial of the refractometer.

(3) Use of refractometer. According to the rule that pure sucrose solution has refractive index and its refractive index increases with the increase of concentration, the weight percentage of sucrose can be obtained by measuring the refractive index of sugar solution. The impure sugar solution contains non-sugar impurities, and the result is the weight percentage of the refractive solid solution. This value is an approximation of the weight percent of true solid solution. The influence of non-sugar impurities on the determination result is smaller than that of the density method when the refractive index is used for determination, the method is more accurate and quick. It is widely used in production and scientific research.

Refractometer measure the concentration of sugar solution, it can be read directly on the scale of the instrument, or the temperature can be corrected by looking up the refractive index table. The weight percentage in the sugar solution obtained from this is the refractive index Brix. For impure sugar solution, it is called the approximate value of the weight percentage of true solid solution.

With the continuous development of the electronic industry, the refractometer is now an electronic digital display, and the data can be read directly. Is being used more and more widely. The WAY-2S digital Abbe refractometer is shown in Fig. 2-6.

Fig. 2-6　The WAY-2S digital Abbe refractometer

The WAY-2S digital Abbe refractometer has the advantages of high precision, automatic correction of the influence of temperature on the Brix value, and display of the sample solution temperature.

1) Structural diagram of WAY-2S digital Abbe refractometer is shown in Fig. 2-7.

The specifications are as follows:

Measuring range: refractive index nD　1.300 0~1.700 0

Brix: Bx-TC　0~95%

Brix: Bx　0~95%

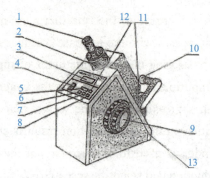

Fig. 2-7 Structural diagram of WAY-2S digital Abbe refractometer

1—Eyepiece; 2—Dispersion correction handwheel; 3—Display window; 4—"POWER" waveform power switch;
5—"READ" reading display key; 6—"Bx–TC" Brix display key; 7—"nD" refractive index display key;
8—"Bx" without temperature correction display key; 9—Adjustment handwheel; 10—Spotlight illumination component;
11—Refractive prism part; 12—"TEMP" temperature display key; 13—RS233 socket

Measurement accuracy: refractive index nD　±0.0002

Brix: Bx–TC　±0.1%

Brix: Bx　±0.1%

Temperature: display range　0~50 ℃

Corrected Bx temperature range　15~45 ℃

2) Operation steps and usage.

① Press the "POWER" waveform power switch, the lighting lamp in the component is on, and the display window displays "00000" (sometimes the display window displays "— —" first, and then displays "00000" after a few seconds).

② The refractive prism part is opened, and the lens paper is removed (the lens paper only needs to be used in a single layer, placed between the two prisms when the instrument is not in use, so as to prevent the working surface of the prism from being damaged by fine hard particles left on the prism when the prism is closed).

③ Check the surfaces of the upper and lower prisms and carefully clean them with water or alcohol. The surfaces of the two prisms should also be carefully cleaned after the measurement of each sample, because a small amount of the sample left on the prism will affect the measurement accuracy of the next sample.

④ Place the sample to be measured on the working surface of the refractive prism below. If the sample is liquid, use a clean dropper to suck 1~2 drops of liquid sample onto the working surface of the prism, and then cover the light inlet prism above. If the sample is a solid, the solid sample must have a flat, polished surface. Before measurement, the polished surface shall be wiped clean, and 1~2 drops of transparent liquid (such as bromonaphthalene) with refractive index higher than that of the solid sample shall be dropped on the working surface of the refractive prism below, and then the polished surface of the solid sample shall be placed on the working surface of the refractive prism to make good contact. It is not necessary to cover the upper light inlet prism when measuring solid samples.

⑤ Rotate the rotating arm and the condenser lens barrel of the spotlight illumination component to uniformly illuminate the light inlet surface (for measuring liquid samples) of the light inlet prism above or the light inlet surface (for measuring solid samples) in front of the solid samples.

⑥ Observe the field of view through the eyepiece while rotating the adjustment handwheel so that the cut-off falls in the crosshair field of view. If the field of view is dark from the eyepiece, turn the adjustment handwheel counterclockwise. If the field of view is bright, turn the adjusting handwheel clockwise. The bright region is at the top of the field. The eyepiece can be rotated in the case of a bright field of view to adjust the visibility to see the clear crossing line.

⑦ Rotate the dispersion correction handwheel in the square notch of the eyepiece and adjust the position of the condenser. That the light and dark parts in the field of view have good contrast and the boundary line of light and dark has minimum dispersion.

⑧ Rotate the adjusting hand wheel to make the light and shade dividing line accurately align with the intersection point of the crossing line, as shown in the Fig. 2-8.

Fig. 2-8 Make the light and shade dividing line accurately align with the intersection point of the crossing line

⑨ Press the "READ" reading display key 5, 00000 in the display window will disappear, and "— —" will be displayed. After a few seconds the refractive index of the tested sample will be displayed. Getting the Brix value of the sample, you can press the "Bx" Brix display key without temperature correction or the "Bx-TC" Brix display key with temperature correction (press ICUMSA) .The "nD", "Bx-TC" and "Bx" keys are used to select the measurement mode. After selection, press the "READ" key again, and the display window will be displayed according to the pre-selected measurement mode. Sometimes press the "READ" key to display "— —". After a few seconds, the display window is completely dark, and there is no other display, reflecting that the instrument may be faulty. At this time, the instrument cannot work normally and needs to be checked and repaired.When the selected measurement mode is "Bx-TC" or "Bx", if the rotation of the adjusting hand wheel exceeds the Brix measurement range (0-95%) , press the "READ" key, and the display window will display "•".

⑩ To detect the sample temperature, press the "TEMP" temperature display key, and the display window will display the sample temperature. If pressing the "READ" key, when the display window displays "— —", pressing the "TEMP" key is invalid,it can not be determined.

In other cases, the temperature of the sample can be detected. When the temperature is displayed, press the "nD", "Bx-TC" or "Bx" key again, and the original refractive index or Brix will be displayed. In order to distinguish whether the displayed value is temperature or Brix, a " ┝ " symbol is added before the temperature, a " ⊏ " Symbol is added before the "Bx-TC" Brix, and a " ┣ " symbol is used before the "Bx" Brix.

⑪ After the sample (sugar solution) measurement, it must be carefully cleaned with alcohol or water.

⑫ The refractive prism of the instrument has a constant temperature water structure. If the refractive index of the sample at a specific temperature needs to be measured, the instrument can be externally connected with a thermostat to adjust the temperature to the required temperature before measurement.

⑬ The computer can be connected to the instrument with an RS232 cable. First, Send an arbitrary character and wait for the message to be received (parameters: baud rate 2400, data bit 8, stop bit 1, total byte length 18).

3) Instrument calibration. The instrument may also be calibrated on a regular basis, or when there is doubt about the measurement data. Distilled water or glass standard block for calibration. If there is an error between the measured data and the standard, the hexagon socket wrench can be used to carefully rotate the screw inside through the small hole in the dispersion correction handwheel, so that the cross line on the reticle moves up and down, and then the measurement is carried out until the measured data meets the requirements. When the sample is a standard block, the measured number shall conform to the data marked on the standard block. If the sample is distilled water, the measured number shall conform to the following Table 2-4.

Table 2-4 Refractive index of distilled water

Temperature	Refractive Index /nD	Temperature	Refractive Index /nD
18	1.333 16	25	1.332 50
19	1.333 08	26	1.332 39
20	1.332 99	27	1.332 28
21	1.332 89	28	1.332 17
22	1.332 80	29	1.332 05
23	1.332 70	30	1.331 93
24	1.332 60		

[Example 2-3] The Brix of the sample was observed to be 19.5 °Bx at a temperature of 18 °C, and the corrected refractive Brix of the sample was calculated.

[Solution] Check the temperature correction at 18 °C from the refraction Brix correction table = 0.14 °Bx,

Corrected Brix = 19.5−0.14 = 19.36 (°Bx).

[Example 2-4] Assume that the Brix of the syrup after quadruple dilution is 16.30 °Bx at 23 °C, and calculate the corrected refractive index of the syrup.

[Solution] The temperature correction at 23 °C obtained from the refraction Brix correction table is 0.22 °Bx,

Corrected Brix = (16.30 + 0.22)×4 = 66.08 (°Bx).

4. Use of Baume Meter

In addition to using a Brix spindle to measure the Brix, a Baume meter is also used in sugar mills to measure the concentration of lime milk. The shape of the Baume meter is the same as the brix spindle, which is a glass tube with a thick lower part and a thin upper part and a scale. The Baume meter is measured in degrees Baume (°Bé). The scale of the Baume meter is formed by taking distilled water as 0 °Bé, taking the mass fraction (weight percent concentration) of salt as 15% as the scale of 15 °Bé, and then dividing equally in 0–15 and extending downward. If the result of the measurement is 5 °Bx′, it means that the density of the solution is equivalent to that of a 5% salt solution. Because there is no special equipment for measuring the concentration of lime milk, and the Baume meter is a general equipment, the Baume degree is approximately equal to the weight percentage of calcium oxide in lime milk (the content of calcium oxide in 10 °Bx′ lime milk is 8.74%), so it is customary to use it to measure the concentration of lime milk in sugar mills.

The relationship between the content of calcium oxide in the lime milk and the Baume degree can be expressed by the following formula:

Content of calcium oxide in lime milk (%) = 0.891 8Be−0.114 1

Be—Baume degree in lime milk.

The use of a Baume meter is the same as that of a brix spindle.

In the general case, 1 °Bx′ is approximately equal to 1.8 °Bx.

Task 2.2 Determination of pol and apparent purity of Semi-product

Enterprise case

The director takes back part of the intermediate cane juice from the production line, and then arranges Xiaoming to determine the pol and purity of the above samples. What instrument should Xiaoming choose and what method should he use to determine it?

Mission objectives

Through the study of this task, students can achieve the following goals:

(1) Understand and master the concepts of pol, sucrose content, apparent purity and gravity purity.

(2) Master the usage of polarimeter.

(3) Grasp the method and calculation of determination of pol and apparent purity.

Quality objectives

Develop the good habit of comparative analysis of learning.

Task description

Determination of pol and apparent purity of semi-product by polarimeter.

Implementation conditions

(1) Ingredients: Sugar juice.

(2) Equipment and instruments: Hammermeter, polarimeter, 200 mm polarimeter tube, thermometer, 250 mL conical flask, measuring cylinder, beaker, filter paper.

(3) Reagent: Basic lead acetate.

Procedures and methods

(1) Sample processing: Molasses, massecuite, massecuite and other samples containing crystals are diluted by 6 times to ensure that all crystals are dissolved.

(2) Measure the Brix and temperature of the sample solution, and record the data.

(3) Determination of pol: Pour about 100 mL of the sample into a dry 250 mL conical flask (if it is not dry, it must be washed with the sample solution first), add a proper amount of alkaline lead acetate powder (the minimum amount with clarification effect is appropriate), shake well, filter, wash the container (small beaker) with the initial filtrate, collect the filtrate after pouring, and measure its optical rotation reading with a 200 mm polarimeter tube. The polarimeter tube must first be washed 2~3 times with filtrate.

(4) Data processing and calculation:

1) Calculation.

Brix of sample solution (%) = Observed optical rotation × Brix factor

Where　Brix factor—obtained by looking up the Brix factor key table based on the observed Brix.

$$\text{Apparent purity (\%)} = \frac{\text{Pol}}{\text{Brix}} \times 100\%$$

2) Data and result processing.

Project	Data and results
Brix of sample solution/°Bx	
Temperature of sample solution/°C	
Brix of sample solution/°Bx	
Observed optical rotation of sample solution/°Z	
Pol of sample solution/%	
Apparent purity of sample solution/%	
Sample pol/%	
Sample apparent purity/%	

Reflection

In recent years, in order to protect the environment and promote the development of ecological civilization, the near-infrared polarimeter has gradually replaced the ordinary optical polarimeter. What are the advantages of the near-infrared polarimeter in environmental protection? Please consult the information extensively and give the answer.

Task-related knowledge—determination of pol and sucrose content by polarimetry

1.Basic principles of polarimetry

(1) Basic knowledge. Polarimetry is one of the simplest and fastest methods to determine sucrose, which is based on the optical activity of sucrose and the fact that the optical rotation of sucrose solution is proportional to its concentration.

Polarimetry can be divided into primary polarimetry and secondary polarimetry. The results of secondary polarimetry are more accurate, but the determination process is more complex and the time is longer. For some special samples (with high purity or low sucrose content), the results of the two methods are not very different or the accuracy of the results is not required to be high in production, so the primary polarimetry can be used instead of the secondary polarimetry.

Polarimeter is the main instrument to determine the sucrose content of sucrose solution in sugar factories. Polarimeter is a kind of instrument to detect optical rotation. Its working principle is based on the optical rotation of sugar solution. It uses the vibration direction of rotating polarized light to detect the size of optical rotation. Before introducing the polarimeter, it is necessary to have a certain understanding of the basic knowledge of the concept of sucrose, polarized light and optical rotation.

1) Sucrose degree (Pol). The weight percentage of sucrose content in the sample measured by single polarimetry.

$$\text{Pol (\%)} = \frac{\text{Sucrose content}}{\text{Sample weight}} \times 100\%$$

2) Sucrose. The percentage by weight of the sucrose content in the sample as determined by the specified method. Generally, samples with high purity (such as white granulated sugar) or low sucrose content (such as filter cake) are determined by primary polarimetry, and other intermediate products are determined by secondary polarimetry.

$$\text{Sucrose (\%)} = \frac{\text{Sucrose content}}{\text{Sample weight}} \times 100\%$$

3) Purity. It refers to the weight percentage of sucrose content in the dry substance of the sample. It can be divided into gravity purity and apparent purity according to different determination methods.

① Gravity Purity (GP).

$$\text{Gravity purity (\%)} = \frac{\text{Sucrose}}{\text{Brix}} \times 100\%$$

② Apparent purity (AP).

$$\text{Apparent purity (\%)} = \frac{\text{Pol}}{\text{Brix}} \times 100\%$$

4) Natural light and polarized light (plane of polarization). Light is an electromagnetic wave, a transverse wave. That is, the direction in which the light travels is perpendicular to the direction in which the light wave vibrates.

Natural light: The light has innumerable vibration planes perpendicular to the direction of the light. If a beam of light is emitted from the paper and directed at eyes, we can see that the vibration plane of the light is as shown in Fig. 2-9 (a). There are innumerable vibration planes, and each double arrow represents a vibration plane.

Polarized light: Because some crystals only allow light in one direction of vibration to pass through, when light passes through these special crystals, it will become light with only one plane of vibration. If a beam of light is emitted from the paper surface and directed to us after passing through a special crystal, we can see that the vibration plane of the light is as shown in Fig. 2-9 (b). There is only one vibration plane, and the light of this single vibration plane is called "polarized light". Its plane of vibration is called the "plane of polarization". This kind of special crystal can not only produce polarized light, but also detect the rotation angle of polarized light, that is, to determine the size of optical rotation.

Fig. 2-9 Natural light and polarized light
(a) Natural light; (b) Polarized light

5) Optically active substances. When light passes through some substances, the vibration plane of light will rotate a certain angle, which is called "optical activity". Any substance that can rotate the vibration plane of polarized light by a certain angle is called "optical activity quality". Sugar solutions have optical activity, among which sucrose and glucose can rotate the vibration plane of polarized light to the right, which is called dextrorotation, represented by (+), and these substances are called dextrorotation substances; fructose can rotate the vibration plane of polarized light to the left, which is called levorotation, represented by (−), and these substances are called levorotation substances.

6) Optical rotation. When the temperature, the wavelength of the light source, and the distance of light passing through the optically active substance are specified values, the angle of rotation of the plane of vibration after light passes through the optically active substance is called optical rotation.

(2) Polarimeter. An instrument for measuring the angle of rotation of the plane of vibration of light passing through an optically active substance is called a polarimeter. Fig. 2-10 shows the WZZ-2SS digital sugar polarimeter.

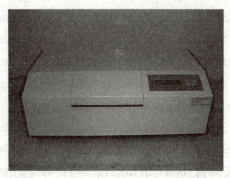

Fig. 2-10 The WZZ-2SS digital sugar polarimeter

In order to compare the angle of rotation of the plane of vibration of the light, the polarimeter must use polarized light with only one plane of vibration.

There are many types of polarimeter, but the basic principle is the same. It is determined by using the optical activity of sucrose solution. When the thickness of the liquid layer of the sucrose solution, the temperature and the wavelength of the light source are certain values, the rotation angle of the polarization plane of the sucrose solution is directly proportional to the concentration of sucrose, and the concentration of sucrose in the solution can be obtained by measuring the rotation angle of the polarization plane of the sugar solution, that is:

$$\alpha = K \cdot C \cdot L$$

Where α—the angle of rotation of the plane of polarization;

C—concentration of sugar solution;

L—the distance of the light travels through the solution, i.e., the thickness of the sugar solution (length of polarimeter tube);

K—coefficient.

K can be determine by measuring that optical rotation α_0 of a sugar solution have a known concentration C_0 using the relationship $K = \alpha_0 / (C_0 \cdot L)$.

1) Basic parts of a polarimeter. No matter what kind of polarimeter it is, it has the following components:

① Light source. It is required to produce light with stable wavelength. Because the same optically active substance has different optical rotation under different wavelengths of light, the wavelength of the light source is required to be stable (also known as high color purity). Sodium light bulbs are often used as light sources, which can produce yellow monochromatic light. For example, the WZZ-2SS automatic polarimeter uses a 20 W sodium lamp as the light source, and a simple point light source is composed of a small aperture diaphragm and an objective lens.

② Polarizer. Its function is to change the monochromatic light produced by the light source into polarized light with only one plane of vibration. A "Nichol" prism composed of two calcite crystals is generally used as a polarizer. Calcite crystal has birefringence. When light passes through the crystal, it can be decomposed into two beams of light with perpendicular vibration direction and different forward direction. By properly selecting the angle of the bonding surfaces of the two crystals and the bonding material, one of the light rays can be totally reflected at the bonding surface to the black side of the prism and absorbed there. The other beam of light can pass through the prism almost undeflected, and the beam of light coming out of the prism becomes polarized light with only one plane of vibration.

③ Polarimeter tube. It is used to hold the sample solution to be tested. This is a long, thin glass tube with a glass cover fixed at both ends by a screw cap with a hole in the middle to allow light to pass through. The shape is shown in the Fig. 2-11. When measuring, put the sample into the polarimeter tube and place it in the light path of polarized light. If the sample is optically active, the plane of vibration of the polarized light will be rotated by a certain angle after the polarized light passes through the polarimeter tube. Since the sample rotates the plane of vibration by an angle proportional to the path of light through the optically active substance, the length of the polarimeter tube must be accurate. Polarimeter tubes shall be 200 mm in specified length, 100 mm in half specified length and 400 mm in double length.

Fig. 2-11 Polarimeter tube

④ Analyzer. The structure of the analyzer is the same as that of the polarizer. It is used to measure the rotation angle of the vibration plane of polarized light. When the two crystals are in the same direction (the vibration planes that produce polarized light are the same) the polarized light produced from the polarizer can pass through the analyzer completely, and the light is brightest from the back of the analyzer (called the "field of view"). When the two crystals are perpendicular to each other, at the zero point of the instrument, the polarized light from the polarizer cannot pass through the analyzer at all (the "field of view" of the analyzer is the darkest). When an optically active substance is placed in the light path, part of the polarized light can pass through the analyzer because the vibration plane of the polarized light is rotated by a certain angle, and the "field of view" of the analyzer should become bright. If the polarizer or analyzer is rotated so that the vibration of the polarized light entering the analyzer is perpendicular to the direction of the analyzer crystal, the "field of view" of the analyzer will become dark again. The rotation angle of the polarizer or analyzer in this process is the optical rotation of the measured substance.

2) Working principle of polarimeter.

① Zero. As shown in Fig. 2-12, the natural light passes through the polarizer to generate polarized light in the vertical direction. The clean polarimeter tube is placed in the polarimeter after distillation, and the direction of the analyzer is adjusted to be perpendicular to the polarizer, that is, only the polarized light in the horizontal direction can pass through the analyzer, and the light can not pass through the polarizer and the analyzer crystal at the same time. At this time, no light passes through after observing the analyzer. The field of vision in the polarimeter tube is dark.

② After placing the sample. As shown in Fig. 2-13, the polarimeter tube filled with sugar solution is placed in the polarimeter, that is, between the polarizer and the analyzer crystal. The natural light passes through the polarizer to generate polarized light in the vertical direction. After the polarized light passes through the polarimeter tube filled with sugar solution, the vibration plane is rotated by the sugar solution for a certain angle. The light after rotation can be decomposed into horizontal part and vertical part.The light energy in the horizontal direction passes through the crystal of the analyzer. At this time, the light seen from the polarimeter tube is weaker than the original one, that is, the field of view in the polarimeter tube is slightly brighter.

③ Measurement of optical rotation angle. As shown in Fig. 2-14, rotate the analyzer crystal to make the axial plane of the analyzer perpendicular to the vibration plane of the polarized light coming out of the polarimeter tube. Even if the polarized light cannot pass through the analyzer, the field of view in the lens barrel will become dark again. At this time, the rotation angle of the analyzer crystal is the angle at which the sample rotates the polarization plane.

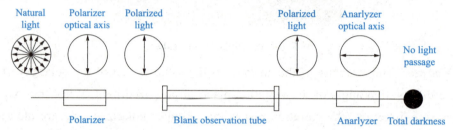

Fig. 2-12 The condition when the polarimeter tube is not filled with sugar solution

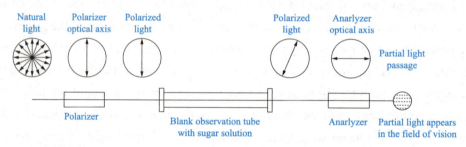

Fig. 2-13 The condition of the polarimeter tube after filling with sugar solution

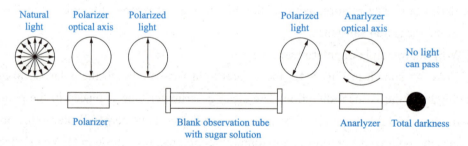

Fig. 2-14 The condition of detector after adjustment of optical axis

3) The method of dividing the scale of polarimeter. The rotation angle of the sample can be measured by the polarimeter. For the sake of convenience, as a special equipment for sugar industry, the rotation angle of the polarization plane can be directly converted into the sucrose concentration of the sample solution. The international pol (°Z) is uniformly used as the unit for the reading of the polarimeter.

① Scale division of polarimeter.

a. Take 20 °C as the standard temperature of the sample solution to be measured, and the length of the polarimeter tube is 200 mm.

b. Put the distilled water into a clean polarimeter tube, place it on the polarimeter for measurement, and set the reading as the "0" point of the instrument. Indicates that the concentration of sucrose in the sample solution is "0", that is, the weight percentage of sucrose content in the sample is "0".

c. Dissolve 26.000 g (one normal weight) of pure sucrose and transfer it into a 100 mL volumetric flask, add water to the marked line (normal weight of dilution), shake well, put the sugar solution into a 200 mm polarimeter tube (which has been washed several times with the sugar solution for determination) and place it on the polarimeter for determination, and the reading obtained is set as "100" of the instrument. The weight percentage indicating the sucrose content in the original sample (note: not the sample solution to be tested) is "100".

d. Divide equally between "0" and "100" to obtain the scale of the polarimeter.

② International Brix (°Z). The reading obtained on a polarimeter for sugar manufacturing is called the "international pol" and is represented by the symbol "°Z".

It can be seen from the division method of the polarimeter scale that when the sample is diluted with the normal weight, if the reading measured on the polarimeter is 100 °Z, the mass percentage of sucrose content in the original sample is "100". The sucrose content of the tested sample solution was 26 g/100 mL. Since the rotation angle of the sugar solution is directly proportional to the concentration of sucrose in the sample solution, 1 °Z means that the content of sucrose in the sample solution to be tested is 0.26 g/100 mL.

4) That relationship between optical rotation read and sucrose. Assuming that the polarimetric reading of a sample is P, it does not necessarily directly indicate the pol (or sucrose content) of the sample. Different dilution methods are used for samples, and the results of polarimetric determination have different meanings.

① The sample is diluted with the normal weight. The weight percentage of sucrose content in the original sample (before dilution) is P (i.e., the sucrose content of the original sample is divided into P); or the sucrose content of the sample solution to be tested (the sample solution obtained after dilution of the sample) is 0.26 g/100 mL.

② The sample is diluted by multiple dilution (including undiluted). Indicates that the sucrose content of the test sample solution (sample solution obtained by diluting the sample) is $0.26P$ g/100mL. In the sugar mill, the sucrose content is generally expressed by mass percentage, and the conversion method is as follows:

$$\text{Mass percentage of sucrose content in the tested sample solution} = \frac{0.26P}{100d} \times 100\% = \frac{26P}{100d} (\%)$$

Where P—optical rotation reading (°Z);

D—density of tested sample solution (g/mL).

Note: d can be found in the relevant table according to the observed Brix of the tested sample solution or calculated according to the following formula:

Apparent density of sugar solution to be measured $=0.9982+0.0037B+0.00001816B^2$

Where B—observed Brix of sample solution (°Bx).

In practical application, for the sake of convenience, $26/100\ d$ is made into a table, which is called Pol (polarization) factor table, and is directly retrieved from the observed Brix of the sample solution. It can also be calculated according to the following formula:

$$\text{Pol (polarization) factor} = (0.5107 - 0.001B)^2$$

Where B—observed Brix of tested sample solution (°Bx).

The mass percentage of the sucrose content of the sample solution to be tested can be calculated according to the following formula:

Mass percentage of sucrose content in the sample solution to be tested $= P \times$ pol (polarization) factor

The mass percentage of the sucrose content of the original sample can be calculated according to the following formula:

Mass percentage of sucrose content in the original sample = mass percentage of sucrose content in the sample solution to be tested $\times$ dilution times

[Example 2-5] When the temperature is close to 20°C, the observed Brix of the sugar solution is measured to be 20.00 °Bx, and the reading of the optical rotation is measured to be 71.20 °Z with a 200 mm polarimeter tube. Find the Brix of the sugar solution.

[Solution] Solution 1:

Check the relevant table according to the observed Brix of the tested sample solution or calculate according to the following formula get d:

Apparent density of sugar solution to be measured $d=0.9982+0.0037B+0.00001816B^2$
$$=0.9982+0.0037\times20+0.00001816\times20^2$$
$$=1.0796$$

$$\text{Pol of the tested sugar solution (\%)} = \frac{0.26P}{100d} \times 100\% = \frac{26P}{100d} \times 100\% = 17.14\%$$

Solution 2:

According to the Brix of the sample solution, directly look up the Pol (polarization) factor table to obtain the Pol (polarization) factor or calculate it according to the following formula:

Pol (polarization) factor $= (0.5107 - 0.001B)^2 = (0.5107 - 0.001\times20)^2 = 0.24076$

Pol (%) of the sugar solution to be tested $= P \times$ Pol (polarization) factor $\times 100\% = 71.2 \times 0.24076 \times 100\% = 17.14\%$

5) Factors affecting the determination of optical rotation.

① Wavelength of light. For the same substance, the optical rotation angle measured at different wavelengths is also different. For saccharides, the measured rotation angle increases as the wavelength decreases. The wavelength of light is related to the temperature of the light source, so before using the polarimeter, it must be preheated until the wavelength of the light emitted by the bulb is stable.

② Temperature. When the temperature is high, the influence of temperature on the determination of optical rotation is small, while when the temperature is low, the influence of temperature on the determination of optical rotation is large. The optical rotation angle of sucrose decreases with the increase of temperature, but the optical rotation angle of reducing sugar increases with the increase of temperature (the optical rotation of reducing sugar is opposite to that of sucrose, which is left-handed, and the reading of its optical rotation angle is negative, and the decrease of left-handed angle is the increase of the value of optical rotation). In the case of high requirements for the results, the results of polarimetry should be corrected for temperature.

③ Impurities. There are two kinds of impurities that affect the determination of optical rotation in sugar factory samples, one is the substances with optical rotation, mainly reducing sugar, and the other is the substances that affect the determination of sugar rotation, mainly alkali metals, alkaline earth metals and various salts.

a. Reducing sugar. Reducing sugars in sugar mills are mainly glucose (+ dextrose) and fructose (-levulose). The total optical rotation of the two sugars is levorotatory, that is, the optical rotation is negative, which can reduce the optical rotation of the sucrose sample. This is also one of the main reasons why the pol of the same sample is usually smaller than that of sucrose in sugar factories.

b. Alkali metal, alkaline earth metal. Hydroxides of these two metals and carbonates of alkali metals can react with sucrose to form sucrose salts with less dextrorotation, which reduces the optical rotation angle of sucrose. When the solution is dilute, the measurement error caused by alkali will be smaller due to the larger degree of hydrolysis of sucrose salt.

Alkali metal chlorides, nitrates, sulfates, acetates and citrates, alkaline earth metal chlorides, magnesium sulfate and many other salts can reduce the optical rotation of sucrose. The degree of decrease increases with the increase of the amount of salt and the decrease of the molecular weight of salt.

④ Clarifying agent and acid. The clarifier can precipitate fructose and increase the optical rotation reading. Under the condition that the determination method can not avoid the influence of reducing sugar on the results, in order to reduce the error and make the determination results of different batches of the same sample comparable, the dosage of clarifier must be strictly in accordance with the regulations.

Inorganic acid can affect the optical rotation of reducing sugar. When the sample contains more reducing sugar, the influence of inorganic acid on the determination of optical rotation should be corrected. The effect of organic acids on the determination of optical rotation is negligible.

⑤ Mutarotation. The reducing sugar has optical activity. When these substances are dissolved, their optical rotation will change. The change is rapid at first, then gradually becomes slower, and after a few hours, the optical rotation is basically stable. When the sugar solution is heated, the mutarotation can be quickly balanced.

2. Use of polarimeter

(1) Application method.

1) Insert the power plug of the instrument into the 220 V AC power supply and turn on the power switch. At this time, the sodium lamp should be turned on. It is necessary to preheat the sodium lamp for 5 min to make it emit light stably.

2) Switch the light source switch to DC, so that the sodium lamp is lit under DC. If the sodium lamp does not light up after switching to DC, switch back to AC, and then switch to DC after increasing the warm-up time.

3) Turn on the measurement switch, and the nixie tube shall have digital display.

4) Put the test tube filled with distilled water or other blank solvents into the sample chamber. After the reading is stable, press the reset button or adjust the zero adjustment knob to make the count of the instrument zero.

5) Take out the test tube of the blank solvent, inject the sample to be tested into the polarimeter tube, put it into the sample chamber, and the digital display window of the instrument will display the optical rotation of the sample.

6) Press the retest button one by one, repeat reading for several times, and take the average value as the determination result of the sample.

7) After the instrument is used, turn off the measurement, light source and power switch in turn.

(2) Attention to problems.

1) The instrument should be placed in a dry and ventilated place to prevent moisture erosion. It should be used in a working environment with a relative humidity of not more than 85% at 20 °C as far as possible. The instrument should be handled with care to avoid vibration.

2) When zeroing or measuring, there shall be no bubbles in the polarimeter tube. If there are bubbles, they shall be removed from the middle funnel first. If there are foggy water drops at both ends of the transparent surface, it should be wiped with a soft cloth.

3) When the sodium lamp cannot be used due to the failure of the DC power supply system, the instrument can also be tested under the AC power supply of the sodium lamp, but the performance of the instrument may be slightly reduced.

(3) Polarimetric polarimeter tube.

The polarimeter tube is used to hold the sample liquid and is one of the most frequently touched parts in the polarimeter. Polarized light is the phenomenon of optical rotation in the process of passing through the sample liquid in the polarimeter tube, so its quality directly affects the accuracy of optical rotation determination.

1) Specification. The polarimeter tube used for sugar detection should have a funnel-shaped measuring tube. The length is divided into three types: 400 mm, 200 mm and 100 mm, and the length must be strictly verified.

The polarimeter tube of 200 mm is used to measure the pol (polarization) of various sugar solutions, which is the most commonly used polarimeter tube in sugar factories. The polarimeter tube of 100 mm is used to measure the pol (polarization) of darker sugar solutions, and the polarimeter tube of 400 mm is generally used to measure bagasse pol (polarization).

The polarimeter tube slot of most domestic automatic polarimeters can only accommodate 200 mm polarimeter tube. In this case, the conventional method is used to measure the pol (polarization) of sugar solution with dark color and the pol (polarization) of bagasse. Attention should be paid to the treatment of the measurement readings.

2) Usage and precautions.

① The polarimeter tube must be clean. If there is lead acetate powder or scale in the tube, it can be washed with 10% ~ 20% acetic acid (dilute hydrochloric acid if necessary), and then washed with distilled water.

② Rinse the polarimeter tube with distilled water during measurement, rinse it with sample solution for 2~3 times, and then fill it with sample solution.

③ The sample liquid in the polarimeter tube shall not be mixed with bubbles to avoid measurement errors.

④ The screw cap of the polarimeter tube should not be screwed too tightly, so as to avoid stress on the cover glass and affect the measurement results. The tightness should not leak liquid.

⑤ The sample liquid (or water drops) adhered to the tube body, the cap at both ends and the cover glass in the cap must be wiped dry with a soft cloth, and then measured, so as to avoid blurring the observation field and causing errors.

⑥ Do not hold the glass tube by hand (hold the screw cap) to prevent the sample liquid from being heated and causing errors due to temperature rise.

⑦ After use, pour out the sample solution, rinse it with distilled water, dry it, and place it in a small box covered with soft cotton cloth.

⑧ The cover glass of the polarimeter tube must be regularly checked with distilled water for optical activity due to dirt, and the cover glass must be cleaned or corrected.

3) Verification method. Inspection of the length of the polarimeter tube: remove the screw caps at both ends of the polarimeter tube, measure the two ends of the glass tube with a vernier caliper (it is necessary to measure several times at different positions), and record its length. The allowable error of its length is shown in the Table 2-5.

Table 2-5 Length error of polarimeter tube

Nominal length /mm	100	200	400
Grade A/μm	±10	±20	±40
Grade B/μm	±200	±400	±800

If the length error of the polarimeter tube exceeds the above figures, it is better not to use it. If necessary, the length correction number can be used to calculate the correct rotation reading.

Example: The measured length of 200 mm polarimeter tube is 199.8 mm.

Then: Length correction = 200/199.8 = 1.001.

Assume that the optical rotation degree of a white granulated sugar measured by the tube is 99.60 °Z, then the correct optical rotation reading of the white granulated sugar is:

$$99.60 \times 1.001 = 99.70.$$

3. Application of single polarimetry

The method of determining the sucrose content in a sample by measuring the optical rotation reading of the sample is called the single optical rotation method. In addition to the main component of sucrose, the products of sugar mills also contain a small amount of other optically active substances (such as glucose, fructose, etc.) , which affects the accuracy of the determination of optical rotation, so the results measured by the single optical rotation method are only approximate to the weight percentage of sucrose for impure sugar solution, call it "Pol" (also known as "rotation degree"). However, there are few impurities in white granulated sugar, which have little interference on the determination of optical rotation, so the results of single polarimetry can be used as the sucrose content of the sample. For bagasse and filter cake with little pol, the absolute value of sucrose content is very small, so large measurement error can be allowed, and the result of single polarimetry can also be regarded as the sucrose content of the sample.

(1) Determination of pol of mixed juice, syrup and massecuite.

1) Main equipment and reagents.

① Equipment and instruments: polarimeter, 200 mm polarimeter tube, industrial balance or scale

② Reagent: distilled wat or deionized water, absolute ethyl alcohol and alkaline lead acetate

2) Assay step.

① Sample processing. No dilution is required for samples with concentrations below 25 ° Bx. Samples with higher concentrations can be diluted with distilled water. Syrup shall be diluted by 4 times, and molasses, massecuite, massecuite and other samples containing crystals shall be diluted by 6 times to ensure that all crystals are dissolved.

② Brix determination. Determine the observed Brix of the sample (or diluted sample solution) .

③ Clarification and filtration. About 100 mL of the sampling solution is put into a dry and clean 250 mL conical flask (if it is not dry, it must be washed with the sample solution for 2~3 times) . Add appropriate amount of basic lead acetate powder (the minimum amount with clarifying effect is appropriate) , shake well, filter; wash the container (usually a small beaker) with the initial filtrate, then pour out, and collect the filtrate in the container.

Note: The amount of clarifier should be determined according to the actual situation of the sample. If the amount is too small, the filtrate will be turbid, the optical rotation reading will be unstable, or even the determination can not be carried out, and the filtration rate will be significantly reduced. However, if the dosage is too much, it will increase the reading of optical rotation and make the error of the determination result too large.

④ Determination of optical rotation: Take a 200 mm polarimeter tube, wash it with filtrate for 2~3 times, fill it with filtrate, put it into a polarimeter, and determine its optical rotation reading P (°Z).

3) Calculation.

$$\text{Pol of sample solution (\%)} = \frac{26P}{100d} \times 100\%$$

Where P—optical rotation reading of the sample solution to be measured (°Z);

d—apparent density of the sample solution to be measured (look up the table according to the observed Brix or calculate according to the following formula).

$$\text{Apparent density of tested sample liquid } (d) = 0.998\ 2 + 0.037B + 0.000\ 018\ 16B^2$$

Where B—observed Brix of sample solution (°Bx).

For the convenience of use, $26/(100d)$ can be made into a table according to different observed Brix, which is called the Pol (polarization) factor table.

It can also be calculated according to the following formula based on the observed Brix:

$$\text{Brix (RPM) factor} = (0.510\ 7 - 0.001B)^2$$

Where B—observed Brix of tested sample solution (°Bx).

$$\text{Brix of sample solution (\%)} = P \times \text{RPM factor} \times 100\%$$

or

$$\text{Pol of sample solution (\%)} = P(0.510\ 7 - 0.001B)^2 \times 100\%$$

Note: If the sample is diluted multiple times during the determination, then:

Pol of original sample = pol of tested sample solution × dilution factor

[Example 2-6] After the syrup is diluted by four times, the observed Brix of the sugar solution is measured to be 15.35 °Bx, and the optical rotation reading is measured to be 54.48 °Z with a 200 mm polarimeter tube. Find the pol of the syrup.

[Solution] Calculate the Brix (polarization) factor from the observed Brix by the formula:

$$\text{Brix (polarization) factor} = (0.510\ 7 - 0.001 \times 15.35)^2 = 0.245\ 4$$

$$\text{Brix of diluted sugar solution (\%)} = 54.48 \times 0.245\ 4 \times 100\% = 13.37 \times 100\%$$

Since the sample was diluted four times, than:

$$\text{Pol of syrup (\%)} = 13.37 \times 4 \times 100\% = 53.47\%$$

(2) Determination of pol of bagasse. The pol of bagasse is determined by boiling bagasse with water to make the sugar exudate, and finally the pol of the internal solution of bagasse is basically equal to that of the external exudate. Because the sugar mill regards the cane material as consisting of juice and fiber, sucrose and all soluble substances are included in the juice, while the other insoluble parts are regarded as fiber. As long as the fiber content of the sample is calculated in advance, the content of sugarcane juice can be calculated. The pol of the bagasse can be calculated according to the pol of the cane juice.

1) Main equipment and reagents.

① Sugar meter (polarimeter). The sugar meter shall be calibrated according to the international sugar scale and the pol (°Z), and the measuring range shall be from (−30 °Z) − (+120) °Z.

② Bagasse digester. The bagasse digester comprises a water bath pot, a bagasse cup and a bagasse press.

③ 1 000 W electric furnace.

④ 12.5 °Bx sodium carbonate solution.

⑤ Basic lead acetate.

2) Assay step.

① Determine that fiber content of the sample. The fiber content of the sample can be determined directly or estimated as specified.

Bagasse fibre fraction F (%) is defined as follows:

$F = 60$ for $R \leqslant 1.0$

$F = 55$ for $R = 1.05-2.0$

$F = 50$ for $R > 2.0$

R refers to the optical rotation reading of the cooking liquor measured with a 200 mm polarimeter tube.

② Cooking. First weigh the weight of bagasse cup and cup press, then quickly weigh 100.0g of bagasse, add 500 mL of water containing 5 mL of 12.5 °Bx sodium carbonate solution at 70 °C, cover with cup press, gently flatten the bagasse, and put it into a boiling water bath for cooking. After 0.5 h, press the sugar for the first time, and then press it every 15 min for three times, so that the sugar can fully exude. The total cooking time is 1 h. After cooking, put the cup in a cold water bath to cool to room temperature, wipe dry and weigh, and record the weight of bagasse and solution.

③ Clarification and filtration. Squeeze out the solution as much as possible with a cup, inject it into a 200 mL conical flask, add a proper amount of alkaline lead acetate powder, shake it up and filter it, and discard the initial part of the filtrate after washing the container. Filtrate was collected.

④ Polarimetry. After washing the 200 mm polarimeter tube with the filtrate for 2~3 times, fill the filtrate and place it on the polarimeter to measure the polarimetric reading.

3) Calculation.

$$\text{Pol of bagasse (\%)} = \frac{\text{Weight of sucrose in bagasse}}{\text{Weight of bagasse}} \times 100\%$$

Weight of sucrose in bagasse = Percolate weight × pol of leachate

Percolate weight = Bagasse weight with solution − Fiber weight

Fiber weight = sample weight × sample fiber content

Since the concentration of the exudate is very low, its density can be taken as 1.

$$\text{Pol of exudate (\%)} = \frac{26P}{100d} \times 100\% = 0.26R \times 100\%$$

$$\text{Pol of exudate (\%)} = \frac{0.26R(W-F)}{100} \times 100\%$$

Where R—optical rotation reading (°Z) measured with 200 mm polarimeter tube;

W—weight of bagasse with solution (g);

F—sample fiber content (selected as required according to R).

[Example 2-7] Assume that the total weight of the bagasse cup and the bowl pressure is 855.5 g, the total weight of the bagasse cup, the bowl pressure and the solution after cooking is 1 301.5 g, and the optical rotation reading measured with a 200 mm polarimeter tube is 2.2 (°Z), and then calculate the pol of bagasse.

[Solution] Weight of bagasse and solution = 1 301.5−855.5 = 446 (g)

Select the sample fiber according to R, $F = 50$

then:

$$\text{Pol of bagasse(\%)} = \frac{0.26 \times 2.2 \times (446-50)}{100} \times 100\% = 2.27\%$$

(3) Determination of pol (sucrose content) of filter cake. The pol of filter cake is determined by the primary polarimetry. Although the filter cake contains more non-sugar, which has a certain impact on the polarimetry determination, the filter cake contains less sucrose, and the absolute value of the difference between sucrose and pol is not large, so the requirement for the determination result is low. The pol of filter cake can also be regarded as sucrose content in the calculation.

The pol of filter cake is determined by diluting the normal weight. Because not all of the filter cake can be dissolved in water, some insoluble substances occupy part of the volume of the volumetric flask, resulting in less than 100 mL of liquid in the volumetric flask after dilution. According to the requirements of quantitative dilution, the volume of diluted sample solution should be 100 mL. This volume error should be corrected.

Assuming that the moisture content of the filter cake is 50%, the dry substance content is also 50%, and the density of the dry substance in the filter cake is about 2.9 g/cm^3. When 26 g of sample is weighed, 13 g of which is dry substance, the volume occupied by the dry substance shall be 13/2.9 = 4.86 (mL). A volumetric flask of 104.86 mL shall be used according to the requirements for quantitative dilution. If a volumetric flask of 100 mL is used, the amount of sample shall be reduced appropriately.

$$26 : 104.86 = X : 100$$

$$X = 26 \times 100/104.86 = 24.79 \text{ (g)}$$

Since this is only an approximate estimate, for the convenience of weighing, the normal weight of filter cake is 25 grams.

1) Main equipment and reagents.

① Sugar meter (polarimeter). The sugar meter shall be calibrated according to the international sugar scale and the pol (°Z) , and the measuring range shall be (−30) °Z − (+120) °Z.

② Porcelain evaporating dish.

③ Volumetric flask 100 mL ± 0.08 mL.

④ Alkaline lead acetate solution (54 °Bx).

⑤ For carbonated sugar plants, the burette must be added. Burette calibration 0.1mL, 20 ℃ standard, maximum allowable error 0.05 mL.

2) Assay step.

① Sulfitation sugar mill. Weigh 25.0 g of sample and put it into a porcelain evaporating dish, add 2~7 mL of alkaline lead acetate solution (54 °Bx) and a small amount of distilled water, use a round-headed glass rod or porcelain rod to carefully grind it into a uniform paste completely free of particles, then wash it with distilled water into a 100 mL wide-mouth volumetric flask, add water to the marked line, fully shake it up and filter it. An initial portion of that filtrate was was the vessel and discarded. Filtrate was collected. After washing the 200 mm polarimeter tube with the filtrate for 2-3 times, fill the filtrate and place it on the polarimeter to measure the polarimetric reading.

② Carbonated sugar mill. Weigh 25.0 g of sample and put it into a porcelain evaporating dish, add 2~7 mL of alkaline lead acetate solution (54 °Bx) and a small amount of distilled water, use a round-headed glass rod or porcelain rod to carefully grind it into a uniform paste completely free of particles, then wash it with distilled water into a 100 mL wide-mouth volumetric flask, drop 2 drops of phenolphthalein indicator, shake it up and slowly drop it with dilute acetic acid until the red color disappears. Add water to the marked line, shake well and filter. An initial portion of that filtrate was was the vessel and discarded. Filtrate was collected. After washing the 200 mm polarimeter tube with the filtrate for 2~3 times, fill the filtrate and place it on the polarimeter to measure the polarimetric reading.

Note: As the filter cake of the carbonated sugar mill contains part of the combined sugar, that is, the insoluble calcium salt formed by sucrose and lime, the sample needs to be neutralized with dilute acetic acid before determination, so that the sucrose calcium salt is decomposed and this part of sucrose is released.

3) Calculation. Since the sample has been diluted with the normal weight, the reading of the polarimeter is the pol (sucrose content) of the mud.

Task 2.3 Determination of sucrose content and gravity purity

Enterprise case

The director takes back part of the intermediate cane juice from the production line, and then arranges Xiaoming to determine the sucrose content and gravity purity of the above samples. What instruments and methods should Xiaoming choose to determine?

Mission objectives

Through the study of this task, students can achieve the following goals:

(1) Further understanding of concepts such as sucrose and gravity purity.

(2) Understand the difference between primary and secondary polarimetry.

(3) Master the method and calculation of determining sucrose content and gravity purity.

Quality objectives

Develop the good habit of comparative analysis of learning.

Task description

Determination of sucrose (%) and gravity purity of semi-products using a polarimeter.

Procedures and methods

(1) Ingredients: Sugar juice.

(2) Equipment and instruments: Brix spindle, polarimeter, 200 mm polarimeter tube, thermostatic water bath, analytical balance, thermometer (0~100 ℃), mercury thermometer (0~50 ℃, division of 0.1 ℃), 200 mL conical flask, 100 mL volumetric flask, 50 mL pipette, measuring cylinder, beaker, filter paper.

(3) Reagent: Alkaline lead acetate powder, 24.85 °Bx hydrochloric acid solution [specific gravity d (20 ℃) = 1.102 9], sodium chloride solution (231.5 g/L), zinc powder.

(4) Steps.

1) Sample preparation: Various sugar juices can be determined with the original sample, and other samples with higher concentration can be prepared with the dilution method or the normal weight of solution method.

2) Measure the Brix and temperature of the sample solution, and record the data.

3) Determination of sucrose content. Pour about 200 mL of sample solution into a conical flask, add about 2 g of alkaline lead acetate powder, shake up quickly, filter, suck two 50 mL filtrates with a pipette, and transfer them into two 100 mL volumetric flasks respectively. Add 10 mL of sodium chloride (231.5g/L) solution into one of the bottles, and then add water to the scale. Shake it up. If it is turbid, filter it. Measure the observed polarization of the filtrate with a 200 mm polarimeter tube. Multiply this number by 2 to get the direct optical rotation reading P

and record the temperature of the sugar solution when reading. Add 20 mL of distilled water and 10 mL of 24.85 °Bx hydrochloric acid into another bottle, insert the thermometer, heat it to 60 °C in the water bath, keep it at this temperature for 10 min (shake it continuously during the first 3 min), take it out and immerse it in cold water, and cool it rapidly to the temperature close to the temperature when reading the direct polarization. Spray a small amount of water with a washing bottle, wash the sugar solution attached to the thermometer into the bottle, take out the thermometer, add water to the scale (if the solution is dark, add a small amount of zinc powder), shake well, and filter if it is turbid. Use a 200 mm polarimeter tube to measure the observed optical rotation, multiply this number by 2 to obtain the conversion optical rotation reading P', and use a 0.1 °C scale thermometer to measure the temperature t of the sugar solution at the time of reading (The difference between the sugar solution temperature when measuring P and P' shall not exceed 1 °C).

(5) Data processing and calculation.

Project	Data and results
Observation Brix of sample solution/°Bx	
Temperature of sample solution/°C	
Brix of sample solution/° Bx	
Direct optical rotation reading P/°Z	
Liquid temperature during determination of direct optical rotation/°C	
Conversion rotation reading P'/°Z	
Liquid temperature when measuring conversion optical rotation/°C	
Sucrose content of sample solution/%	
Gravity purity of sample solution/%	
Sucrose content of sample/%	
Gravitational purity of the sample/%	

Reflection

What is the relationship between pol (%) and sucrose (%), apparent purity and gravity purity? In what scenarios are they used? Please use the philosophy of "concrete analysis of concrete situations" to think.

Task-related knowledge—secondary polarimetry

Polarimetry is one of the most commonly used methods for determining sucrose content. It is based on the principle that sucrose solution has optical activity and can rotate the vibration plane of light passing through the solution by a certain angle, and the rotation angle is proportional to the concentration of the sucrose solution. However, for samples containing impurities, because some impurities also have optical activity, these impurities will also have an impact on the angle of optical activity. The determination of optical rotation of such samples is actually a comprehensive effect of various optically active substances on the optical rotation. The optical rotation produced by sucrose is only part of the result. In order to separate the optical rotation produced by sucrose alone from the sample containing impurities, a new determination method is needed, that is, the secondary optical rotation method.

1. Basic principle of secondary polarimetry

In addition to the optical activity of the sugar solution, the second optical rotation method also makes the following assumptions: Sucrose can be completely converted into reducing sugar under appropriate conditions, and the optical activity of all impurities in the sample does not change in the process of sucrose conversion. If this assumption is true, a test with a pure sucrose solution of a specified concentration (26.000 g/100 mL) will give the following results:

$$C_{12}H_{22}O_{11} + H_2O \rightarrow C_6H_{12}O_6 + C_6H_{12}O_6$$

      Sucrose   Glucose   Fructose

That is to say, sucrose is hydrolyzed into equal amounts of glucose and fructose under the action of an invertase (hydrochloric acid or invertase), and the mixture of these two sugars is called invert sugar.

If the sugar solution before and after conversion is measured separately on the polarimeter, the polarimetric reading shall be:

Before transformation: $P = 100\ °Z$

After transformation: $P' = -32.1\ °Z$

Then the ratio of the change in optical rotation before and after conversion to the optical rotation of sucrose is:

$$\frac{P-P'}{P} = \frac{100-(-32.1)}{100} = 1.321$$

The above is the result measured by a pure sugar solution with a specified concentration. Since the optical rotation is proportional to the concentration of the sugar solution, the concentration of the sugar solution changes, and P and P' change at the same time in the same proportion, so the above formula is applicable to a pure sugar solution with any concentration. That is to say, for pure sucrose solution, the ratio of the change of optical rotation before and after sucrose conversion to the optical rotation of sucrose is a fixed value of 1.321.

For impure sugar solution, the P and P' obtained by polarimetric determination are the comprehensive results of the optical rotations of various optically active substances in the sample. For any sample, if the optical rotation produced by the pure sucrose portion of the sample is α_s. The sample was converted under appropriate conditions, and the optical rotation of the invert sugar produced after sucrose conversion was α_r. The optical rotations of the impurities in the sample are $\alpha_1, \alpha_2, \alpha_3, \alpha_4 \ldots \alpha_n$, since it has been assumed that the optical rotation of the impurity does not change during the conversion of sucrose, we have:

$$P = \alpha_s + \alpha_1 + \alpha_2 + \alpha_3 + \alpha_4 + \cdots + \alpha_n$$
$$P' = \alpha_r + \alpha_1 + \alpha_2 + \alpha_3 + \alpha_4 + \cdots + \alpha_n,$$

Subtract P and P', because the optical rotation of other impurities does not change, the subtraction will completely cancel each other. Therefore, the change of optical rotation of the sample before and after conversion is the change of optical rotation of the pure sucrose part in the sample, namely:

$$P - P' = \alpha_s - \alpha_r$$

α_s and α_r are the pure sucrose part in the sample and the optical rotation produced after the conversion of this part of sucrose, so there should also be the following relationship:

$$\frac{\alpha_s - \alpha_r}{\alpha_s} = \frac{100 - (-32.1)}{100} = 1.321$$

Since $P - P' = \alpha_s - \alpha_r$,

so

$$\frac{P - P'}{P} = \frac{\alpha_s - \alpha_r}{\alpha_s} = \frac{100 - (-32.1)}{100} = 1.321$$

That is, the ratio of the change in optical rotation ($P-P'$) of any concentration of the sample before and after conversion to the optical rotation α_s produced by the pure sucrose in the sample is also a fixed value (1.321). P and P' in the formula are the results determined by the polarimeter, while α_s is the reading of the optical rotation produced by the pure sucrose in the sample, which is a part of the results determined by the optical rotation and cannot be directly measured by the polarimeter, but can be indirectly obtained by calculation.

According to this principle, the direct optical rotation of the sample (the optical rotation before conversion) is determined first, then the sample is converted by an appropriate method, the optical rotation of the converted sample is determined, and finally the optical rotation produced by pure sucrose in the sample is calculated according to the above formula. That is to say, the optical rotation produced by pure sucrose in the sample is not directly measured on the polarimeter, but indirectly obtained by calculation. It is called secondary polarimetry because it is determined twice.

$$\alpha_s = \frac{P-P'}{1.321} = \frac{P-P'}{132.1} \times 100$$

Where α_s—optical rotation generated by pure sucrose in the sample (°Z);

P—direct optical rotation of sample (optical rotation before conversion) (°Z);

P'—optical rotation of the sample after conversion (optical rotation after conversion) (°Z).

2. Kreiger divisor

The ratio of the change in optical rotation before and after sucrose conversion to the optical rotation of pure sucrose is called the Kreiger divisor, which is the denominator term in the above formula. The value of Kreiger's divisor is affected by the way of sucrose conversion (acid or enzymatic), the content of dry substance in the sample, and the temperature. Sugar mills generally adopt the acid conversion method, and the formula for calculating the Kelaijie divisor is as follows:

$$\text{Kreiger divisor} = 132.56 - 0.0794(13-g) - 0.53(t-20)$$

Where g—grams of dry substance in 100 mL of the sample solution to be determined

$$(g = \frac{\text{Apparent Brix}}{2} \times \text{Density of the corresponding apparent Brix}) \text{ (g)};$$

t—the temperature at which the optical rotation of the transformed sample solution is measured (°C).

Note: Because the determination of Brix is carried out before the determination of optical rotation, and it is necessary to add hydrochloric acid to convert sucrose during the determination of optical rotation, usually take 50 mL of the sample solution after the measurement of Brix, add hydrochloric acid to a constant volume of 100 mL, that is, double dilution, so the observed Brix should be divided by 2 when calculating the weight of dry substance in the sample solution to be measured.

3. Basic operation of secondary polarimetry

(1) Scope of application. In sugar mills, the sucrose content of all intermediate products is determined by the secondary polarimetric method, except that the sucrose content of white granulated sugar with few impurities, bagasse and filter cake with little sugar is determined by the primary polarimetric method.

(2) Main instruments and reagents.

1) Main instruments.

① Polarimeter (polarimeter), international sugar unit (°Z).

② Standard quartz tube.

③ 200mm polarimeter tube.

④ Volumetric flask. Standard temperature 20 °C, (200 ± 0.10) mL, (100 ± 0.08) mL.

⑤ Pipette. Standard temperature 20 °C, (50 ± 0.05) mL.

⑥ Precision thermometer. 0–50 °C, 0.1 °C mercury thermometer.

⑦ Ordinary thermometer. 0–100 °C, 1 °C graduation.

⑧ Thermostatic water bath ±1 °C graduation.

⑨ Analytical balance. Sensitivity 0.001 g.

2) Main reagent.

① Basic lead acetate. Chemically pure, the main quality indicators are as follows:

Total Lead (as PbO) 76%;

Basic Lead (as PbO) 33%.

The particle size can pass through the sieve with a pore size of 0.42 mm completely and at least 70% can pass through the sieve with a pore size of 0.12 mm.

② Hydrochloric acid solution (24.85 °Bx, density 1.102 9 g/cm^3).

③ Sodium chloride solution (231.5 g/L).

④ Zinc powder.

⑤ Aluminum hydroxide slurry.

4. Measurement steps and calculation

(1) Sample preparation.

1) Dilution. Generally, the sugar juice can be directly determined by the original sample, and the sample with higher concentration can be diluted by the specified quantitative dilution method or the weight dilution method. Solid samples can be considered to be diluted with the normal weight. For example, brown granulated sugar and tablet sugar can be used to prepare a normal weight of solution, that is, weigh 52.000g of sample and dilute to 200 mL, or weigh 65.000 g of sample and dilute to 250 mL. The final molasses is prepared into 1/3 of the normal weight of solution, that is, 43.333 g of sample is weighed and diluted to 500 mL, because when the normal weight of dilution is used, the reading of the polarimeter (or the result of secondary optical rotation calculation) directly represents the sucrose content of the original sample, which can simplify the calculation. Syrup shall be diluted by 4 times, massecuite and molasses shall be diluted by 6 times, and samples with crystals must ensure that all crystals are completely dissolved.

2) Clarification and filtration. After the sample is diluted, a proper amount of clarifier is added for clarification and filtration. The dosage of clarifier shall be sufficient for clarification, and shall not be excessive, otherwise, precipitation may be generated after adding NaCl solution, which will affect the results of optical rotation determination.

Pour about 200 mL of the sample solution into a conical flask, add about 2 g of alkaline lead acetate powder, shake up quickly, and filter. Discard the first 25 mL of filtrate when filtering the sample solution. Since two 50 mL samples are required for the determination, the determination can be started only after the filtrate volume exceeds 100 mL to ensure that the two samples are completely consistent.

(2) Determination.

1) Brix determination. Measure the Brix and temperature of the prepared sample solution.

Note: The temperature is used for the calculation of gravity purity. If the gravity purity is not calculated, the temperature may not be measured.

2) Measurement of direct optical rotation. Pipette 50 mL of the filtrate into a 100 mL volumetric flask. Add 10 mL of sodium chloride (231.5 g/L) solution, and then add water to the scale. Shake well. If it is turbid, filter it. Measure the observed polarization of the filtrate with a 200 mm polarimeter tube. Multiply this number by 2 to obtain the direct optical rotation reading P, and record the temperature of the sugar solution when reading.

3) measurement of conversion optical rotation. Suck 50 mL of the filtrate with a pipette, transfer it into a 100 mL volumetric flask, add 20 mL of distilled water, add 10 mL of 24.85 °Bx hydrochloric acid, insert a thermometer, heat it accurately to 60 °C in a water bath, keep it at this temperature for 10 min (shake it continuously within the first 3 min), take it out and immerse it in cold water. It cools rapidly to a temperature close to that at which the direct polarization is read. Spray a small amount of water with a washing bottle, wash the sugar solution attached to the thermometer into the bottle, and take out the thermometer. Add water to the scale (if the solution is dark, add a small amount of zinc powder) shake well. If turbidity is found, filter. Use a 200 mm polarimeter tube to measure the observed optical rotation, multiply this number by 2 to obtain the conversion optical rotation reading P', and use a 0.1 °C scale thermometer to measure the temperature t of the sugar solution at the time of reading (the difference between the sugar solution temperature when measuring P and P' shall not exceed 1 °C).

4) Data recording.

① The observed Brix of the diluted sample solution (this data does not need to be determined when the normal weight of dilution is used).

② Direct optical rotation reading P (°Z).

③ Conversion rotatory reading P' (°Z).

④ Temperature t of sample solution at the time of determination of conversion rotation reading P' (required to be accurate to 1/10 °C)

(3) Calculation method.

1) Optical rotation reading S produced by pure sucrose in the diluted sample solution. By measuring the optical rotation readings of the sample before and after conversion, the optical rotation S produced by pure sucrose in the sample can be calculated.

$$S(°Z) = \frac{100(P-P')}{132.56 - 0.0794(13-g) - 0.53(t-20)}$$

Where P — direct optical rotation reading (°Z);

P' — conversion rotation reading (°Z);

g — the weight of dry substance in 100 mL of the sample solution to be determined.

$$g = \frac{\text{Apparent Brix}}{2} \times \text{Density of the corresponding apparent Brix}$$

or:

$$g = \frac{1}{2}(0.998\,2B + 0.003\,7B^2 + 0.000\,018\,16B^3)$$

B—brix of sample solution;

t—the temperature of the sample solution at which the conversion rotation reading P' is determined.

2) Sucrose content of the sample to be tested. The above calculation result S is only equivalent to the optical rotation reading produced by the pure sucrose in the sample solution to be measured. Its exact meaning is also related to the dilution method of the sample. For the sample diluted in the normal weight, S represents the sucrose content (weight percent) of the original sample (undiluted sample). For undiluted or diluted samples, S is only equivalent to the reading of pure sugar solution on the polarimeter. Its unit is °Z, indicating that the sucrose content of diluted samples is $0.26S$ g/100 mL. This is a volume weight percent concentration, which should be converted to weight percent concentration by an appropriate method, that is, converting 100 mL to weight. The specific method is to observe the Brix of the diluted sample solution, look up the table to get the pol (polarization) factor or density, and then calculate.

Sucrose content of diluted sample (%) = S Brix (polarization) factor

The Brix (polarization) factor can be calculated as follows:

$$\text{Pol factor} = (0.510\,7 - 0.001B)^2$$

Where B—observed Brix of diluted sample solution.

or:

$$\text{Sucrose content of diluted samples (\%)} = \frac{26S}{100 \times \text{Density of diluted sample}}$$

3) Sucrose content of original sample.

Sucrose content of original sample (%) = dilution times × sucrose content of sample after dilution

Note: If the sample is not diluted, the dilution factor is 1.

When the sample is diluted with the normal weight, 26.000 g of sample is diluted to 100 mL and the corresponding dilution factor is:

$$\text{Multiple dilutions of the normal weight of dilution} = \frac{100 \times \text{Density of diluted sample}}{26.000}$$

Sucrose content of the original sample $S =$

$$\frac{26S}{100 \times \text{Density of diluted sample}} \times \frac{100 \times \text{Density of diluted sample}}{26.000}$$

That is to say, when the sample is diluted with the normal weight, the optical rotation reading produced by the pure sucrose in the measured sample solution is the sucrose content of the original sample, which can simplify the calculation.

(4) Precautions.

1) Because the conversion of sucrose requires the addition of HCl, which is equivalent to the addition of an impurity in the sample, according to the principle of secondary polarimetry, when determining the direct optical rotation, an impurity NaCl, which has basically the same effect on the optical rotation of sucrose as HCl, with the same amount as HCl. The effect of HCl on the optical rotation of the sugar after conversion can be counteracted by $(P-P')$.

Note: After adding m g of HCl or NaCl to the converted sugar solution, the effect on the optical rotation is:

NaCl $0.540m$

HCl $0.540\ 7m$

The two are very similar, so they can be offset.

2) The content g of dry substance in the sugar solution shall be calculated by the observed Brix. Since the unit of g is the weight of dry substance in 100 mL of the sample solution to be measured, and the unit of Brix is the weight of dried solids in 100 g of the sample solution to be measured, if the Brix is converted into g, the density (converted into volume) corresponding to the observed Brix shall be divided by 100 g of the sample liquid to be measured.

3) Since there is only one temperature correction for the Kreiger divisor, it is required that the temperature difference between the two polarimetric determinations shall not exceed 1 °C, and the temperature for correction shall be the temperature at which the optical rotation of the converted sample solution is measured.

4) Because it is only a hypothesis that sucrose can be completely converted into reducing sugar under appropriate conditions, in fact, it is very difficult to achieve complete conversion. In order to compare the results of each determination, the time and temperature of sucrose conversion should be strictly controlled.

5) Since the sample is diluted twice before the determination of optical rotation (take 50 mL to make a constant volume of 100 mL), the optical rotation reading should be multiplied by 2 to obtain the optical rotation reading of the sample before dilution.

6) If too much clarifier is added, precipitation will occur after adding NaCl solution, which shall be filtered before polarimetric determination (if necessary, aluminum hydroxide slurry can be added as filter aid).

5. Measurement examples

(1) Determination of sucrose content in mixed juice.

1) Brix determination. The Brix and temperature were measured after bagasse was filtered from the mixed juice sample with a sieve.

Note: If the gravity purity calculation is not carried out, the temperature may not be measured.

2) Clarification and filtration of sample solution. Take about 200 mL of mixed juice from

which bagasse is filtered, add about 2 g of alkaline lead acetate powder, shake up quickly, and filter. Discard the first 25 mL of filtrate when filtering the sample solution. Since two 50 mL samples are required for the determination, the polarimetric determination can be started only after the amount of filtrate exceeds 100 mL to ensure that the two samples are completely consistent.

3) Measurement of direct optical rotation. Pipette 50 mL of the filtrate into a 100 mL volumetric flask. Add 10 mL of sodium chloride (231.5 g/L) solution, and then add water to the scale. Shake well. If it is turbid, filter it. Measure the observed polarization of the filtrate with a 200 mm polarimeter tube. Multiply this number by 2 to obtain the direct optical rotation reading P, and record the temperature of the sugar solution when reading.

4) Measurement of conversion optical rotation. Suck 50 mL of the filtrate with a pipette, transfer it into a 100 mL volumetric flask, add 20 mL of distilled water, add 10 mL of 24.85 °Bx hydrochloric acid, insert a thermometer, heat it accurately to 60 °C in a water bath, keep it at this temperature for 10 min (shake it continuously within the first 3 min), take it out and immerse it in cold water. It cools rapidly to a temperature close to that at which the direct polarization is read. Spray a small amount of water with a washing bottle, wash the sugar solution attached to the thermometer into the bottle, and take out the thermometer. Add water to the scale (if the solution is dark, add a small amount of zinc powder) shake well. If turbidity is found, filter. Use a 200 mm polarimeter tube to measure the observed optical rotation, multiply this number by 2 to obtain the conversion optical rotation reading P', and use a 0.1 °C scale thermometer to measure the temperature t of the sugar solution at the time of reading (the difference between the sugar solution temperature when measuring P and P' shall not exceed 1 °C).

5) Data recording. Assume that the measured data of a mixed juice are as follows:

The observed Brix of the mixed juice was 16.10 °Bx and the temperature was 18.5 °C.

Direct optical rotation reading $P = 27.83 \times 2 = 55.66$ (°Z).

Conversion rotatory reading $P' = -9.84 \times 2 = -19.68$ (°Z), Temperature $t = 18.52$ °C

6) Calculation (take the above data as an example).

① Calculate the grams of dry substance in 100 mL of the sample solution to be determined.

The Brix of the juice mixture was observed to be 16.10 °Bx,

By formula

$$g = \frac{1}{2}(0.998\,2B + 0.003\,7B^2 + 0.000\,018\,16B^3)$$

$$g = (0.998\,2 \times 16.10 + 0.003\,7 \times 16.10^2 + 0.000\,018\,16 \times 16.10^3) / 2$$

$$g = 9.6936 \text{ (g/100 mL)}$$

② Calculate the optical rotation reading S produced by the pure sucrose in the mixed juice According to the formula (°Z).

$$S=\frac{100(P-P')}{132.56-0.0794(13-g)-0.53(t-20)}$$

$$=\frac{100[55.66-(-19.68)]}{132.56-0.0794(13-8.5529)-0.53(18.52-20)}=56.650\ (°Z)$$

③ Calculate that sucrose content of the mixed juice.

Sucrose content of mixed juice = optical rotation reading produced by pure sucrose in mixed juice $S \times$ Brix factor $\times 100\%$

The Brix (polarization) factor can be calculated as follows:

$$\text{Sucrose content factor} = (0.5107 - 0.001B)^2 \times 100\%$$

Sucrose content of mixed juice = $56.612 \times (0.5107 - 0.001 \times 16.10)^2 \times 100\% = 13.85\%$

(2) Determination of sucrose content in syrup. The determination of sucrose in syrup is basically the same as that in mixed juice, but the difference lies in the dilution of the sample and the calculation of the result.

1) Sample dilution. Take a proper amount of syrup sample and dilute it four times (for example, take 150 g of syrup and add 450 g of water), and determine the Brix of the diluted sample solution.

The rest of the operation is the same as that of the mixed juice.

2) Result calculation. It is the same as the mixed juice, but it should be noted that the calculated result is the sucrose content of the sample solution to be measured, while the original sample is syrup, so the calculation result should be multiplied by the dilution factor to obtain the sucrose content of syrup.

(3) Determination of sucrose content in final molasses.

1) Preparation of sample solution. Mix the molasses sample evenly, weigh 43.333 g of sample, add a little distilled water, mix evenly with a small glass rod, carefully wash the sample into a 500 mL volumetric flask, add water to the marked line, and shake well.

2) Clarification and filtration. Pour out the prepared sample solution from the volumetric flask to determine the Brix, then take about 200 mL into a conical flask, add about 3.5 g of alkaline lead acetate powder (because the molasses contains more impurities, the amount of clarifier should be increased appropriately), shake up and filter. The initial part of the filtrate is discarded after washing the container. When the amount of the filtrate is enough to absorb two 50 mL samples, the optical rotation determination is carried out.

3) Measurement of optical rotation. The direct optical rotation reading (P) and the conversion optical rotation reading (P') of the filtrate were measured according to the method shown in 3) and 4) of the measurement of the sucrose content of the mixed juice, and the temperature (t) of the sugar solution at the time of conversion optical rotation was measured with a 1/10 scale thermometer.

Note: If the color of the sugar solution is dark when determining the conversion rotation reading, add a small amount of zinc powder after adding water to the marked line. The zinc powder reacts with the hydrochloric acid in the sugar solution to generate nascent hydrogen with bleaching effect, which lightens the color. The determination can be carried out after filtration.

4) Data recording. Same sucrose content as mixed juice.

5) Calculation of molasses sucrose content. According to the preparation method of sample solution, the amount of sample in 100 mL sample solution is:

Amount of sample in 100 mL sample solution = 43.333/5 ≈ 8.667 (g)

The normal weight for the sugar mill is 26.000 g. While 26.000/3 ≈ 8.667 g, therefore, the concentration of the sample solution prepared according to the above method is 1/3 of the normal weight. Because the optical rotation reading is proportional to the concentration, the measured reading is only 1/3 of the dilution of a normal weight. As long as the reading of optical rotation measurement is multiplied by 3, the optical rotation reading corresponding to a solution with a specified concentration can be obtained. According to the calibration method of the polarimeter, when the sample is diluted according to the normal weight, the reading of the polarimeter indicates the sucrose content (or pol) of the original sample. That is to say, the diluted sample solution is placed on the polarimeter for determination, and the reading displayed on the instrument indicates the sucrose content of the original sample (undiluted). Calculate the sucrose content of the molasses.

[Example 2-8] Take 43.333 g of molasses to a constant volume of 500 mL, and measure the observed Brix to be 15.08 °Bx.

Content of the molasses.

Direct optical rotation reading $P = 3.45 \times 2 = 6.90$ (°Z).

Conversion rotatory reading $P' = -3.23 \times 2 = -6.46$ (°Z), Temperature t = 19.45 °C.

[Solution] After dilution, the weight of molasses per 100 mL of sugar solution is 43.333/5 g. The concentration X (specified concentration) of the diluted sugar solution is:

$$(43.333/5) : X = 26 : 100$$

$$X = 33.33 \text{ (i.e. 1/3 of the normal weight)}$$

Calculate the weight of dry substance in 100 mL of the sample solution to be determined:

According to the observation that the Brix of the prepared sugar solution is 15.08 °Bx, according to the formula

$$g = \frac{1}{2}(0.998\,2B + 0.003\,7B^2 + 0.000\,018\,16B^3)$$

$$g = (0.998\,2 \times 15.08 + 0.003\,7 \times 15.08^2 + 0.000\,018\,16 \times 15.08^3)/2$$

$$g = 7.98 \text{ (g/100 mL)}$$

Since the concentration of the sugar solution is 1/3 of the normal weight, and the optical rotation reading is proportional to the concentration, the measured optical rotation reading can be converted into the reading when the concentration is 1 normal weight after being multiplied by 3.

When the concentration of the sample solution to be measured is 1 normal weight, the reading of the polarimeter indicates the sucrose content of the original sample.

$$\text{The sucrose content of molasses} = \frac{3 \times 100 (P-P')}{132.56 - 0.0794(13-g) - 0.53(t-20)} \times 100\%$$

$$= \frac{3 \times 100 [6.90 - (-6.46)]}{132.56 - 0.0794(13-7.98) - 0.53(19.45-20)} \times 100\% = 26.16\%$$

Task 2.4 Determination of reducing sugar in sugarcane juice by Lane-Eynon method

Enterprise case

The director looked at the report that had just come out and found that the reducing sugar of the clear juice was particularly high. He immediately took back part of the clear juice from the production line and arranged Xiaoming to determine the reducing sugar of the above samples. What instrument and method should Xiaoming choose to determine?

Mission objectives

Through the study of this task, students can achieve the following goals:

(1) Understand the principle of Lane–Eynon method.

(2) Grasp the determination method and calculation of Lane–Eynon method.

Quality objectives

Develop the good habit of comparative analysis of learning.

Task description

Determination of reducing sugars in products in process by the Lane–Eynon method.

Procedures and methods

(1) Ingredients: cane juice.

(2) Equipments and instruments: electric furnace, stopwatch, 250 mL volumetric flask, 50 mL burette, 5 mL pipette.

(3) Reagents: Fehling reagent A and B, 54 °Bx neutral lead acetate solution, deleading agent (phosphate and oxalate mixture), 1% tetramethyl blue indicator.

(4) Steps.

1) Preparation of sample solution. For samples with low concentration (below 26 °Bx), such as various sugar juices, it is not necessary to dilute them and prepare them according to the original concentration of the samples.

Take 50 mL of sample cane juice, transfer it into a 250 mL volumetric flask, add 1mL of 54 °Bx neutral lead acetate solution, shake well, then add 3 mL of deleading agent and shake well, add distilled water to the scale, shake well and filter, discard about 15 mL of the initial filtrate, and the filtrate is the prepared sugar solution. The concentration of reducing sugar in the prepared sugar solution should be kept between 2.50 and 4.00 g/L.

2) Titration.

① Pre-test: Use two 5mL pipettes to suck 5mL of Fehling's solution B into a 250 mL conical flask, then suck 5 mL of solution A into solution B, and mix well. Add 15 mL of sugar solution from the burette into the conical flask, shake it up, heat it on the electric stove covered with asbestos net, and boil it accurately for 2 min (controlled by stopwatch), add 3–4 drops of tetramethyl blue indicator, and continue to add sugar solution until the blue color disappears, which is the end point. This operation should not exceed 1 min, so that the total time of the whole boiling and dropping operation is controlled within 3 min. Record the volume of prepared sugar solution consumed for titration.

② Retest: Suck 5 mL of Fehling test solution B and 5 mL of Fehling test solution A into a 250 mL conical flask according to the above sequence, add 1 mL less prepared sugar solution than the amount consumed in the pre-test from the burette, and shake well. The titration procedure is the same as the pre-test, and the boiling time should also be accurately controlled to 2 min, and the titration to the end point should not exceed 1 min. When titrating, the conical flask should be shaken gently, but it should not be separated from the heat source to keep the solution boiling, so as to avoid errors caused by the oxidation of tetramethyl blue caused by air entering the flask.

(5) Data processing and calculation.

Brix of sample: Sucrose content:

Project	Data and results		
Volume of prepared sugar solution consumed for titration			
Reducing sugar/%			

Reflection

What is the relationship between sucrose and reducing sugar?

Determination of reducing sugar

There is a certain amount of reducing sugar (RS) in all kinds of materials in sugar mills. The reducing sugar in sugar reflects the maturity of sugar, that is, the quality of sugar. The content of reducing sugar in semi-finished products is an important basis for formulating process conditions. The content of reducing sugar in finished sugar is related to the storage performance of products. Therefore, the determination of reduction sugar is an important item in chemical management of sugar mills.

The reducing sugar in the intermediate products of sugar mills was determined by the Lane-Eynon method or the Lane-Eynon constant volume method, and the reducing sugar in white granulated sugar was determined by the Offner method. All these methods are based on the fact that Cu^{++} (Cu^{2+}) ions can quantitatively react with reducing sugar to produce Cu^+ ions under certain conditions, so as to determine the amount of reducing sugar. However, different methods have different quantitative methods for reducing sugar.

1. Determination of reducing sugar by Lane–Ainon method

The Lane–Eynon constant volume method is a volumetric analysis method based on the redox reaction of Cu^{++} ions with reducing sugars in Fehling's reagent. Fehling's reagent is an alkaline solution of copper salts consisting of copper sulfate, potassium sodium tartrate, and sodium hydroxide. Copper sulfate provides the Cu^{++} ions required for the reaction, and sodium hydroxide provides the alkaline conditions required for the reaction. Under alkaline conditions, copper sulfate can form copper hydroxide precipitate. Due to the presence of potassium sodium tartrate, copper hydroxide can form a soluble complex with potassium sodium tartrate, so that Cu^{++} ions can also exist stably in the solution under alkaline conditions.

(1) Basic principles. In the boiling state, a certain amount of Fehling reagent is titrated with the sample solution containing reducing sugar. In the quantitative Fehling reagent, the amount of Cu^{++} ions is certain and can only react with a considerable amount of reducing sugar. If the reducing pol of the sample solution is high, the volume consumed during titration is small, and vice versa. Therefore, the amount of reducing sugar can be calculated according to the consumption of the sample solution. In order to determine the end point, tetramethyl blue, which has a weaker oxidation ability than Cu^{++}, was added. During titration, Cu^{++} with strong oxidation ability was completely reduced by reducing sugar, and then the excess reducing sugar reacted with tetramethyl blue to make it change from blue to colorless, indicating the arrival of the end point. The main chemical reactions in the determination process are as follows:

$$\text{Reducing sugar} + CuSO_4 \text{ (blue)} \rightarrow \text{organic acid} + Cu_2O \text{ (brick red)}$$

$$\text{Reducing sugar} + \text{tetramethyl blue (blue)} \rightarrow \text{tetramethyl blue (colorless)}$$

The above reaction is reversible, and when the colorless tetramethyl blue combines with oxygen in the air, it will turn blue. In order to prevent this from happening, the solution must be kept boiling during titration, and the rising steam is used to prevent air from entering the conical flask.

1) Fehling's reagent. The Cu^{++} ions during the titration were provided by Fehling's reagent. Fehling's reagent consists of two solutions A and B stored separately.

Solution A: $CuSO_4$ solution, providing Cu^{++} ions required for the reaction. Weigh 69.278 g of crystalline copper sulfate ($CuSO_4 \cdot 5H_2O$), dissolve it with water, transfer it into a 1 000 mL volumetric flask, add water to the marked line, shake well and filter.

Solution B: Contains potassium sodium tartrate and sodium hydroxide to provide the conditions required for the reaction. Dissolve 346 g of sodium potassium tartrate ($NaKC_4H_4O_6 \cdot 4H_2O$) in about 500 mL of water, and dissolve 100 g of sodium hydroxide in about 200 mL of water, mix them and transfer to a 1 000 mL volumetric flask, and add water to the marked line. Let stand for 2 d. If the liquid level drops, add more water to the marked line. Shake well and filter.

Because Cu^{++} ions can only react with reducing sugar under alkaline conditions, $CuSO_4$

will slowly react with potassium sodium tartrate under alkaline conditions to produce Cu_2O precipitate, which will change the concentration of $CuSO_4$ solution. The concentration of $CuSO_4$ solution is the key to the quantification of reducing sugar, so the solution A and B can only be mixed before use, and should be used immediately after mixing.

Potassium sodium tartrate in solution B can form a stable soluble complex with Cu^{++}, which is beneficial to the reaction. When the Fehling reagent is transferred into the conical flask, the solution B shall be transferred first, and then the solution A. Because solution B is moved in first, there is a large amount of potassium sodium tartrate in the conical flask, which can ensure that when solution A is moved in, Cu^{++} ions will immediately form a stable soluble complex without forming copper hydroxide precipitate. If the liquid A is moved first, when the liquid B is moved into the moment of contact with the liquid A, because the environment is dominated by liquid A, the amount of potassium sodium tartrate is very small, Cu^{++} ions may generate copper hydroxide precipitation, once this precipitation is formed, it can not be guaranteed to be completely dissolved in a short time. The reaction rate of Cu^{++} ions in the precipitated state with reducing sugar is much lower than that of Cu^{++} ions in the dissolved state. In the same reaction time, some Cu^{++} may not have time to react with reducing sugar, thus affecting the determination results.

2) Fehling's reagent concentration. The determination of reducing sugar is to use the sample to titrate a certain amount of Fehling reagent, so the amount of Cu^{++} ions in Fehling reagent is the key to the quantitative determination of reducing sugar. The amount of Cu^{++} ion is related to the concentration and amount of Fehling reagent and reaction conditions. How much reducing sugar can react with a certain amount of Cu^{++} ion is not a fixed value. It is related to the conditions of the reaction. It can not be calculated by a simple chemical reaction formula, but can only be determined by experimental methods. Therefore, it is necessary to titrate a certain amount of Fehling reagent with standard invert sugar under certain conditions before determination, and then calculate the amount of reducing sugar equivalent to Fehling reagent according to the consumption of standard invert sugar solution. At present, it is uniformly stipulated that in the absence of sucrose, if 10 mL of Fehling reagent (5 mL of solution A and solution B) consumes 25.64 mL of 2.000 g/L inverted sugar solution (experimental data), its concentration is set as an equivalent, which is equivalent to 51.28 mg of reducing sugar. Otherwise, the concentration correction factor shall be calculated according to the consumption of Fehling reagent.

$$\text{Concentration correction factor of Fehling reagent } K = V/25.64$$

Where V—2g/L standard invert sugar solution actually consumed during calibration (mL).

3) Reducing sugar factor. There are many factors affecting the reaction between reducing sugar and Cu^{++} ions, including the electric furnace used and the specification of the conical flask (related to the time used to heat to boiling), the content of sucrose in the sample (because sucrose can also react with Cu^{++} ions), and the content of reducing sugar in the sample (because of the different content of reducing sugar, the consumption of prepared sugar solution during titration is

also different, after mixing with a fixed amount of Fehling's reagent, the volume is different, and the concentration of Cu^{++} ions in the conical flask is also different, when the consumption of sugar solution is large, the total volume will be large, and the concentration of Cu^{++} ions will decrease, otherwise, it will increase, and the oxidation ability of Cu^{++} ions is also different). Therefore, the concentration of Fehling reagent can only be determined after calibration with standard invert sugar solution. The calibration conditions are required to be completely consistent with the test conditions. The weight of reducing sugar equivalent to 10 mL of Fehling reagent (5 mL of solution A and solution B) is called the reducing sugar factor. It varies slightly with the content of reducing sugar and sucrose in the sample solution. The weight of reducing sugar corresponding to 10 mL of Fehling's reagent can be obtained through experiments under different sucrose concentrations and when the sample consumption is between 15 and 50 mL, and a "reducing sugar factor table" can be made for calculation.

The factor of reducing sugar can be obtained by referring to attached list 8 from the weight of sucrose in 100 mL of prepared sugar solution G and the volume of prepared sugar solution consumed by titration V, or can be calculated according to the following formula:

$$F=(50.62-0.548\,8V+0.305\,6G)/(1-0.011\,33V+0.015\,98G-0.000\,1G^2)$$

Where F—factor of reducing sugar;

G—weight of sucrose in 100 mL of prepared sugar solution (g);

V—volume of prepared sugar solution consumed for titration (mL).

(2) Measurement steps.

Main equipment and reagents:

① Alkaline burette graduation 0.1 mL, 20 ℃ standard, maximum allowable error 0.05 mL.

② 250 mL conical flask.

③ 5 mL pipette.

④ Electric furnace, stopwatch, etc.

⑤ Standard invert sugar solution, mass concentration: 2.000 g/L.

⑥ 54 °Bx neutral lead acetate solution.

⑦ Fehling's reagent.

⑧ Tetramethyl blue indicator.

⑨ Lead remover (mixed solution of disodium hydrogen phosphate and potassium oxalate).

(3) Assay step.

1) Preparation of sample solution.

① Dilution. Samples below 26 °Bx may not be diluted, syrup is diluted by 4 times, and massecuite and molasses are diluted by 6 times. The Lane–Eynon method requires that the content of reducing sugar in the prepared sugar solution used for titration should be in the range of 250–400 mg per 100 mL of the prepared sugar solution. The dilution multiple of the sample shall be adjusted according to the pre-test results. If the pre-test consumes too much prepared sugar solution (more than 50 mL), the dilution multiple shall be reduced; if too little will increase the analysis error, the dilution multiple shall be increased.

For samples with higher concentration, sugar solution can also be prepared by direct weighing. For example, syrup can be weighed about 12.500 g; molasses, massecuite and massecuite can be weighed about 1.500–12.500 g (accurate to three decimal places), dissolved in distilled water and transferred to a 250 mL volumetric flask, and then diluted by volume.

② Clarification and deleading. Take 50 mL of the diluted sample solution and filter it with a fine copper mesh, then transfer it into a 250 mL volumetric flask, add a proper amount (about 1 mL) of 54 °Bx neutral lead acetate solution for clarification, shake it well, then add the lead remover (about 3 mL), shake it well, add water to the marked line, filter and collect the filtrate (called prepared sugar solution).

2) Pre-inspection. Use a pipette to suck 5 mL of Fehling's solution B into a 250 mL conical flask, and then use another pipette to suck 5 mL of solution A into solution B, and mix well. Add 15 mL of sugar solution from a burette into a conical flask, shake it up, heat it on an electric stove covered with an asbestos net, boil it accurately for 2 min (controlled by a stopwatch), add 3–4 drops of tetramethyl blue indicator, keep the sugar solution boiling, and continue to add the prepared sugar solution until the blue color disappears. This operation should not exceed 1 min, so that the total time of the whole boiling and titration operation is controlled within 3 min. Record the volume of prepared sugar solution consumed for titration, V_1.

3) Re-inspection. Suck 5 mL of Fehling test solution B and 5 mL of Fehling test solution A into a 250 mL conical flask according to the above sequence, add 1 mL less prepared sugar solution from the burette according to the volume of prepared sugar solution V_1 consumed in the pre-inspection, and shake up. The titration procedure is the same as the pre-inspection, and the boiling time should also be accurately controlled at 2 min. The time of titration to the end point after adding the indicator should not exceed 1 min. When titrating, the conical flask should be shaken gently, but the heat source should not be removed to keep the solution boiling, so as to avoid errors caused by the oxidation of tetramethyl blue caused by air entering the flask. Record the milliliter T of the prepared sugar solution consumed for titration.

(4) Calculation method. There are many factors affecting the reaction of Fehling's reagent with reducing sugar, such as boiling time, the volume of solution in the conical flask at the end of titration and the content of sucrose. Among these factors, except that the boiling time can be controlled by standard operation, the volume of the solution in the conical flask at the end of titration is affected by the concentration of reducing sugar in the prepared sugar solution. When the concentration of reducing sugar is high, the amount of prepared sugar solution consumed is small, and the volume of the solution in the conical flask at the end of titration is small, otherwise it is large. At the same time, because the sucrose can also consume part of Cu^{++} ions, the used amount of the prepared sugar solution is large, and the sucrose content of the prepared sugar solution is high, when the titration is finished, the more the sucrose in the solution in the conical

flask is, the more the Cu^{++} ions consumed by the sucrose are, and the less the copper ions which can participate in the reducing sugar reaction are. Therefore, it can not be calculated directly according to the reaction equation. The amount of reducing sugar equivalent to 10 mL Fehling reagent should be determined according to the actual situation and the relevant data in the "Reducing Sugar Factor Table".

The equivalent amount of 10 mL Fehling reagent was determined according to the consumption of 100 mL and the weight of sucrose in the prepared sugar solution, that is, the "reducing sugar factor". The consumption of the prepared sugar solution can be directly obtained from the titration results, and the weight of sucrose in 100 mL of the prepared sugar solution is calculated from the sucrose content (or pol) of the sample. The reducing sugar factor obtained by looking up the table is the amount of reducing sugar equivalent to 10 mL of Fehling reagent, that is, the weight of reducing sugar contained in the sample in the conical flask.

1) The data required for the calculation process are as follows:

① Weight of the sample (this data is not available for the sample solution prepared by volumetric method) (g).

② Brix B of the sample (this data is not available for the sample solution prepared by direct weighing method) (°Bx).

③ Sample sucrose content S.

④ Concentration correction factor K of Fehling reagent.

⑤ The milliliter T of the prepared sugar solution is consume for titration.

2) The calculation can be performed as follows:

① Calculate the weight of sample W in 100 mL of prepared sugar solution.

a. Prepare sample solution by volumetric method:

$$W=Vd$$

Where V—volume of sample contained in 100 mL of sugar solution (mL);

D—density of sample solution (refer to the table according to the observed Brix B or calculate according to the following formula: $d=0.998\ 2+0.003\ 7B+0.000\ 018\ 16B^2$).

For example, take 50 mL of sample solution to prepare 250 mL, and the Brix of the sample is 15.32 °Bx, then the volume V of the sample in 100 mL of prepared sugar solution is:

$$V=50\times 100/250=20\ (mL)$$

$$W=20\times(0.998\ 2+0.003\ 7\times 15.32+0.000\ 018\ 16\times 15.32^2)=21.18\ (g)$$

b. Preparing sugar solution by a direct weighing method;

$$W = \text{Sample weight}/2.5$$

② Calculate the weight of sucrose in 100 mL of sugar solution.

$$G=WS/100$$

Where G—100 mL weight of sucrose in the prepared sugar solution (g);

W—100 g of sample in 100 mL of prepared sugar solution W (g);

S—sucrose content of sample (%).

③ Calculate weight I of reducing sugar in 100 mL of prepared sugar solution.

$$I=100FK/T$$

Where I—mg of reducing sugar in 100 mL of prepared sugar solution (mg);
  K—Fehling reagent concentration correction factor;
  T—amount of prepared sugar solution consumed by titration (mg);
  F—factor of reducing sugar (obtained from the of sucrose in 100 mL of prepared sugar solution G and the volume of prepared sugar solution consumed by titration V according to attached list 8) , which can also be calculated according to the following formula:

$$F=(50.62-0.548\,8V+0.305\,6G)/(1-0.011\,33V+0.015\,98G-0.000\,1G^2)$$

Where G—the weight of sucrose in 100 mL of prepared sugar solution;
  V—volurne of prepared sugar solution consumed for titration.

④ Calculate the reducing sugar R of the sampled.

$$R=100I/(1\,000W)=I/(10W)$$

Where R—sample reducing sugar (%).

The rest of the notation is as above.

[Examples 2-9] The observed Brix of the mixed juice was 15.32 °Bx, and the sucrose content was 13.58%. Suck 50 mL of sample and dilute it to 250 mL, the concentration correction coefficient of Fehling reagent is 0.975, 35.00 mL of prepared sugar solution is consumed for reexamination, and the reducing pol of mixed juice is calculated.

[Solution] The volume V_1 containing the sample in 100 mL of the prepared sugar solution is:

$$V_1=50\times 100/250=20\,(mL)$$

The weight W of the sample contained in 100 mL of the prepared sugar solution is:

$$W=20\times(0.998\,2+0.003\,7\times 15.32+0.000\,018\,16\times 15.32^2)=21.18\,(g)$$

The gram of sucrose in 100 mL of that prepared sugar solution is G:

$$G=21.18\times 13.58\%=2.88\,(g)$$

The reducing sugar factor F is obtained from the weight of sucrose in 100 mL of prepared sugar solution G and the volume of prepared sugar solution consumed by titration V according to attached list 8, and can also be calculated according to the following formula:

$$F=(50.62-0.548\,8V+0.305\,6G)/(1-0.011\,33V+0.015\,98G-0.000\,1G^2)$$

$$F=\frac{50.62-0.548\,8\times 35+0.305\,6\times 2.88}{1-0.011\,33\times 35+0.015\,98\times 2.88-0.000\,1\times 2.88^2}=49.78$$

The concentration correction coefficient of Fehling reagent is 0.975, and 10 mL of Fehling reagent is equivalent to the amount of reducing sugar:

10 mL Fehling reagent is equivalent to reducing sugar = $0.975\times 49.78=48.54$ (mg)

That is to say, 10 mL Fehling reagent is equivalent to 48.54 mg of reducing sugar; or 35.00 mL of prepared sugar solution consumed by titration contains 48.54 mg of reducing sugar. The reducing sugar contained in 100mL of prepared sugar solution is:

$$35 : 48.54 = 100 : X$$

$$X = 100 \times 48.54/35 = 138.69 \text{ (mg)}$$

The reducing pol of mixed juice:

$$= \frac{\text{The reducing sugar contained in 100 mL of prepared sugar solution}}{\text{The weight of the sample contained in 100 mL of the prepared sugar solution}} \times 100\%$$

$$= \frac{138.69}{21.18 \times 1\,000} \times 100\% = 0.65\%$$

[Example 2-10] The sucrose content of brown granulated sugar is 88.45%. After diluting with the normal weight, take 50 mL to prepare 200 mL. The concentration correction coefficient of Fehling reagent is 1.028. 32.35 mL of prepared sugar solution is consumed for re-inspection to calculate the reducing pol of brown granulated sugar.

[Solution] The sample solution with the specified concentration contains 26.000 g of sample per 100 mL. Take 50 mL to prepare 200 mL, and then prepare 200 mL of sugar solution containing 13.000 g of original sample.

The weight W of the sample contained in 32.35 mL of the prepared sugar solution is:

$$13 : 200 = W : 32.35$$

$$W = 13 \times 32.35/200 = 2.10 \text{ (g)}$$

The weight G of sucrose contained in 100 mL of prepared sugar solution is:

$$(13 \times 88.45\%) : 200 = G : 100$$

$$G = 13 \times 88.45\% \times 100/200 = 5.75 \text{ (g)}$$

Calculate the reducing sugar factor F according to G and the amount of 32.35 mL sugar solution consumed.

$$F = (50.62 - 0.548\,8V + 0.305\,6G) / (1 - 0.011\,33V + 0.015\,98G - 0.000\,1G^2)$$

$$F = \frac{50.62 - 0.548\,8 \times 32.35 + 0.305\,6 \times 5.75}{1 - 0.011\,33 \times 32.35 + 0.015\,98 \times 5.75 - 0.000\,1 \times 5.75^2} = 47.95$$

10 mL Fehling reagent is equivalent to reducing sugar $= 1.028 \times 47.95 = 49.29$ (mg)

$$\text{The reducing pol of brown granulated sugar} = \frac{49.29}{2.10 \times 1\,000} \times 100\% = 2.35\%$$

(5) Precautions.

1) Fehling reagent A and B should be stored separately. If they are premixed, potassium sodium tartrate slowly reduces Cu^{++} ions to Cu^+ ions, causing a change in the concentration of Cu^{++} ions. When in use, a pipette is used for accurate suction, and the pipette is moved into the conical flask in the order of moving the liquid B first and then the liquid A.

2) The boiling time is accurately controlled to be 2 min, and the time is calculated when the

sugar liquid is fully boiled.

3) The heat source used and the specification of conical flask have a certain influence on the analysis results. Therefore, it is better to use the same set of instruments and appliances when calibrating Fehling reagent and determining samples.

4) When titrating, the solution in the conical flask should be kept boiling, so as to avoid air entering the flask to oxidize the tetramethyl blue and make it blue, which will cause errors in the determination.

5) The tetramethyl basket indicator itself can consume a certain amount of invert sugar, so it must be added according to the normal weight.

6) The Purpose of pre-inspection. One of the purposes of the pre-test is to know whether the dilution ratio of the prepared sugar solution for titration is appropriate. The consumption of sugar solution should be 20~40 mL. If 15 mL of the prepared sugar solution is added to Fehling's reagent, the blue color is completely removed after boiling (all Cu^{++} ions are reduced to Cu^+ ions), Indicating that the prepared sugar solution is too thick; if more than 50 mL of prepared sugar solution is consumed for titration, it indicates that the prepared sugar solution is too thin. In both cases, readjustment of the dilution factor is required. The second is to know the approximate amount of prepared sugar solution consumed for titration, and calculate the amount of prepared sugar solution that should be added in advance according to this amount, so that the final titration operation can be completed within 1 min when the solution is kept boiling during re-examination, thus improving the accuracy of determination.

7) Because reducing sugar can form a complex with calcium and lead, which interferes with the determination, it is necessary to remove calcium and lead.

2. Determination of reducing sugar by Lane-Eynon constant volume method

The Lane-Eynon constant volume method, also known as the constant volume method, is an improvement of the Lane-Eynon method. The Lane-Eynon Method was once the formal method recommended by ICUMSA (International Committee for the Uniform Method of Sugar Analysis). In the new national standard of brown granulated sugar, the Lane-Eynon Constant Volume Method was used to replace the Lane-Eynon Method for the determination of reducing sugars. In 1978, the 17th ICUMSA meeting had revoked the formal method qualification of the Lane-Eynon Method. The Lane-Eynon constant volume method is adopted as a formal method, and the method can overcome the defects that the total volume of the solution in the conical flask is changed at the end of titration due to different reducing pols of samples, so that the oxidation capacity of Cu^{++} ions is different, and the calculation is complicated.

(1) Basic principles. The Lane-Eynon constant volume method is an improvement of the traditional Lane-Eynon method. Its determination principle and main chemical reaction are exactly the same as those of the Lane-Eynon method, but the amount of Fehling reagent used is 20 mL (10 mL for solution A and 10 mL for solution B).

The results show that the oxidation ability of $CuSO_4$ is affected by many factors, such as the concentration of Cu^{++}, temperature, time, the amount of sucrose in the sample and so on. In order to reduce the influence of these factors, the volume of the solution in the conical flask at the end of titration is required to be exactly 75 mL (controlled by adding a proper amount of water in advance) in addition to the amount of Fehling reagent and boiling time, so as to reduce the influence of volume change on the determination and simplify the calculation. So this method is called the constant volume method.

The constant volume method is to add a certain amount of water into the conical flask in advance to achieve the goal that the volume of the solution in the conical flask is exactly 75 mL at the end of titration. However, because it is not known in advance how much sample will be consumed for titration, it is necessary to carry out pre-inspection first, preliminarily determine the consumption of the sample, calculate the amount of water to be added according to the consumption of the sample, and add the water to the conical flask in advance. At the end of the titration, the volume of the solution in the conical flask is essentially 75 mL.

(2) Concentration of Fehling reagent. The Fehling reagent used in the Lane-Eynon constant volume method is the same as that used in the Lane-Eynon method, but the calibration method is slightly different.

Because there are many factors affecting the reaction between reducing sugar and Cu^{++} ions, the concentration of Fehling reagent can only be determined after calibration with standard invert sugar solution, and the calibration conditions are required to be completely consistent with the detection conditions. If the concentration of Fehling reagent is appropriate, in the absence of sucrose, 20 mL of Fehling reagent (10 mL of solution A and 10 mL of solution B) should consume 40 mL of 2.5 g/L standard invert sugar solution, which is equivalent to 100 mg of reducing sugar. If the concentration is not appropriate, the correction factor (K) shall be calculated based on the equivalent amount of reducing sugar during calibration.

$$\text{Fehling reagent concentration correction factor } (K) = V/40$$

Where V—2.5 g/L standard invert sugar solution actually consumed during calibration (mL).

If sucrose is present in the sample, it will consume some of the Cu^{++} ions. The more sucrose in the conical flask, the more Cu^{++} ions were consumed. At this time, according to the weight of sucrose in the conical flask, the correction factor f should be obtained by looking up the relevant table (attached list 9), then the weight of reducing sugar equivalent to 20 mL Fehling reagent (10 mL of solution A and 10 mL of solution B) should be calculated according to the following formula.

$$\text{Weight of reducing sugar equivalent to 20 mL Fehling reagent} = 100fK$$

Where K—concentration correction coefficient of Fehling reagent;

f—sucrose correction factor.

(3) Main instruments and reagents.

1) Main instruments. Basic burette, 300 mL conical flask, 10 mL pipette, electric stove,

stopwatch, etc.

2) Reagent. EDTA solution (40 g/L), tetramethylene blue, Fehling reagent, etc.

(4) Measurement steps.

1) Preparation of sample solution. For samples with low concentration (below 26 °Bx), it is not necessary to dilute various sugar juices and prepare them according to the original concentration of the sample. Samples with high concentration (above 26 °Bx) can be diluted with distilled water (syrup is diluted by 4 times, molasses is diluted by 6 times, and samples containing crystals such as massecuite and massecuite are diluted by 6 times, and the crystals must be completely dissolved).

Take 100 mL of the above diluted sugar solution (sugar juice does not need to be diluted) and transfer it into a 300 mL volumetric flask, add 1-2.4 mL of potassium oxalate solution (50 g/L) to each gram of dry substance, fully shake up, add distilled water to the scale, fully shake up and filter. The potassium oxalate solution is added to remove calcium ions that interfere with the determination of reducing sugar. Discard about 15 mL of the initial filtrate, which is the sugar solution.

Alternatively, add 4 mL of EDTA solution (40 g/L) to each gram of dry substance in the solution, shake well, and add distilled water to the scale. Because the complex formed by EDTA and calcium ions is more stable than the complex formed by reducing sugar and calcium ions, filtration is not needed.

For samples with higher concentration, sugar solution can also be prepared by direct weighing. For example, syrup can be weighed about 12.500 g; molasses, massecuite and massecuite can be weighed about 1.500-12.500 g (accurate to three decimal places), dissolved in distilled water and transferred to a 250 mL volumetric flask, and then diluted by volume.

The concentration of reducing sugar in the prepared sugar solution should be kept between 2.50 and 4.00 g/L. If it is not within this range, the prepared sugar solution consumed during titration will be too much or too little, which should be adjusted by changing the dilution ratio.

2) Determination method.

① Pre-inspection. Use a pipette to suck 10 mL of Fehling reagent solution B into a 300 mL conical flask, and then use another pipette to suck 10 mL of solution A into solution B, and mix well. Add 25 mL of sugar solution from a burette into a conical flask, add 15 mL of distilled water with a measuring cylinder, shake it up, heat it on an electric stove covered with an asbestos net, boil it accurately for 2 min (controlled by a stopwatch), add 3-4 drops of tetramethylene blue indicator, and continue to add the prepared sugar solution until the blue color disappears, which is the end point. This operation should not exceed 1 min, so that the total time of the whole boiling and dropping operation is controlled within 3 min. Record the number of milliliters of prepared sugar solution consumed for titration, V_1.

② Re-inspection. Suck 10 mL of Fehling test solution B and 10 mL of Fehling test solution A into a 300 mL conical flask according to the above sequence, add distilled water $(75-20-V_1)$

mL into a measuring cylinder according to the volume V_1 of the prepared sugar solution consumed in the pre-test, add 1 mL of the prepared sugar solution less than the volume V_1 consumed in the pre-test from a burette, and shake up. The titration procedure is the same as the pre-test, and the boiling time should also be accurately controlled at 2 min. The time of titration to the end point after adding the indicator should not exceed 1 min. When titrating, the conical flask should be shaken gently, but the heat source should not be removed to keep the solution boiling, so as to avoid errors caused by oxidation of tetramethyl blue due to air entering the flask. Record the volume T of prepared sugar solution consumed for titration, and use this data for the calculation of reducing sugar.

If V_1 is far from T due to unskilled operation, the result of reexamination can be V_1, and the determination can be repeated again.

(5) Calculation method.

1) Required data.

① Number of milliliters of prepared sugar solution consumed for titration T.

② Concentration correction factor K of Fehling reagent.

③ Sucrose content S of diluted sample solution.

2) Calculation method. The principle of Lane-Eynon constant volume method is the same as that of Lane-Eynon method, and the ability of Fehling reagent to react with reducing sugar can not be calculated directly according to the reaction equation. Only experimental methods can be used to determine the amount of reducing sugar equivalent to 20 mL Fehling reagent under different conditions, and these data can be made into a "correction coefficient table" for actual determination. However, the Lane-Eynon constant volume method controls the amount of solution in the conical flask at the end of titration to a fixed value, so that the ability of Fehling reagent to react with reducing sugar is not affected by the amount of sugar solution consumed, and only the influence of sucrose on the oxidation ability of Fehling reagent needs to be considered in the calculation, so as long as the weight of sucrose in the conical flask is calculated, the correction coefficient of Fehlin reagent can be found out. Compared with the Lane-Eynon method, it is simpler.

In the absence of sucrose, 20 mL of Fehling's reagent is equivalent to 100 mg of reducing sugar. If sucrose is present, multiply by the corresponding correction factor for Fehling's reagent. The calculation method is as follows:

① Calculate the weight of sucrose in the solution in the conical flask after titration A:

When the volumetric dilution method is used, first calculate the sample volume V_0 in 100 mL of the prepared sugar solution:

$$V_0 = 100 \times 100/250 = 40 \text{ (mL)}$$

The weight of sucrose in the solution in the conical flask $A = \dfrac{V_0 \times d_{(20°C)} \times T \times S}{100 \times 100}$

When the weight dilution method is used, it can be calculated directly according to the weight W of the sample.

$$\text{The weight of sucrose in the solution in the conical flask } A = \frac{W \times T \times S}{100 \times 250}$$

Where W—when the weight dilution method is used, weigh the sample (g).

② Calculate the sucrose correction factor f with the following formula or refer to attached list 9 according to the sucrose content A in the solution in the conical flask:

$$f = \frac{0.994\ 7 + 0.041\ 8A}{1 + 0.067\ 2A}$$

Where A—the number of grams of sucrose contained in the solution in the conical flask (g).

③ Sample reducing sugar:

$$\text{Reducing pol (\%)} = \frac{f \times K \times 100}{T \times \frac{V_0}{100} \times d_{(20℃)} \times 1\ 000} \times 100\% = \frac{10 \times f \times K}{T \times V_0 \times d_{(20℃)}} \times 100\%$$

Where f—sucrose correction factor;

V_0—100 mL of prepared sugar solution contains ml of sample (mL);

T—volume of prepared sugar solution consumed for titration (mL);

K—correction factor of Fehling reagent concentration;

$d_{(20℃)}$—apparent density of sample (20 ℃).

The apparent density $d_{(20℃)}$ of the sample is calculated by the observed Brix of the sample or by the following formula:

$$\text{Apparent density of the sample } d_{(20℃)} = 0.998\ 2 + 0.003\ 7B + 0.000\ 018\ 16B^2$$

Where B—observed Brix of sample.

Note: For syrup, massecuite and other samples that can only be determined after dilution, the reducing pol of the original sample shall be obtained by multiplying the calculation result by the dilution factor.

[Example 2-11] The sucrose content of the mixed juice was 13.98%, the Brix was observed to be 16.55 °Bx, 100 mL of the sample was absorbed and prepared to 250 mL, the concentration correction coefficient of Fehling reagent was 1.028, and 33.35 mL of the prepared sugar solution was consumed for re-examination to calculate the reducing pol of the mixed juice.

[Solution] The volume V of the sample contain in 100 mL of the prepared sugar solution is:

$$V = 100 \times 100/250 = 40 \text{ (mL)}$$

The weight W of the sample contained in 100 mL of the prepared sugar solution is:

$$W = 40\ (0.998\ 2 + 0.003\ 8 \times 16.55 + 0.000\ 018\ 16 \times 16.55^2) = 42.58 \text{ (g)}$$

The solution in the conical flask (33.35 mL) contains sucrose weight A:

$$A = 33.35 \times 42.58 \times 13.98 / (100 \times 100) = 1.99 \text{ (g)}$$

Calculate the correction factor of Fehling reagent as follows:

$$f = \frac{0.994\ 7 + 0.041\ 8A}{1 + 0.067\ 2A}$$

$$f = \frac{0.994\ 7 + 0.041\ 8 \times 1.99}{1 + 0.067\ 2 \times 1.99} = 0.950\ 7$$

Amount of reducing sugar equivalent to 20 mL Fehling reagent R:

$$R = 100 \times 1.028 \times 0.950\ 7 = 97.73\ (\text{mg})$$

The weight W of the sample in the conical flask after titration is:

$$W = 33.35 \times 42.64/100 = 14.22\ (\text{g})$$

$$\text{Reducing pol of mixed juice} = \frac{R}{W} \times 100\% = \frac{97.73}{14.22 \times 1\ 000} \times 100\% = 0.69\%$$

[Example 2-12] The sucrose content of brown granulated sugar is 88.45%, and the reagent correction coefficient of Fehling reagent is 0.986. The brown granulated sugar is diluted according to 1/2 of the normal weight to determine the reducing sugar, and the titration consumption is 31.30 mL, so as to calculate the reducing sugar of red granulated sugar.

[Solution] Because the sample is diluted according to the semi-normal weight, every 100 mL contains 13 g of sample.

The solution in the conical flask (31.30 mL) contains sucrose weight A:

$$A = 31.3 \times 13 \times 88.45/(100 \times 100) = 3.60\ (\text{g})$$

Calculate the correction factor of Fehling reagent according to the formula:

$$f = \frac{0.994\ 7 + 0.041\ 8A}{1 + 0.067\ 2A}$$

$$f = \frac{0.994\ 7 + 0.041\ 8 \times 3.60}{1 + 0.067\ 2 \times 3.60} = 0.922$$

Calculate the weight W of the sample according to the weight A of sucrose in the solution in the conical flask:

$$W = 3.6/0.884\ 5 = 4.07\ (\text{g})$$

Amount of reducing sugar equivalent to 20 mL Fehling reagent R:

$$R = 100 \times 0.986 \times 0.922 = 90.91\ (\text{mg})$$

$$\text{Reducing pol of brown granulated sugar} = \frac{R}{W} \times 100\% = \frac{90.91}{4.07 \times 1\ 000} \times 100\% = 2.23\%$$

[Example 2-13] The sucrose content of final molasses is 34.32%, each 100 mL of prepared sugar solution contains 1.733 g of final molasses, the concentration correction coefficient of Fehling reagent is 0.988 5, the titrated consumption of prepared sugar solution is 28.65 mL, and the reducing pol of final molasses is calculated.

[Solution] The solution in conical flask (28.65 mL) contains sucrose weight A as follows:

$$A = 28.65 \times 1.733 \times 34.32/(100 \times 100) = 0.170\ 4\ (\text{g})$$

Calculate the correction factor of Fehling reagent according to the formula:

$$f = \frac{0.9947 + 0.0418A}{1 + 0.0672A}$$

$$f = \frac{0.9947 + 0.0418 \times 0.1704}{1 + 0.0672 \times 0.1704} = 0.990$$

Amount of reducing sugar equivalent to 20 mL Fehling reagent R:

$$R = 100 \times 0.9885 \times 0.990 = 97.86 \text{ (mg)}$$

The weight W of the sample in the conical flask after titration is:

$$W = 28.65 \times 1.733/100 = 0.497 \text{ (g)}$$

$$\text{Reducing sugar of final molasses} = \frac{R}{W} \times 100\% = \frac{97.86}{0.497 \times 1\,000} \times 100\% = 19.69\%$$

(6) Precautions. The total volume of the solution at the end of the titration must be kept constant (75 mL). Others are the same as the Lane–Eynon method.

Task 2.5　Determination of sulphur strength of neutralized juice

Enterprise case

The factory director found that the quality of the clear juice today was very poor and quite turbid. He estimated that it was related to the recent quality of sugarcane and the intensity of sulfitation in the purification process. He asked the laboratory director and Xiaoming to collect the neutralized juice carefully, and then analyze the intensity of sulfitation to judge whether his judgment was accurate.

Mission objectives

Through the study of this task, students can achieve the following goals:

(1) Understand the definition of sulfitation intensity.

(2) Grasp the determination method of sulfitation intensity intensity.

Quality objectives

Develop the overall consideration of the problem.

Task description

Determination of sulfitation intensity of neutralized juice.

Procedures and methods

(1) Burette: 50 mL.

(2) Erlenmeyer flask: 150 mL.

(3) Measuring cylinder: 10 mL.

(4) 1/64 mol/L I_2 standard solution.

(5) 10g/L starch solution.

(6) Steps. Measure 10 mL of sulfur fumigation juice with a measuring cylinder, pour it into a 150 mL conical flask, and then measure 30 mL of distilled water with the same amount, pour it into the conical flask to dilute the sample, and test it with pH test paper; if the solution is alkaline, add a proper amount of dilute acetic acid (or dilute hydrochloric acid) for neutralization to make it acidic, then add several drops of 10 g/L starch solution, and then titrate with 1/64 mol/L I_2 standard solution until the solution turns blue, and record the volume of iodine standard solution consumed for titration.

1 mL of 1/64 mol/L I_2 standard solution is equivalent to 1 mg of SO_2.

(7) Data processing and calculation.

Sample	Consumed volume of 1/64 mol/L I_2 standard solution
1	
2	
3	

Reflection

Intensity of sulfitation is an important control index of sulfurous acid method. Please think about why this index is so important.

2.5.1 Task-related knowledge — determination of sulphur strength

Sulphitation intensity is the quantity of sulfur dioxide absorbed by cane juice, which is an important process condition directly affecting the purification effect in sulfitation sugar mills. Through the determination of sulfitation intensity, the neutralization process of sulfitation can be understood and controlled to obtain good purification effect.

1. Basic Principles

The method is based on the principle of iodometry, and the basic reaction is as follows:

$$I_2 + 2e^- = 2I^-$$

I_2 is a weak oxidant, and SO_2 with strong reducibility can be directly titrated by using I_2 standard solution. The chemical equation of the reaction is as follows:

$$I_2 + SO_2 + 2H_2O = 2I^- + SO_4^{2-} + 4H^+$$

Starch can be used as an indicator for this reaction. Starch reacts with I_2 to form a blue complex, which indicates the end point according to the appearance of blue.

2. Main instruments and reagents

(1) Burette: 50 mL;

(2) Erlenmeyer flask: 150 mL;

(3) Measuring cylinder: 10 mL;

(4) 1/64 mol/L I_2 standard solution;

(5) 10g/L starch solution.

3. Measurement steps

Measure 10 mL of sulfur fumigation juice with a measuring cylinder, pour it into a 150 mL conical flask, and then measure 30 mL of distilled water with the same amount, pour it into the conical flask to dilute the sample, and test it with pH test paper; if the solution is alkaline, add a proper amount of dilute acetic acid (or dilute hydrochloric acid) for neutralization to make it acidic, then add several drops of 10 g/L starch solution, and then titrate with 1/64 mol/L I_2 standard solution until the solution turns blue, and record the volume of iodine standard solution consumed for titration.

1 mL of 1/64 mol/L I_2 standard solution is equivalent to 1 mg of SO_2.

4. Calculation method

Sulfur intensity = volume of 1/64 mol/L I_2 standard solution used for titration.

If the iodine solution consumed for titration is 18 mL, that is, the sulfitation intensity is 18 mL, the content of SO_2 in the neutralized juice is 1.8 g/L.

2.5.2 Task-related knowledge——determination of alkalinity and total calcium

Alkalinity refers to the amount of alkali contained in the sugar juice when it is in an alkaline reaction. Alkalinity is generally expressed weight of calcium oxide in 100 mL of solution, or weight of CaO per liter of sample.

By measuring the alkalinity of pre-lime juice, one-carbon juice and one-carbon clear juice, the use of lime and the effect of one-carbon saturation are understood, so as to change the control conditions of saturation and obtain good cleaning effect.

The total calcium content is expressed as the weight of unreacted calcium oxide, calcium oxide that has formed calcium carbonate and some alkali metal equivalent calcium oxide contained in 100 mL of sugar juice. The total calcium content was used for the analysis of one-carbonatation juice in a carbonated sugar mill. The analysis method is basically the same as that of alkalinity.

The determination of alkalinity is based on the principle of acid-base titration reaction. The standard acid solution with known concentration is used to titrate the sample solution to be determined until the acid-base neutralization reaction is completed, and then the content of the component to be determined is calculated according to the concentration and volume of the consumed standard solution and the stoichiometric relationship of the chemical reaction.

1. Main equipments and reagents

(1) Burette: 50 mL, graduation: 0.1 mL.

(2) Conical flask: 150 mL, 250 mL.

(3) Measuring cylinder: 10 mL.

(4) Sulfuric acid solution: 1/56 mol/L, 1/5.6 mol/L.

(5) Phenolphthalein indicator solution: 1%.

(6) Methyl orange indicator solution: 0.1%.

2. Alkalinity determination steps

(1) Determination of alkalinity of preliming juice, one-carbon juice and one-carbon clear juice. The sample was stirred and filtered. Pipette 10 mL of the filtrate into a 150 mL conical flask. Dilute with 20 mL of distilled water in a conical flask. Add 1~2 drops of phenolphthalein indicator. Shake well. The solution is red. Fill 1/56 mol/L H_2SO_4 standard solution with an acid

burette, adjust the zero point, titrate the solution in the conical flask under constant shaking until the red color of the solution in the conical flask just disappears, and record the volume V_0 of the consumed sulfuric acid standard (repeat the determination for three times, and take the average value of V).

(2) Alkalinity calculation.

1) One milliliter of 1/56 mol/L H_2SO_4 standard solution is equivalent to 0.001 g of CaO.

2) The alkalinity of the sample is expressed as the equivalent weight of calcium oxide in 100 mL of sugar juice.

$$\text{Alkalinity} = \frac{0.001 \times V}{10} \times 100$$

Where V—the volume of 1/56 mol/L H_2SO_4 standard solution consumed for titration (mL).

3. Determination of total calcium

(1) Determination procedure of total calcium content in a carbon juice. Measure 10 mL of unfiltered sample with a measuring cylinder, transfer it into a 250 mL conical flask, add 50 mL of distilled water for dilution, drop 2–3 drops of methyl orange indicator, shake it up and titrate it with 1/5.6 mol/L H_2SO_4 standard solution, and keep shaking until the solution appears reddish and does not disappear within 30 s. Record the volume of standard solution consumed V_0 (repeat the determination three times, V is averaged).

(2) Calculation of total calcium. Each milliliter of 1/5.6 mol/L H_2SO_4 standard solution is equivalent to 0.01 g CaO.

$$\text{Total calcium} = \frac{0.01 \times V}{10} \times 100$$

Where V—volume of 1/5.6 mol/L H_2SO_4 standard solution consumed for titration.

Note: The sample for the determination of total calcium shall not be filtered, otherwise the accuracy of the analysis results will be affected.

Task 2.6　Determination of sugar juice pH

Enterprise case

Today's production report reflects that the loss of sucrose is very large. The director of the laboratory immediately arranges Xiaoming to collect various intermediate products for pH determination. At this time, what instrument should he choose and how to detect it?

Mission objectives

Through the study of this task, students can achieve the following goals:

Understand and master the principle and method of pH measurement of sugar juice.

Quality objectives

Develop the habit of comparative analysis of learning.

Task description

Determination of the acidity of the work-in-process using a pH meter.

Procedures and methods

(1) Ingredients: Sugarcane juice.

(2) Equipments and instruments: pH meter (measuring range pH=0~14, accuracy from pH=0.01~0.1), glass electrode, calomel electrode, electromagnetic stirrer.

(3) Reagents: potassium hydrogen phthalate buffer solution (pH= 4.00), borax buffer solution (pH = 9.22), mixed phosphate buffer (pH = 6.88).

The pH changes of the above three buffer solutions at different temperatures are shown in Table 2-6.

Table 2-6　The pH changes of the above three buffer solutions at different temperatures

Temperature /°C	Potassium hydrogen phthalate buffer solution 0.05 mol/L	Borax buffer solution 0.01 mol/L	Mixed phosphate buffer solution 0.025 mol/L
5	4.00	9.39	6.95
10	4.00	9.33	6.92
15	4.00	9.27	6.90
20	4.00	9.22	6.88
25	4.01	9.18	6.86
30	4.01	9.14	6.85
35	4.02	9.10	6.84
40	4.03	9.07	6.84
45	4.04	9.04	6.83
50	4.06	9.01	6.83
55	4.08	8.99	6.84
60	4.10	8.96	6.84

(4) Steps.

1) Preparation of sample solution. When determining the pH of solid (except white granulated sugar) and viscous samples, the sample shall be diluted 1 ∶ 1, and the pH of other products in process in the sugar factory can be determined at the original concentration.

2) Calibration of pH meter. Before making the measurement, calibrate the pH meter with a suitable buffer solution so that the pH indicated on the instrument scale is exactly that of the buffer solution.

3) Determination. Adjust the temperature regulator of the instrument so that the indicated temperature is the same as the temperature of the solution to be measured. Immerse the electrode rinsed with the solution to be measured into the solution, start the stirrer to stir the solution, press the measurement switch, and record the reading when the instrument stably displays the pH of the solution. The specific operation steps are carried out according to the instruction manual of the instrument.

(5) Precautions.

1) The glass electrode used for the first time or reused for a long time should be soaked in distilled water for more than 24 h to stabilize its asymmetric potential. When in use, the bulb part shall be completely immersed in the solution to be tested; after use or short-term use, it can be put back into distilled water for soaking.

2) The calomel electrode should not be immersed in distilled water frequently. A few potassium chloride crystals should be retained. The potassium chloride solution should be kept saturated frequently. There should be no bubbles in the solution in the electrode.

3) The pH of the buffer solution used to calibrate the acidometer should preferably be close to the pH of the solution being measured. The buffer solution should not be used if precipitates are produced.

4) When using the calomel electrode, pull out the small rubber plug at the upper end where the potassium chloride solution is added, so as to keep enough liquid level difference in the capillary to prevent the sample solution from entering the capillary and affecting the determination result.

5) The glass film of the glass electrode is brittle and easily damaged, so care should be taken when using it. When installing the glass electrode, the bulb at the lower end should be slightly higher than the ceramic core end of the calomel electrode to avoid damage. If there is greasy dirt on the bulb glass membrane, it can be immersed in ethanol first, then in ether or carbon tetrachloride, and finally in ethanol, and then washed with distilled water.

Reflection

The pH meter can be put back into distilled water for soaking after being used or not used for a short time? Please develop a good habit of taking good care of instruments and equipment and maintaining them carefully.

Task 2.7 Determination of phosphate value of sugarcane juice

Enterprise case

Today's production report reflects that a lot of phosphoric acid has been consumed recently when the product is stable. The factory director judges that it is possible to use too much phosphoric acid, and then immediately asks the laboratory director to arrange Xiaoming to collect sugarcane juice for the determination of phosphate. What should he do at this time?

Mission objectives

Through the study of this task ,students can achieve the following goals:

(1) Understand the determination principle of phosphate value of sugarcane juice.

(2) To master the determination method and calculation of phosphate value of sugarcane juice.

Quality objectives

Develop the habit of seeing through the phenomenon to the essence.

Task description

Determination of phosphate in sugarcane juice.

Procedures and methods

(1) Ingredients: cane juice.

(2) Equipments and instruments: one set of photoelectric colorimeter or spectrophotometer, one set of cuvette, 25 mL colorimetric tube, pipette and 100 mL volumetric flask.

(3) Reagent: potassium dihydrogen phosphate standard solution, 25 g/L ammonium molybdate sulfuric acid mixed solution, 50 g/L stannous chloride glycerol solution.

(4) Steps.

1) sugarcane juice preparation solution: Suck 5 mL of cane juice with a pipette, transfer it into a 100 mL volumetric flask, add water to the marked line, shake well, filter, discard the initial filtrate, suck 5 mL of filtrate, transfer it into another 100 mL volumetric flask, add water to about 95 mL, accurately add 2 mL of 25 g/L ammonium molybdate sulfuric acid mixed solution, and then add 5 drops of 50 g/L stannous chloride glycerin solution. Then add water to the line and shake well. After standing for 5 min, the prepared solution was a 400-fold dilution of sugarcane juice.

2) Preparation of phosphoric acid value standard color solution: Suck 0.5 mL, 1.0 mL, 1.5 mL, 2.0 mL, 2.5 mL, 3.0 mL, 3.5 mL and 4.0 mL of potassium dihydrogen phosphate standard solution into 8 volumetric flasks of 100 mL size by pipette, and dilute with distilled water to constant volume. The content of P_2O_5 in each volumetric flask is 50 mg/L, 100 mg/L,

150 mg/L, 200 mg/L, 250 mg/L, 300 mg/L, 350 mg/L, 400 mg/L respectively. The above solution was also diluted 400 times, and the mixed solution of ammonium molybdate and sulfuric acid and the solution of stannous chloride in glycerol were added. After standing for 5 min, the solution became a blue standard color solution with the color from light to dark.

3) Colorimetric method. The phosphate value of cane juice can be determined more quickly and accurately by using spectrophotometer or photoelectric colorimeter. However, the phosphate value curve must be determined and drawn first. Since the colour is proportional to the absorbance, the curve is a straight line.

① Drawing of phosphate value curve: Put distilled water in the cuvette, and correct the zero point with a red filter (wavelength is about 660 nm). Then the absorbency is respectively measured by the standard color liquid with the known phosphoric acid value. The regression method of mathematical statistics is used to draw a more accurate curve of phosphoric acid value according to the regression equation.

② Determination of phosphate value of cane juice: Fill the blank sample solution without any reagent in the cuvette, and correct the zero point with a red filter (wavelength is about 660 nm). Then the absorbance of the sugarcane juice preparation solution is measured, and the corresponding phosphoric acid value is searched from the phosphoric acid value curve or substituted into the regression equation to calculate the phosphoric acid value.

4) Draw a graph.

Task-related knowledge — analysis of sugarcane juice composition

The various components in cane juice have an important influence on the sugar production process, energy saving, sugar recovery and the quality of finished sugar. Through the analysis of sugarcane juice composition, the basis for the establishment of the best clarification process conditions was provided.

1. Determination of phosphoric acid value

(1) Purpose. The amount of soluble phosphate present in the juice is called the phosphate value, expressed as P_2O_5. Generally, sugar mills only analyze the natural phosphate value of cane juice, that is, the part of phosphate that can quickly produce phosphate ions and precipitate calcium phosphate in the process of extraction. Natural phosphoric acid value is an important factor for effective clarification of sugarcane juice. Generally, the content of P_2O_5 in sugarcane juice should not be less than 300 mg/L. If the content of P_2O_5 in sugarcane juice is less than 300 mg/L, an appropriate amount of phosphoric acid or superphosphate clear liquid can be used to help the normal clarification. Therefore, the purpose of determining the acid phosphorus value of sugarcane juice is to provide the basis for formulating the optimal extraction process conditions.

(2) Common methods. Ammonium molybdate colorimetry is a common method for the

determination of phosphoric acid value.

(3) Ammonium molybdate colorimetric method.

1) Basic principles. At a certain acidity, the phosphate reacts with the added ammonium molybdate to form a yellow crystalline ammonium phosphomolybdate precipitate. The reaction formula is:

$$PO_4^{3-}+3NH_4^{+}+12MoO_4^{2-}+24H^{+}=(NH_4)_3PO_4 \cdot 12MoO_3 \cdot 6H_2O+6H_2O$$

Ammonium molybdate is easily reduced by a suitable reducing agent (such as stannous chloride) to form a dark blue phosphomolybdate blue [$MoO_2 \cdot 4MoO_3 \cdot H_3PO_4$], The shade of blue is directly proportional to the content of phosphate, so the content of soluble phosphate in sugarcane juice can be measured by colorimetry. The reaction formula is:

$(NH_4)_3PO_4 \cdot 12MoO_3+11H^{+}+4Sn^{2+}=(MoO_2 \cdot 4MoNO_3)_2 \cdot H_3PO_4+2MoO_2+4Sn^{4+}+3NH_4^{+}+4H_2O$

2) Main instruments, equipment and reagents.

① Equipments and instruments: One set of photoelectric colorimeter or spectrophotometer, one set of cuvette, 25 mL colorimetric tube, pipette and 100 mL volumetric flask.

② Reagents: potassium dihydrogen phosphate standard solution, 25 g/L ammonium molybdate sulfuric acid mixed solution, 50 g/L stannous chloride glycerol solution.

3) Determination method.

① Sugarcane juice preparation solution. Suck 5 mL of cane juice with a pipette, transfer it into a 100 mL volumetric flask, add water to the marked line, shake well, filter, discard the initial filtrate, suck 5 mL of filtrate, transfer it into another 100 mL volumetric flask, add water to about 95 mL, accurately add 2 mL of 25 g/L ammonium molybdate solution, then add 5 drops of 50 g/L stannous chloride solution, and then add water to the marked line. Shake well. After standing for 5 min, the prepared solution was a 400-fold dilution of sugarcane juice.

② Preparation of phosphoric acid value standard color solution. Suck 0.5 mL, 1.0 mL, 1.5 mL, 2.0 mL, 2.5 mL, 3.0 mL, 3.5 mL and 4.0 mL of potassium dihydrogen phosphate standard solution into 8 volumetric flasks of 100 mL size, and dilute with distilled water to constant volume. The content of P_2O_5 in each volumetric flask is 50 mg/L, 100 mg/L, 150 mg/L, 200 mg/L, 250 mg/L, 300 mg/L, 350 mg/L and 400 mg/L respectively. The above solution was also diluted by 400 times, and ammonium molybdate solution and stannous chloride solution were added. After standing for 5 min, it became a blue standard color solution with color from light to dark.

③ Colorimetric method. The phosphate value of cane juice can be determined more quickly and accurately by using spectrophotometer or photoelectric colorimeter. However, the phosphate value curve must be determined and drawn first. Since the colour is proportional to the absorbance, the chart is a straight line.

Drawing of phosphate value curve: Put distilled water in the cuvette, and correct the zero point with a red filter (wavelength is about 660 nm). Then the absorbency is respectively measured by the standard color liquid with the known phosphoric acid value. The regression method of mathematical statistics is used to draw a more accurate curve of phosphoric acid value according to the regression equation.

Determination of phosphate value of cane juice: Fill the blank sample solution without any reagent in the cuvette, and correct the zero point with a red filter (wavelength is about 660 nm). Then the absorbance of the sugarcane juice preparation solution is measured, and the corresponding phosphoric acid value is looked up from the phosphoric acid value curve or substituted into the regression equation to calculate the phosphoric acid value.

④ Precautions. When drawing the phosphoric acid value curve and determining the phosphoric acid value of the cane juice sample, the cuvette thickness shall be selected so that the transmittance reading of the instrument is 20%~80%, and the matching cuvette shall be used. The transmittance difference of the matching cuvette at the same optical path shall not be greater than 0.2%. If the cuvette of another specification is used, the curve shall be made again.

2. Determination of calcium and magnesium salt content

(1) Purpose. The existence of calcium and magnesium salts in sugar juice will form a product in the heating and evaporation equipment, which will affect the heat transfer efficiency and increase the ash content of finished sugar. When the content of magnesium salt is high, brown granulated sugar will have a bitter taste. Therefore, its content should be analyzed in order to take appropriate measures to remove it in the purification process and reduce its impact.

(2) Basic principles. This method is based on the principle of complexometry. The sample was titrated with complexing agent as standard solution to form complex compounds with calcium and magnesium ions in sugarcane juice. Ethylenediaminetetraacetaic acid disodium salt (EDTA) is a carboxylic complexing agent, which can form stable complexes with cations such as calcium and magnesium salts under different pH conditions. Most of the calcium and magnesium salts in sugarcane juice form purplish red complexes with Ca^{2+} and Mg^{2+}. When all Ca^{2+} and Mg^{2+} have been reacted, the end point is reached when the indicator changes to its original blue color. The reaction formula is as follows:

$$M^{2+} + HIn^{2-} \longrightarrow MIn^- + H^+$$
$$\text{(Blue)} \quad \text{(Violet Blue)}$$

When titrating to the endpoint:

$$MIn^- + H_2Y^{2-} \longrightarrow MY^{2-} + HIn^{2-} + H^+$$
$$\text{(Violet Blue)} \quad \text{(Blue)}$$

In the above reaction formula, M^{2+} represents calcium and magnesium ions, HIn^{2-} represents eriochrome black T indicator, and H_2Y^{2-} represents EDTA.

(3) Main instruments, equipments and reagents.

1) Burette: 50 mL.
2) Conical flask: 250 mL.
3) 0.01 mol/L EDTA solution.

4) 150 g/L sodium hydroxide solution.

5) 20 g/L hydroxylammonium hydrochloride solution.

6) Calcium indicator.

7) Eriochrome black T indicator.

8) Ammonium salt buff solution

(4) Determination method.

1) Determination of total calcium and magnesium salts. Accurately suck 10 mL of sugarcane juice sample, transfer it into a 250 mL conical flask, add about 100 mL of distilled water and 5 mL of ammonium salt buffer solution, mix well, and then add 3 drops of 20 g/L hydroxylamine hydrochloride and about 0.1 g of Eriochrome Black T indicator. After mixing and dissolving, the solution is purplish red. Titrate with 0.01 mol/L EDTA solution until the solution suddenly turns from wine red to blue. In addition, 100 mL of distilled water was used for the blank test according to the above method. The difference in the number of milliliters of EDTA consumed before and after titration is the number of milliliters of EDTA consumed to titrate the total amount of calcium and magnesium salts.

2) Determination of calcium salt content. Accurately suck 10 mL of sugarcane juice sample, transfer it into a 250 mL conical flask, add about 100 mL of distilled water and 2 mL of 150 g/L sodium hydroxide solution, mix well, stand for about 3 min to precipitate magnesium hydroxide, then add 3 drops of 20 g/L hydroxylamine hydrochloride solution and about 0.1 g of calcium indicator, after mixing and dissolving, the solution is grape red. Titrate with 0.01 mol/L EDTA solution until pure blue appears. In addition, 100 mL of distilled water was used for blank test according to the above method. The difference in the number of milliliters of EDTA consumed before and after titration is the volume of EDTA consumed for titration of calcium salt.

The content of magnesium salt can be calculated by the difference between the amount of EDTA consumed for titrating the total amount of calcium and magnesium salt and the amount of EDTA consumed for titrating the calcium salt.

(5) Calculation. The content of calcium salt (or magnesium salt) in the sample is calculated according to next formula, and is expressed by the weight of CaO (or MgO) contained in 100g of sugarcane juice:

The content of calcium salts (or magnesium salt) in the sample is expressed weight of CaO (or MgO) contained in 100 g of cane juice:

$$CaO = \frac{C \times V_2 \times 0.05608 \times 100}{W}$$

$$MgO = \frac{C \times (V_1 - V_2) \times 0.04032 \times 100}{W}$$

Where V_1—the volume of EDTA consumed for titration of total calcium and magnesium salts (mL);

V_2—the volume of EDTA consumed for titration of calcium salt (mL);

C—concentration of EDTA standard solution (mol/L);

W—the weight of sugarcane juice sample (g);

0.05608—the mass of 1 mmol CaO (g);

0.04032—the mass of 1 mmol MgO (g).

(6) Precautions.

1) Juice samples for determination of calcium and magnesium salts must be collected before preliming. The content of calcium salt in prelimed cane juice has changed, which affects the authenticity of the analysis results.

2) Mercuric salt should not be added to the sample as a preservative, because mercury ions can consume EDTA, which will make the analysis results higher.

3) Because cane juice or other sugar juice has its own color, it is difficult to show blue at the end point of titration, and it is generally grey–green, so it should be carefully observed when judging the end point. At the same time, the amount of indicator should not be too much, so as to avoid too dark color and affect the judgment of the key points.

4) Some ions in sugarcane juice can interfere with the analysis, among which iron, aluminum and manganese are the main ones. If necessary, potassium cyanide can be added to remove iron, triethanolamine to remove aluminum, and hydroxylamine hydrochloride to remove manganese to reduce their interference.

Sugarcane daily analysis

Moisture determination

Determination of colour and turbidity of WIP

Near infrared assay intermediate

Module 3

Analysis of physical and chemical index of finished sugar

Task 3.1　Determination of sucrose content in white granulated sugar

Enterprise case

Some customers reported that the sucrose (%) of the latest product was not up to the standard, so the director arranged Xiaoming to determined the sucrose (%) of the batch of white granulated sugar. What instrument should Xiaoming choose and what method should he use to determine it?

Mission objectives

Through the study of this task, students can achieve the following goals:

(1) Further familiarize yourself with the use of the saccharometer.

(2) To master the method of determining sucrose content in white granulated sugar by single polarimetry and its calculation.

Quality objectives

Develop the habit of comparative analysis of learning.

Task description

The sucrose (%) of white granulated sugar was analyzed by polarimeter.

Procedures and methods

(1) Ingredients: White granulated sugar

(2) Equipment and instruments: Polarimeter, 200 mm polarimeter tube, analytical balance, mercury thermometer (0 ~ 50 ℃, division of 0.1 ℃) , 100 mL volumetric flask, measuring cylinder and beaker

(3) Reagents: Distilled water or deionized water, absolute ethyl alcohol

(4) Steps: Weigh the sample of white granulated sugar [(26.000 ± 0.002) g] into a dry and

clean small beaker, add 40~50 mL of distilled water, and stir it with a thin glass rod to dissolve it completely. Pour it into a 100 mL volumetric flask, wash the beaker and glass rod with a small amount of distilled water for 3~5 times, pour the washing water into the volumetric flask, shake up the solution in the flask after pouring the washing water each time until the distilled water is added below the marked line of the volumetric flask, place it for at least 10 min to reach the room temperature, and then add distilled water to about 1mm below the marked line of the volumetric flask to ensure that the neck of the flask has been cleaned. Be careful not to entrain air bubbles in the solution. If there are air bubbles, they can be eliminated with ethanol or ether. Hold the top of the neck of the volumetric flask vertically so that the line of the volumetric flask is level with the operator's eyes. Observe it in the bright background, add water to the marked line with a long dropper, dry the inner wall of the bottle neck with clean filter paper, plug the plug tightly and shake it well. If turbidity is found, filter it with filter paper, cover the funnel with surface glass, discard the first 10 mL of solution, and collect 50~60 mL of the filtrate.

Flush the polarimeter tube with the solution to be tested at least twice, and then fill the polarimeter tube with the solution, taking care not to entrain air bubbles in the polarimeter tube. Place the polarimeter tube in the sugar meter, determine the temperature of the solution in the polarimeter tube (accurate to 0.1 °C) immediately after determining the optical rotation reading, and record the above data.

(5) Data processing and calculation.

Project	Data and results
Observe the Brix reading/°Z	
Observe the temperature of sugar solution at P_1/°C	
Sucrose content/%	

3.1.1 Task-related knowledge — determination of sucrose content in white granulated sugar

The optical activity of sucrose is used to determine the concentration of sucrose in the solution by measuring the optical activity of the aqueous solution of sugar sample with a sugar meter with a scale of 100 °Z on the international sugar scale under specified conditions. The purity of white granulated sugar is very high, and it contains few impurities, so it has little interference to the determination of polarimetry, and the sucrose content of white granulated sugar can be determined by one-step polarimetry. However, since the optical activity of the sugar solution is affected by temperature, the measurement results must be corrected for temperature.

1. Instruments and equipment

(1) Sugar meter. The saccharometer shall be calibrated according to the international saccharinity scale and the saccharinity (°Z), with the measuring range (−30)~ (+120) °Z, and

approved by the standard quartz tube. Three types can be selected:

1) Sugar analyzer (disk polarimeter) is equipped with an adjustable analyzer (analyzer), using monochromatic light source (wavelength between 540 nm and 590 nm), usually using a green mercury light or yellow sodium light.

2) Quartz wedge sugar meter:

① With monochromatic light source (wavelength between 540 nm and 590 nm);

② Equipped with an incandescent lamp as the light source, while separating the light with an effective wavelength of 587 nm by means of a suitable colour filter;

3) Sugar meter is equipped with Faraday coil as compensator, using monochromatic light source (wavelength between 540 nm and 590 nm).

Note: The old °S scale can still be used, but the °S reading must be converted to °Z by multiplying it by a factor of 0.999 71.

(2) Volumetric flask. Capacity, (100.00±0.02) mL, is corrected by weighing with water at (20.00±0.1) °C respectively. If the capacity of the volumetric flask is within the range of (100.00 ± 0.01) mL, it can be used without correction; if the capacity exceeds this range, it can be used after correction with the correction factor corresponding to 100.00 mL.

(3) Polarimetric polarimeter tube. Length: (200.00±0.02) mm, the qualification certificate shall be issued by the legal measuring agency, or the polarimeter tube with the certificate shall be used for comparison and inspection.

(4) Analytical balance: Sensitivity 0.1 mg.

2. Reagent

Distilled water: Free of optically active substances.

3. Calibration of sugar meter

The sugar meter shall be calibrated with a standard quartz tube that has been verified by a statutory metrological institution.

(1) Temperature correction of optical rotation of quartz tube. The temperature at which the quartz tube is read using a sugar meter (without a quartz compensator) shall be determined to an accuracy of 0.2 °C. The temperature of the environment and the sugar solution at which the optical rotation is determined shall be as close as possible to 20 °C and shall be within the range of 15–25 °C. If this temperature is greater than ± 0.2 °C in relation to 20 °C, correct the optical rotation of the standard quartz tube for temperature using the following formula:

$$\alpha_t = \alpha_{20} [1+1.44 \times 10^4 (t-20)]$$

Where α_t—optical rotation value of standard quartz tube at t (°C), in unit of international pol (°Z);

α_{20}—optical rotation value of standard quartz tube at 20 (°C), in unit of international pol (°Z);

T—the temperature of the quartz tube at the time of reading, in degrees Celsius (°C).

(2) Conversion factor of quartz tube reading (°Z) at different wavelengths. The pol reading of the quartz tube is based on the green mercury light (wavelength 546 nm) at different wavelengths. When other light sources are used, the conversion shall be made by dividing by the corresponding coefficient in the Table 3-1.

Table 3-1 Conversion coefficient of quartz tube readings at different wavelengths

Light source	Wavelength/nm	Conversion factor
Filtered incandescent light	587	1.001 809
Yellow sodium light	589	1.001 898
Helium/Neon laser	633	1.003 172

4. Preparation of solution

Weigh 26.000 g of sample (white granulated sugar) into a dry and clean small beaker, and add 40~50 mL of distilled water to dissolve it completely. Transfer to a 100 mL volumetric flask, wash the beaker and glass rod with a small amount of distilled water for at least 3 times, transfer the washing water to the volumetric flask, shake up the solution in the flask, and add distilled water to the vicinity of the volumetric flask marking. Allow to equilibrate to room temperature for at least 10 min, then add distilled water to about 1 mm below the mark of the volumetric flask. When there are bubbles, they can be eliminated with ether or ethanol. Add distilled water to the line and shake well.

If the solution is found to be turbid, filter it with filter paper and cover the funnel with a watch glass to reduce the evaporation of the solution. Discard the first 10 mL of filtrate and collect 50~60 mL of the remaining filtrate.

5. Determination of optical rotation

Flush the optical rotation polarimeter tube with the solution to be tested at least twice and then fill the polarimeter tube, and note that there is no bubble in the polarimeter tube. Place the optical rotation polarimeter tube in the sugar detector, and visually check the sugar detector for 5 times, and the reading shall be accurate to 0.05 °Z; if the automatic sugar detector is used, there shall be enough time for the instrument to be stable before the measurement.

Immediately after determining the optical rotation reading, determine the temperature of the solution in the polarimeter tube and record to 0.1 °C.

6. Calculation and result presentation

The temperature of the environment and the sugar solution when determining the optical rotation shall be as close as possible to 20 °C and shall be within the range of 15~25 °C. If this temperature correlates with 20 °C by more than ± 0.2 °C, a temperature correction should be made.

According to different polarimeters, different temperature correction formulas can be used, and the calculation results are expressed in %, with one decimal place.

Sugar meter using quartz wedge compensator:
$$P=P_t [1+0.000\ 32\ (t-20)]$$

Sugar meter without quartz wedge compensator:
$$P=P_t [1+0.000\ 19\ (t-20)]$$

Where P—sucrose content (%);

P_t—observed optical rotation reading in unit of international Brix (°Z);

t—the temperature of the sugar solution at the time of observation, in degrees Celsius (°C).

7. Allowable error

The difference between the two measured values shall not exceed 0.05% of the average value.

Task 3.2 Determination of reducing sugar in white granulated sugar

Enterprise case

A customer responded that a batch of white granulated sugar was damp in a short period of time. The director estimated that the reducing sugar (%) of the product exceeded the standard. He arranged Xiaoming to determine the reducing sugar (%) of the batch of white granulated sugar. What instrument and method should Xiaoming choose to determine it?

Mission objectives

Through the study of this task, students can achieve the following goals:

(1) Understand the determination principle of Offner method.

(2) Grasp the determination method and calculation of Offner method.

Quality objectives

Develop the habit of comparative analysis of learning.

Procedures and methods

(1) Ingredients: White granulated sugar.

(2) Equipment and instruments: 250 mL iodine flask, 50 mL burette, 25 mL pipette, 300 mL conical flask, electric furnace.

(3) Reagents: Offner reagent (copper solution), 0.032 3 mol/L $Na_2S_2O_3$ solution, 0.016 15 mol/L I_2 solution, 1 mol/L HCl solution, 10 g/L starch solution, glacial acetic acid.

(4) Steps: Weigh 10.00 g of white granulated sugar, dissolve it in a 300 mL conical flask with 50 mL of distilled water. The sugar solution contains no more than 20 mg of invert sugar. Then add 50 mL of Offner reagent, mix well, cover it with a small beaker upside down, heat it on an electric stove, make it boil within 4~5 min, and continue to boil accurately for 5 min (the time when boiling beginsnot from the occurrence of bubbles at the bottom of the bottle, but from the occurrence of a large number of bubbles at the surface of the liquid). Remove and cool to room temperature in a cold bath (do not shake). Take it out, add 1 mL of glacial acetic acid, add accurately measured iodine solution under constant shaking, which adds 5~30 mL depending on the amount of reduced copper, and the amount of which is subject to ensure that the iodine solution is excessive. Add 15 mL of 1 mol/L hydrochloric acid along the wall of the conical flask with a measuring cup. Immediately cover the small beaker, place it for 2 min, shake the solution from time to time, and then titrate the excessive iodine with $Na_2S_2O_3$ solution. When the solution is light yellow-green, add 2~3 milliliters of starch indicator (dark blue at this time), and continue to titrate until the blue color has just faded. Record the volume consumed for titration.

(5) Data processing and calculation.

Project	Data and results
Concentration of sodium thiosulfate solution for titration/(mol · L^{-1})	
Volume of sodium thiosulfate solution for titration/mL	
Volume of 0.032 3 mol/L sodium thiosulfate solution for titration/mL	
Concentration of iodine solution added/(mol · L^{-1})	
Volume of iodine solution added/mL	
Volume of 0.016 15 mol/L iodine solution added/mL	
Sample reducing sugar/%	

Reflection

What are the similarities and differences in the determination of reducing sugars (%) in intermediate and final products? Please use the comparative analysis method to answer. What philosophical principles does the method of contrastive analysis embody?

3.2.1 Task-related knowledge — determination of reducing sugar in white granulated sugar

The reducing pol of white granulated sugar was determined by Offner method. Offner method is suitable for the determination of reducing sugar in the presence of a large amount of sucrose. In the current national and industrial standards, the determination of reducing sugar in all kinds of finished sugar is designated by Offner method.

1. Basic Principles

The Offner method is also a copper reduction method, but its method for quantifying reducing sugar is different from the Lane-Eynon method. The method of Offner firstly uses the copper ion in the excessive Offner reagent to react with the reducing sugar in the sample to produce cuprous oxide, and then uses the iodometric method to quantify the amount of cuprous oxide (Cu_2O) produced, so as to determine the content of reducing sugar in the sample. Because of its weak alkalinity, the Offner reagent used in this method can be mixed with copper salt and potassium sodium tartrate to form a solution, which can still be stored for a long time, and potassium sodium tartrate will not reduce divalent copper. In addition, due to the weak alkalinity of the reagent, the reaction process between reducing sugar and copper salt is slow, and the amount of sucrose oxidized is greatly reduced, so the determination results are more accurate. At present, it is used for the determination of reducing sugar in white granulated sugar in sugar factories.

Firstly, a certain amount of sample was heated with excessive offner reagent, and the reducing sugar in the sample reacted with copper ions to form Cu_2O precipitate, and the amount of Cu_2O precipitate was proportional to the amount of reducing sugar in the sample.

$$\text{Reducing sugar (quantitative)} + Cu^{2+} \longrightarrow Cu_2O \text{ (precipitation)}$$

After the reaction was completed, acetic acid was added to make the excessive Cu^{++} ions form a copper acetate complex and lose the oxidation ability, thus preventing the sucrose from being oxidized.

The Cu_2O precipitate was dissolved with hydrochloric acid:

$$Cu_2O + 2HCl \longrightarrow Cu_2Cl_2 + H_2O$$

Then accurately adding excessive iodine standard solution to oxidize Cu_2Cl_2. Because the added iodine has been accurately measured, the amount of iodine reacting with cuprous ions can be known by trying to find the amount of excess iodine.

$$Cu_2Cl_2 + 2KI + I_2 \longrightarrow 2CuI_2 + 2KCl$$

Finally, the excess iodine in the reaction was titrated with sodium thiosulfate standard solution, and starch solution was used as an indicator.

$$I_2 \text{ (excess)} + 2Na_2S_2O_3 \longrightarrow Na_2S_4O_6 + 2NaI$$

The amount of Cu_2O formed with the reducing sugar can be determined from the amount of iodine consumed, and the amount of the reducing sugar can be determined.

2. Offner reagent

The Offner reagent, also known as a copper solution, provides the copper ions needed for the reaction. The Offner reagent consists of copper sulfate, potassium sodium tartrate, anhydrous sodium carbonate and disodium hydrogen phosphate (or anhydrous sodium chloride). The alkalinity of Offner reagent is weak, so it is not necessary to prepare it as A solution and B solution stored separately. Because there is no need to measure Cu^{2+} ions in the determination process, the Offner reagent can be used directly after preparation without calibration. However, it is required to heat and sterilize in a boiling water bath for 2 h after preparation and store in a brown bottle.

3. Main instruments and equipment

(1) Iodine flask (250 mL).

(2) Burette (50 mL), one for acid type and another one for basic type.

(3) 25 mL pipette.

(4) Electric furnace.

4. Main reagents

(1) Offner reagent (copper solution). Weighing 5 g of crystalline copper sulfate ($CuSO_4 \cdot 5H_2O$), 300 g of potassium sodium tartrate ($NaKC_4H_4O_6 \cdot 4H_2O$), 10 g of sodium carbonate (Na_2CO_3) and 50 g of disodium hydrogen phosphate (Na_2HPO_4), dissolving with 900 mL of cold water, heating and sterilizing in a boiling water bath for 2 h, cooling and diluting to 1 000 mL, adding a small amount of activated carbon or refine diatomite, filtering, and storing that filtrate in a colored

reagent bottle.

(2) 0.032 3 mol/L sodium thiosulfate solution. Accurately weigh 8.000 0 g of sodium thiosulfate ($Na_2S_2O_3$), dissolve it in water, transfer it into a 1 000 mL volumetric flask, and calibrate it with potassium dichromate standard solution after constant volume.

Preparation and calibration method of potassium dichromate standard solution. Accurately weigh about 1.58 g (accurate to 0.000 2 g) of standard potassium dichromate dried to constant weight at 120 °C, dissolve it in about 100 mL of water, transfer it to a 1 000 mL volumetric flask for constant volume, shake it up, accurately suck 25 mL of the solution with a pipette, and inject it into a 250 mL iodine flask. Add 2 g of potassium iodide and 15 mL of sulfuric acid with a concentration of 2 mol/L, cover the bottle tightly, shake it gently, and then place it in a dark place to react for 5 min. Add 100 mL of water, use sodium thiosulfate to be calibrated to drip to light yellow, add 2 mL of 5 g/L starch indicator (turn into blue-black), and continue to use sodium thiosulfate to drip until the solution turns into bright green. This calibration shall be carried out for several times until the relative error of two times is within 0.2%. At the same time, a blank test was conducted.

The concentration of sodium thiosulfate standard solution is calculated according to the following formula:

$$C = \frac{25m}{49.03(V-V_1)}$$

Where C—concentration of sodium thiosulfate (mol · L^{-1});
  M—reference weight of potassium dichromate (g);
  V—sodium thiosulfate consumption during calibration (mL);
  V_1—volume of sodium thiosulfate consumed in blank test (mL)
  25—conversion factor
  49.03—molar mass of potassium dichromate (1/6 mol) (g/mol)

(3) 0.016 15 mol/L iodine solution. Accurately weigh 4.100 0 g of chemically pure iodine and 20 g of potassium iodide, and dissolve in a small amount of water. Then transfer it into a 2 000 mL volumetric flask, add water to the marked line, shake it up, calibrate it with standard sodium thiosulfate solution and store it in a brown bottle.

Because iodine is almost insoluble in water, but can be dissolved in potassium iodide solution, a certain amount of potassium iodide must be added when preparing. In addition, because iodine is easy to volatilize, it is difficult to weigh accurately. Generally, it is prepared as a stock solution with high concentration, and then diluted and calibrated before use.

(4) 10 g/L starch solution.

(5) Glacial acetic acid.

(6) 1 mol/L hydrochloric acid.

5. Measurement procedure

Weigh a certain amount of samples (refined white granulated sugar, white granulated sugar,

sugar cube 10.00 g, white crystal sugar 5.00 g, yellow crystal sugar 1.00 g) and dissolve them in a 300 mL conical flask with 50 mL of distilled water. Sugar solution should not contain more than 20 mg of invert sugar, otherwise the amount of sample should be reduced appropriately. Then add 50 mL of Offner reagent, mix well, cover it with a beaker upside down, heat it on the electric stove, make it boil in 4~5 min, and continue to boil for 5 min accurately (the time of boiling is not from the time when bubbles occur at the bottom of the bottle, but from the time when a large number of bubbles appear on the liquid surface). Remove and cool to room temperature in cold water (do not shake to prevent oxidation of Cu^+ ions). Add 1 mL of glacial acetic acid, add accurately measured iodine solution under constant shaking, which adds 5~30 mL depending on the amount of reduced copper, and the amount of which shall be subject to ensure that the iodine solution is excessive (the yellow-brown solution is excessive). 15 mL of hydrochloric acid with a concentration of 1 mol/L was added along that wall of the conical flask with a measuring cup. Cap the flask and place it in a dark place to react for 2 min, shaking the solution occasionally. Excess iodine was then titrated with $Na_2S_2O_3$ solution. When the solution is light yellow-green, add 2~3 mL of starch indicator (dark blue at this time) and continue to titrate until the blue color has just faded. Record the milliliters of $Na_2S_2O_3$ solution consumed for titration.

6. Calculation method

(1) Required data.

1) Amount of the added iodine solution A_1 (mL);
2) Concentration of iodine solution M_a (mol · L^{-1});
3) Amount of sodium thiosulfate solution consumed during titration B_1 (mL);
4) Concentration of sodium thiosulfate M_b (mol · L^{-1}).

(2) Calculation method. In the absence of sucrose, each milliliter of 0.032 3 mol/L standard iodine solution reacted with Cu_2O is equivalent to 1 mg of reducing sugar in the sample. If there is sucrose, since the sucrose will also consume part of the iodine solution, the consumption of the standard iodine solution shall be calculated by the following formula (applicable to 10 g of sucrose) or a correction value shall be obtained by looking up the attached list 4, and the correction value shall be subtracted from the result.

$$\text{Correction value of 10 g sucrose } C = 1.025 + 0.069\,5\,X - 0.001\,24X^2$$

Where X—consumption of standard iodine solution (mL).

For example, for the determination of reducing sugar in white granulated sugar, the sample weighed is 10.00 g, and the determination results are as follows:

The amount of iodine solution added is A_1 (mL), the concentration of iodine solution is M_a (mol · L^{-1});

The amount of sodium thiosulfate solution consumed during titration is B_1 (mL), and the concentration of sodium thiosulfate is M_b (mol · L^{-1}).

When calculating the content of reducing sugar in the sample, first convert the iodine solution and sodium thiosulfate solution into milliliters corresponding to the standard

concentration:
$$A = A_1 M_a / 0.016\ 15\ (\text{mL})$$
$$B = B_1 M_b / 0.032\ 3\ (\text{mL})$$

Amount of iodine solution consumed: $(A-B)$ mL

Substitute $X = (A-B)$ into the calculation formula of sucrose correction value (or refer to the attached list 4) to obtain the correction value C of 10 g of sucrose.

The reducing sugar of the sample is:

$$\text{Reducing sugar (\%)} = \frac{\text{Weight of reducing sugar of the sample}}{\text{Weight of sample}} \times 100$$

$$= \frac{0.001 \times (A-B-C)}{10} \times 100\% = 0.01(A-B-C)\%$$

For sucrose weights other than 10 g, the sucrose correction value may be calculated as follows:

$$\text{The sucrose correction value} = 0.214\ 7\sqrt{X} + 0.113\ 7Y - 0.299\ 8$$

Where X—iodine solution consumption, $(X = A-B)$ (mL);

Y—weight of sucrose (g).

[Example 3-1] Weigh 10 g of white granulated sugar, determine the reducing sugar by using the Offner method, add 25.00 mL of 0.016 3 mol/L iodine solution, titrate 17.25 mL of 0.037 2 mol/L sodium thiosulfate solution, and calculate the reducing pol of white granulated sugar.

[Solution] First, convert the solution to the volume corresponding to the standard concentration.

$$A = 25 \times 0.016\ 3/0.016\ 15 = 25.23\ (\text{mL})$$
$$B = 17.25 \times 0.037\ 2/0.032\ 3 = 19.87\ (\text{mL})$$
$$X = A - B = 25.23 - 19.87 = 5.36\ (\text{mL})$$

Calculate the correction value C for 10 g of sucrose according to the formula:

Correction value of 10 g sucrose $C = 1.025 + 0.069\ 5X - 0.001\ 24X^2$

$$C = 1.025 + 0.069\ 5 \times 5.36 - 0.001\ 24 \times 5.36^2 = 1.36$$

Reducing sugar of white granulated sugar $= 0.01 \times (25.23 - 19.87 - 1.36) = 0.040\ (\%)$

7. Precautions

(1) Be sure to cool the solution to room temperature before adding the iodine solution into the conical flask, cover it after adding the iodine solution, and place it in a dark place to react for 2 min. And prevent iodine from volatilizing at a higher temperature and decomposing under strong light to cause errors.

(2) The amount of hydrochloric acid added should be able to completely dissolve the Cu_2O precipitate, and slightly excessive to make the solution slightly acidic. Under alkaline conditions, copper sulfate in the Offner reagent can react with potassium iodide in the iodine solution to

precipitate iodine. In addition, the reaction of iodine with sodium thiosulfate is also required to be carried out under neutral or slightly acidic conditions, otherwise side reactions will occur, which will complicate the process and make it impossible to calculate. However, the solution should not be too acidic, because sodium thiosulfate will decompose in strong acidic solution. Therefore, the amount of acid added should be well controlled.

(3) Because of the spiral structure of starch, iodine molecules may enter the structure and make it difficult for these molecules to participate in the reaction, so the starch indicator should not be added too early, so as not to produce a blue color that is not easy to disappear.

(4) Sodium thiosulfate can only be redone after excessive titration, and iodine solution can not be used for back titration, because sodium thiosulfate will decompose in strong acidic solution.

(5) The concentration of sodium thiosulfate solution is unstable, and it is easily decomposed by bacteria, CO_2 and O_2 in air and distilled water. Light irradiation can accelerate the decomposition, so it should be calibrated at least once every 30 days.

Task 3.3 Determination of loss on drying of white granulated sugar

Enterprise case

The director looked at a certain batch of samples that had just been collected and felt a little damp. He thought the product was highly watered. He arranged Xiaoming to measure the loss on drying of the above samples. What instrument should Xiaoming choose and what method should he use to measure it?

Mission objectives

Through the study of this task, students can achieve the following goals:

(1) Understand the principle of determining the moisture content of white granulated sugar by atmospheric drying method.

(2) Grasp the method and calculation of determining the moisture content of white granulated sugar by atmospheric drying method.

Quality objectives

Develop good habits of patience and meticulousness.

Task description

Determination of moisture content in white granulated sugar by atmospheric drying method

Procedures and methods

(1) Ingredients: White granulated sugar

(2) Equipment and instruments: Constant temperature drying oven, glass dryer with thermometer, glass weighing bottle (diameter 6–10 cm, depth 2–3 cm), analytical balance

(3) Reagent: Distilled water or deionized water (the conductivity of the water used to prepare all solutions should be less than 15 μs/cm).

(4) Steps: The oven was preheated to 130 °C. Put the dry and clean empty weighing bottle with the cover opened and its cover into the drying oven, dry for 30 min, and then cover the weighing bottle, take it out of the drying oven, and put it into the dryer to cool to room temperature. Weigh the weighing bottle and take 9.5~10.5 g of sample (accurate to ± 0.1 mg), spread the sample in the weighing bottle, then put the weighing bottle containing the sample and its cap into the drying oven preheated to 130 °C, dry accurately for 18 min, cover the weighing bottle, take it out of the drying oven, and put it into the dryer to cool to room temperature. Weighing shall be accurate to ± 0.1 mg.

5. Data processing and calculation.

Project	Data and results
Weight of weighing bottle or container/g	
Weight of sample in weighing bottle or container/g	
Weight of the sample added to the weighing bottle or container after drying/g	
Loss on drying/%	

Reflection

Why should the atmospheric drying method be used to determine the moisture content of white granulated sugar? Is there a better way?

Task-related knowledge — determination of loss on drying of white granulated sugar

The constant temperature drying oven is used to heat and evaporate at a higher temperature to lose water and volatile substances to achieve the purpose of drying. Because the evaporation loss is not entirely water, the water measured by the drying method is also called "loss on drying". The loss on drying of white granulated sugar is one of the main indexes of product quality control.

1. Main instruments and equipment

(1) Drying oven: During the measurement, the temperature at (2.5 ± 0.5) cm above the weighing bottle shall be kept at (130 ± 1) °C.

(2) Dryer.

(3) Flat weighing bottle: Diameter 6~10 cm, depth 2~3 cm.

(4) Analytical balance: Sensitivity 0.1g.

2. Assay steps

The determination method is divided into two methods: the arbitration method, and the conventional method.

(1) Preheat the drying oven to 105 °C (the arbitration method) or 130 °C (the conventional method).

(2) Put the uncapped dry and clean empty weighing bottle and its cover into the drying oven, dry for 30 min, then cover the weighing bottle, take it out of the drying oven, and put it into the dryer to cool to room temperature.

(3) The mass W of the weighing bottle.

(4) Weigh the sample of 20.000 0~30.000 0 g (the arbitration method) or 9.500 0~10.500 0 g (the conventional method) with a weighing bottle as soon as possible (the accuracy shall be $\pm$ 0.1 mg), and the total mass of the sample and the weighing bottle is W_1.

(5) Flatten the sample in the weighing bottle, and then put the weighing bottle containing the sample and its cap into a drying oven preheated to 105 °C (the arbitration method) or 130 °C (the conventional method), and dry it accurately for 3 h (the arbitration method) or 18 min (the conventional method).

It is not necessary to dry to constant weight. However, it must be ensured that there is no visible loss of the sample at any stage of the determination, and the vessel must be held by a dry and clean crucible clamp.

(6) Cover the weighing bottle, take it out of the drying oven, and put it into the dryer to cool to room temperature.

(7) Weigh the mass W_2 of the weighing bottle (with the lid closed) and the sample, which has been dried and cooled to room temperature, to an accuracy of $\pm$ 0.1 mg.

Note: In order to speed up the determination in sugar factories, microwave drying is also used for white granulated sugar. After being weighed, the sample is directly heated in a microwave oven for 1~2 minutes and weighed after being cooled. However, this is not the method stipulated by the national standard, so it can only be used for internal index control.

There is a complete set of imported equipment for determining the moisture content of white granulated sugar. When the sample is placed on the tray of the equipment and pushed into the instrument, the weighing and drying process can be completed automatically. The equipment uses infrared drying technology, the whole determination process takes no more than 2 min, and it can automatically display and print the results.

3. Calculation

The loss on drying is expressed as a percentage, and the result is calculated to two decimal places. The difference between the two measured values shall not exceed 15% of the average value.

$$\text{Loss on drying (\%)} = 100\% (W_1 - W_2) / (W_1 - W)$$

Where W—weight of weighing bottle or container (g);

W_1—weight of sample adding weighing bottle or container (g);

W_2—weight of weighing bottle or container after adding sample and drying (g).

Task 3.4 Determination of conductivity ash of white granulated sugar

Enterprise case

After inspecting the production line, the factory director found that the calcium salt content of the intermediate products was very high through the production report. He judged that there might be some equipment scaling and high conductivity ash content of white granulated sugar. If you are the director of the laboratory, what methods and instruments should you use to determine the conductivity ash?

Mission objectives

Through the study of this task, students can achieve the following goals:

(1) Understand the principle of determining the ash content of white granulated sugar by conductivity method.

(2) To master the method and calculation of determining the ash content of white granulated sugar by conductivity method.

Quality objectives

Develop the habit of looking at things from a connected philosophical point of view.

Task description

The conductivity ash of white granulated sugar was determined by conductivity method, and the results were calculated by appropriate method.

Procedures and methods

(1) Ingredients: White granulated sugar.

(2) Equipment and instruments: Conductivity meter (DDS-11C or DDS-11A), Abbe refractometer, analytical balance, 100 mL volumetric flask, beaker.

(3) Reagents: Distilled water or deionized water (the conductivity of the water used to prepare all solutions shall be less than 15 μS/cm), and 0.01 mol/L potassium chloride solution [take analytically pure potassium chloride, heat it to 500 °C (dark red and hot), dehydrate for 30 min, weigh 745.5 mg, dissolve it in a 1 000 mL volumetric flask, and add water to the marked line].

(4) Steps: Weigh (31.3 ± 0.1) g of white granulated sugar sample and put it into a dry and clean small beaker, add about 40~50 mL of distilled water, and stir it with a thin glass rod to dissolve it completely. Then carefully pour the sugar solution into a 100 mL volumetric flask, wash the beaker and glass rod several times with a small amount of distilled water, merge the washing water into the volumetric flask, and add water to the marked line (this sample solution is 28 °Bx). Shake well and pour into a dry and clean small beaker specially used for conductivity

measurement (the beaker shall be rinsed with sample solution for 2~3 times before pouring) for measurement. Fill another small clean beaker with distilled water to dissolve the sample. Measure the conductivity of the sample solution and distilled water respectively with a conductivity meter, and record the reading and the temperature at the time of reading. The temperature range for measurement should preferably not exceed (20 ± 5) ℃.

(5) Data processing and calculation.

Project	Data and results
Conductivity of sample solution /($\mu S \cdot cm^{-1}$)	
Temperature of sample solution/℃	
Conductivity of distilled water /($\mu S \cdot cm^{-1}$)	
Temperature of distilled water/℃	
Conductivity ash/%	

Reflection

What substance is determined by the conductivity ash method? Please learn to adopt to the philosophy of connection to study the quality control of the whole plant.

Task-related knowledge — determination of conductivity ash in white granulated sugar

The ash in the sample consists of soluble and insoluble ash, which constitutes the insoluble matter of the weight of the ash and can be reflected in the determination of the insoluble matter of the sample. The conductivity of the solution measured by the conductivity method can reflect the soluble substances in the sample (such as salts, free acids and plasma substances), and can better reflect the quality of the sample. The conductivity method is simple, pollution-free and reproducible.

1. Principle of measurement

Conductance is the ability of a substance to carry an electric current. In liquids, the reciprocal of resistance, conductance, is often used to measure its ability to conduct electricity. The international unit of conductance is "Siemens" (S). In actual use, μS ($1S = 10^6 \mu S$) is generally used. Because pure water is almost non-conductive, when there is an ionic electrolyte in the water, the solution has a certain conductivity. When the concentration and type of electrolyte in the solution are different, the conductance of the solution is also different, that is, the conductance can reflect the degree of electrolyte in the water. The amount of electrolyte in the solution can be analyzed by measuring the conductivity of the solution. This is the basic principle of the conductivity meter.

The conductivity of the solution is proportional to the area of the electrode (A) and inversely proportional to the distance of the electrode (L). The conductance measured when the electrode area and distance are both one unit is called conductivity. A commonly used unit of conductivity is microsieverts/centimeter ($\mu S/cm$). It represents the conductivity of the solution when the area of the electrode is 1 cm^2 and the distance is 1 cm.

The pure sucrose solution is almost non-conductive. The solution conducts electricity when the sample contains ionic non-sugars, which are the main components of the ash in the sample. Therefore, the conductivity of the sample can be used to directly reflect the amount of ionic non-sugar in the sample, which is called conductivity ash.

2. Main equipment

(1) Conductivity meter. The instrument used to measure conductance in the laboratory is a conductometer. Commonly used models are DDS-11A, DDS-11C, etc.

Measuring range: $0 \sim 10^5$ $\mu S/cm$.

Measurement error: not more than 0.5% of full scale.

The DDS-11A conductivity meter and the matched electrode are as shown in the Fig. 3-1.

Fig. 3-1　The DDS-11A conductivity meter

(2) Reagent.

1) Distilled water or deionized water. Refined white granulated sugar must be treated with double distilled water (distilled twice) or deionized water with a conductivity of less than 2 $\mu S/cm$. For ordinary white granulated sugar, distilled water with a conductivity of less than 15 $\mu S/cm$ is allowed.

2) 0.01 mol/L potassium chloride solution. Take analytically pure potassium chloride, heat it to 500 °C, dehydrate it for 30 min, weigh 0.745 5 g after cooling, dissolve it and transfer it into a 1 000 mL volumetric flask, and add water to the marked line.

3) 0.002 5 mol/L potassium chloride solution. Pipette 50 mL of 0.01 mol/L potassium chloride solution into a 200 mL volumetric flask, and dilute with water to the marked line. The conductivity of this solution is 328 $\mu S/cm$ at 20 °C.

3. Measurement steps

(1) Preparation of sample solution. To determine the ash content of the sample by the

conductivity method is actually to determine the conductivity of the sample solution. The conductivity of the sample solution is affected by the concentration. The sample solution has the maximum conductivity in a certain concentration range. In the vicinity of this concentration range, the change of concentration has little effect on the conductivity. In order to reduce the influence of concentration change on the determination results, the concentration of the sample solution is generally adjusted to the concentration corresponding to the maximum conductivity when the ash content is determined by the conductivity method. In the vicinity of this concentration range, the concentration has little effect on the conductivity, so a coarse balance can be used for weighing, which is convenient for operation. According to the experimental data, the concentration should be 5 g/100 mL for the intermediate products of the sugar mill, and 28 g/100 g (i.e. 28 °Bx) for the white granulated sugar.

1) Preparation of 5 g/100 mL sample solution. It is suitable for the determination of ash content of products in process, brown granulated sugar, brown sugar, raw sugar and final molasses in sugar mills.

Firstly, the brix of the sample is determined by Abbe refractometer, and the brix of various sugarcane juices can be directly determined. Samples with higher concentration and containing undissolved sucrose crystals (such as massecuite, molasses, etc.) are diluted with a known amount of water (usually 1 : 1 by weight), and then the refractive index is determined. Calculate the weight of the weighed sample according to the following formula, add water to dissolve it, transfer it into a 100 mL volumetric flask, add water to the scale, shake it up and wait for testing.

$$\text{Sample weight (g)} = 5 \times 100/\text{sample brix}$$

When the original concentration of the sample is lower than 5 g/100 mL, the ash content can be determined directly after the refractive index of the sample is determined.

2) Preparation of 31.3 g/100 mL sample solution. It is suitable for the conductivity ash determination of white granulated sugar, sugar cube, crystal sugar, etc.

Weigh (31.3 ± 0.1) g of sample and put it into a dry and clean small beaker, add water to dissolve it, transfer it into a 100 mL volumetric flask, add water to the scale, and shake it up for determination (the concentration of the sample solution is 28 °Bx).

(2) Determination. Select the electrode constant, temperature compensation and measuring range of the conductivity meter as required, use the prepared sample solution to be measured, wash the conductivity electrode and a clean small beaker for conductivity measurement twice, then pour in the sample solution, and measure its conductivity with the conductivity meter. Pour the distilled water used for dilution into a small beaker and stir it with a glass rod. The stirring time should be close to the time of dissolving the sample (because the stirring process will cause carbon dioxide in the air to dissolve into the distilled water, which will increase the conductivity), and then determine the conductivity of the distilled water used for diluting the sample.

(3) Calculation.

1) Data recording.

① Temperature of the sample (T_1);

② Distilled water temperature T_2 (do not record when the conductivity meter is equipped with temperature compensation);

③ Sample conductivity C_{1t} ($\mu S/cm$);

④ Conductivity of distilled water C_{2t} ($\mu S/cm$).

2) Calculation.

① Temperature correction of sample solution conductivity

$$C_1 = C_{1t} [1+0.026(T-20)]$$

Note: $C_1 = C_{1t}$ when the conductivity meter is provided with temperature compensation.

② Conductivity and temperature correction of distilled water:

$$C_2 = C_{2t} [1+0.022(T-20)]$$

Note: $C_2 = C_{2t}$ when the conductivity meter is provided with temperature compensation.

③ Calculation of conductivity ash content of white granulated sugar (31.3 g/100 mL):

$$\text{Conductivity ash content (\%)} = 6 \times 10^{-4} (C_1 - 0.35 C_2)$$

Note: When determining C_2, the unit volume is completely filled with water, and the determination result indicates the content of ionic substances in the unit volume. However, when determining the sample solution, the proportion of water in the unit volume decreases due to the fact that sugar occupies part of the volume, so the ionic substances brought by water are also relatively reduced. When sucrose is dissolved in water, the viscosity of the solution increases. In order to compensate for these two changes, the conductivity of distilled water in the calculation formula should be multiplied by a correction factor of 0.35.

④ Conductance ash calculation of other products (5 g/100 mL):

Conductivity ash content (%) = $18 \times 10^{-4} (C_1 - 0.9 C_2)$

[Example 3-2] The observation refractive index of the mixed juice is 15.40 °Bx. According to the formula, the weight of the sample amount required for preparing the solution with a concentration of 5 g/100 mL: $5 \times 100/15.4 = 32.47$(g). After the constant volume is 100 mL, use the conductivity meter with temperature correction to measure the conductivity of the sample solution: 277 $\mu S/cm$. The conductivity of distilled water was 0.85 $\mu S/cm$, and the conductivity ash content of the mixed juice was calculated.

[Solution] Since the instrument has been corrected for temperature, the calculation can be performed directly.

Conductivity ash (% by weight of sample) = $18 \times 10^{-4} (277 - 0.9 \times 0.85) = 0.50$

4. Precautions

(1) The newly purchased electrode can be used only after the electrode constant is calibrated, and the electrode constant should be calibrated regularly.

(2) In order to ensure the measurement accuracy, the electrode shall be rinsed twice with distilled water (or deionized water) less than 0.5 µS/cm before use, and then rinsed three times with the test sample before use.

(3) The electrode socket shall not be affected with moisture to avoid unnecessary errors.

(4) Distilled water should be measured immediately after pouring out, otherwise it will absorb carbon dioxide in the air and increase conductivity.

(5) When measuring the conductivity of distilled water, it should be consistent with the operation for measuring the conductivity of dissolving sugar, and it should also be stirred, moved into the volumetric flask, shaken and so on, and the time of each step should be basically the same. Otherwise, a large error will occur. Because the dissolution of white granulated sugar needs to be stirred for a long time, the conductivity of distilled water will increase a lot after a longer time of stirring.

5. Use of conductivity meter

(1) Use of the instrument. In order to ensure the accuracy of measurement and the safety of instruments, the following points must be used:

1) Before powering on, check whether the meter hand points to zero. If it does not point to zero, adjust the adjusting screw of the meter head to make the meter hand point to zero.

2) When the plug of the power cord is inserted into the power hole of the instrument (on the back of the instrument), turn on the power switch, and the light will be on. It can work after preheating.

3) Pull the range selector to the required measuring range. If you do not know the size of the object to be measured, adjust it to the maximum range position first to avoid bending the meter needle due to overload, and then change it to the required range step by step.

4) Insert the electrode plug into the socket, align the groove of the plug with the groove of the socket, and then press the top of the plug with your index finger to insert it. When pulling out, hold the lower part of the plug and pull it up.

5) When it is measured as low conductivity (less than 5 µS/cm), there are fewer ions in the solution. In order to reduce the decrease of conductivity caused by the adsorption of ions on the electrode, a bright electrode with a smaller surface area should be selected. When the conductance of the measured liquid is above 5 µS/cm, the current passing through is larger, the electrolysis is obvious, and the polarization of the electrode is obvious, so the platinum black electrode with larger surface area should be selected to reduce the current density and avoid polarization.

6) Adjusting the electrode constant knob to the position of the electrode constant.

7) Adjusting the temperature compensation knob to the temperature of the solution to be measured, and if temperature compensation is not needed, adjusting the temperature compensation knob to the position of 20 °C.

8) Turn the calibration measurement shift switch to "calibration" and adjust the calibration regulator to make the pointer indicate the full scale.

9) Turn the switch to "measure", and multiply the reading in the indicating ammeter by the multiplying factor on the range selector to obtain the conductivity of the solution to be measured. When reading, if the range switch points to the red point, read according to the red scale; if it points to the black point, read according to the black scale.

10) When the conductivity of the sample solution to be tested is greater than 10 000 μS/cm, the DJS-10 platinum black electrode can be selected, and the electrode constant knob is adjusted to 1/10 of the electrode constant. For example, if the constant of the electrode is 9.8, the electrode constant knob should be adjusted to 0.98, and the measured reading should be multiplied by 10, which is the conductivity of the measured solution.

11) During the measurement, it is necessary to frequently check whether the "calibration" is changed, that is, whether the pointer still indicates the full scale when the switch is turned to "calibration".

(2) Calibration of electrode constant. The electrode constant is an important data, which directly affects the results of the determination, so the electrode constant of the newly purchased electrode should be calibrated before use. The method is as follows:

1) Drying the potassium chlorate above the analytical pure grade at 200 °C for 2 h, and then dehydrating at 500 °C for 30 min.

2) Take 715.5 mg of dehydrated potassium chlorate, dissolve it in a 1 000 mL volumetric flask with distilled water whose conductivity is lower than 2 μS/cm, and add water to the marked line. The concentration of this solution is 0.01 mol/L.

3) Absorb 250 mL of 0.01 mol/L potassium chlorate solution, diluting to 1 000 mL, where in that concentration of the solution is 0.002 5 mol/L, measure the temperature of the solution, and calibrate the electrode with the standard solution.

4) Adjust the electrode constant knob of the conductivity meter to 1.0 (if there is a temperature compensation knob, adjust the temperature compensation knob to the temperature corresponding to the solution).

5) Measuring the conductivity S of the standard potassium chlorate solution by using the electrode to be calibrated.

6) Checking a standard potassium chlorate solution conductivity meter to obtain the conductivity K of the potassium chlorate.

7) Calculate the constant Q of the electrode using the following formula:

$$Q = K/S$$

Where Q—electrode constant;
S—the conductivity of potassium chlorate solution measured by the electrode to be calibrated (μS/cm);
K—the conductivity of potassium chlorate from Table 3-2.

Table 3-2　Conductivity of standard potassium chlorate solution (μS/cm)

Temperature/°C	Consistence/(0.1 mol·L^{-1})	Consistence/(0.01 mol·L^{-1})	Consistence/(0.002 5 mol·L^{-1})
15	10 480	1 147	194
16	10 720	1 173	300
17	10 950	1 198	307
18	11 190	1 225	314
19	11 430	1 251	321
20	11 670	1 278	328
21	11 910	1 305	335
22	12 150	1 332	342
23	12 390	1 359	348
24	12 640	1 386	355
25	12 880	13 413	362

Task 3.5 Determination of colour and turbidity of white granulated sugar

Enterprise case

The color of the recent white granulated sugar samples is relatively high, and some customers are not satisfied. The director asked Xiaoming to test the color valve and turbidity of the corresponding products. Please help Xiaoming choose the appropriate instrument and method.

Mission objectives

Through the study of this task, students can achieve the following goals:

(1) Understand the determination principle of international sugar colour and turbidity of white granulated sugar and Semi-product.

(2) Grasp the determination method and calculation of colour and turbidity.

Quality objectives

Develop a meticulous work habit

Task description

The colour and turbidity of a white granulated sugar sample were determined.

Procedures and methods

(1) Ingredients: White granulated sugar.

(2) Equipment and instruments: 721 spectrophotometer (723OG spectrophotometer), cuvette, Abbe refractometer, pH meter, membrane filter, microporous membrane (pore size 0.45 μm), suction filter bottle, vacuum pump

(3) Reagents: 0.5 mol/L HCl, 0.5 mol/L NaOH, 0.1 mol/L NaOH, triethanolamine-hydrochloric acid buffer solution.

1) Triethanolamine-hydrochloric acid buffer solution preparation: Weigh 14.920 g of triethanolamine [$(HOCH_2CH_2)_3N$], dissolve it with distilled water and fix the constant volume in a 1 000 mL volumetric flask, then transfer it into a 2 000 mL beaker, add about 800 mL of 0.1 mol/L hydrochloric acid solution. Stir well and continue to adjust to pH 7.0 with 0.1 mol/L hydrochloric acid (measure the pH value by immersing the electrode of acidity juice in this solution) and store in a brown glass bottle.

2) 0.1 mol/L hydrochloric acid solution preparation: Pipette 9.0 mL of concentrated hydrochloric acid (specific gravity of 1.19) into a 1 000 mL volumetric flask containing an appropriate amount of distilled water, and dilute to volume.

(4) Steps: Weighing 100.0 g of white granulated sugar sample in a 200 mL beaker, adding 135 mL of triethanolamine-hydrochloric acid buffer solution, stirring until the solution is completely dissolved, measuring the refractive index and temperature of the solution, leaving a

part of the solution (used for measuring the absorbance of the sugar solution before filtration), pouring the rest into a microporous membrane filter with a pore size of 0.45 μm, and filtering under vacuum. The first 50 mL of filtrate was discarded. The collected filtrate shall not be less than 50 mL, and the refractometer and thermometer shall be used to measure the refractive brix and temperature of the filtrate respectively. The solution before and after the filtration was contained in cuvette respertively, and the absorbance was measured on a spectrophotometer at 420 nm wavelength. The filtered triethanolamine-hydrochloric acid buffer solution shall be used as the reference standard for the zero-point colour.

(5) Data processing and calculation.

Name	Project	Data and results
White granulated sugar	Brix of observed refraction of sample solution before filtration/°Bx	
	Temperature of sample solution before filtration/°C	
	Refractive Brix of sample solution before filtration/°Bx	
	Brix of observed refraction of filtered sample solution/°Bx	
	Temperature of filtered sample solution/°C	
	Refractive Brix of filtered sample solution/°Bx	
	Absorbance of sample solution before filtration	
	Absorbance of filtered sample solution	
	Colour/IU	
	Turbidity/MAU	

Reflection

How many voids of the membrane should be used to filter the sample to determine the colour value?

Task-related knowledge — determination of colour and turbidity of white granulated sugar

Colour and turbidity are important indicators to measure the effect of sugar production process, especially the removal rate of colored substances, colloids and suspended particles, and are also one of the indicators that have the greatest impact on the grade of products in sugar mills at present. In sugar mills, the colour and turbidity of products are usually determined by spectrophotometry.

1. Basic principle of spectrophotometry

When parallel monochromatic light passes through a uniform and non-scattering solution, part of the light is absorbed, and the intensity of part of the light passing through the solution will be reduced due to the absorption of energy. The degree of light intensity reduction has a certain proportional relationship with the concentration of solute. By using a spectrophotometer, the degree of light intensity reduction can be measured, so as to determine the content of a certain component. This method is called spectrophotometry. The determination error of spectrophotometry is usually 1%~5%, and its accuracy is lower than that of volumetric analysis and gravimetric analysis in the determination of components with higher content, but for the determination of trace components, such accuracy can meet the requirements.

The change of light intensity can be expressed by transmittance T or absorbance A (also called extinction E). If the intensity of the incident light is I_0 and the intensity of the transmitted light is I, the transmittance T is defined as:

$$T = \frac{I}{I_0}$$

On the scale of the instrument, it is usually expressed by percentage $T(\%)$, that is:

$$T(\%) = \frac{I}{I_0} \times 100\%$$

The concentration C of the solution is proportional to the negative logarithm A of the light transmittance, that is:

$$A = -\lg T = -\lg \frac{I}{I_0} = kcl$$

Where A—the absorbance;

  k—absorption coefficient, which is a characteristic constant of a substance at a certain wavelength;

  c—concentration of solution;

  l—the thickness of the liquid layer through which the light passes.

Transmittance $T(\%)$ and absorbance A can be used to express the measurement results of general spectrophotometers. On the pointer instrument, the scale of light transmittance is equally divided, while the scale of absorbance is not evenly spaced.

The solutions of sugar mill's products in process and finished white granulated sugar are not non-scattering solutions, but contain suspended particles that cause light scattering. When these solutions are determined, the determination results are the sum of light attenuation caused by absorption and scattering. In order to determine the simple colour, the solution must be effectively filtered to eliminate the influence of suspended particles in the solution on the determination.

2. Determination of colour and turbidity of white granulated sugar

(1) Main equipment and reagents.

1) Spectrophotometer.

① Measuring range: Transmittance 0–100%.

② Wavelength range: 360~800 nm.

③ Wavelength error: The wavelength error is not ± 1 nm at 420 nm.

2) Abbe refractometer. Mass fraction of sucrose Brix (°Bx): 0~95, minimum division value: 0.2.

3) Microporous membrane filter. The filter membrane shall be uniform in thickness, with symmetrical, uniform and highly penetrating micropores distributed on the membrane surface, with the pore size of 0.45 μm and the porosity of 80%. The pore channels are linear and do not interfere with each other. The filter membrane shall be used together with a sugar filter with a diameter of 150 mm.

4) Triethanolamine-hydrochloric acid buffer solution. Weigh 14.92g of triethanolamine [(HOCH$_2$CH$_2$)$_3$N], dissolve it with distilled water and fix the constant volume in a 1 000 mL volumetric flask, then move it into a 2 000 mL beaker, add about 800 mL of 0.1 mol/L hydrochloric acid, put it on a stirrer, and put the acidometer electrode. Adjust to pH 7.00 ± 0.02 with 0.1 mol/L hydrochloric acid under constant stirring. Store in a brown glass bottle.

(2) Measurement steps.

1) Preparation of sample solution. Weigh 100.0 g of white granulated sugar sample into a 200 mL beaker, add 135 mL of triethanolamine-hydrochloric acid buffer solution, and stir until it is completely dissolved.

2) Determination of absorbance before filtration. Take part of the prepared sugar solution to measure the refractive index of the filtrate with a refractometer, put it in a cuvette, measure its absorbance with a spectrophotometer at 420 nm wavelength, and use unfiltered triethanolamine-hydrochloric acid buffer solution as the reference of zero point.

3) Determination of absorbance after filtration. Pour the rest of the sample solution into a microporous membrane filter with a pore size of 0.45 μm, filter under vacuum, discard the initial 50 mL of filtrate, collect the filtrate which should not be less than 50 mL, measure the refractive brix of the filtrate with a refractometer, put the filtered sugar solution into a cuvette, and measure its absorbance with a spectrophotometer at 420 nm wavelength. Filtered triethanolamine-hydrochloric acid buffer solution was used as the reference standard for the zero point.

(3) Calculation method.

1) Data logging.

① Observe the Brix B_1 of the sample solution before filtration.
② Observe the Brix B_2 of the filtered sample solution.
③ Cuvette thickness.
④ Absorbance A_1 of the sample solution before filtration.
⑤ Absorbance of filtered sample solution A_2.

2) Calculation method.

① Colour.

$$C = \frac{A_2}{bc} \times 1\,000$$

Where C—colour, in International Sugar Colour Unit (IU);

A_2—absorbance of filtered sample solution;

b—cuvette thickness (cm);

c—concentration of sample solution (g/ml) (multiply the brix of refraction corrected to 20 °C by a correction factor of 0.986 2, and then obtain by looking up the Table 3-3).

② Turbidity. According to the national standard, turbidity also uses the international colour unit MAU (attenuation unit). The calculation result is taken to the unit digit.

$$\text{MAU} = \left(\frac{A_1}{b_1 c_1} - \frac{A_2}{b_2 c_2}\right) \times 1\,000$$

Where A_1—absorbance of sample solution before filtration;

b_1—thickness of cuvette when measuring A_1 (cm);

c_1—sample solution concentration during determination of A_1 (g/mL);

A_2—absorbance of filtered sample solution;

b_2—thickness of cuvette when measuring A_2 (cm);

c_2—sample solution concentration during determination of A_2 (g/mL).

$$c = \text{Brix} \times \text{corresponding apparent density}/100$$

Note: When triethanolamine is used as pH buffer solution, because the buffer solution will bring in dry substance, the measured refractive brix should be multiplied by a correction factor of 0.986 2 before being used for calculation and table lookup to correct the influence of triethanolamine on brix.

For white granulated sugar, the concentration c (g/mL) of the sample solution can be directly checked for the brix of sucrose solution and the grams of sucrose per milliliter (in the air) in Table 3-3.

It can also be calculated by the following formula:

$$\text{Sample concentration } c \text{ (g/ml)} = 0.014\,1B - 0.094\,16$$

Where B—Brix of the calibrated sample solution (°Bx).

Table 3-3 Comparison between brix of sucrose solution and sucrose in grams per milliliter (in Air)

Refractive Brix /°Bx	g sucrose/mL (in the air)	Refractive Brix /°Bx	g sucrose/mL (in the air)	Refractive Brix /°Bx	g sucrose/mL (in the air)
40.0	0.470 2	41.7	0.493 8	43.4	0.517 8
40.1	0.471 5	41.8	0.495 2	43.5	0.519 2
40.2	0.472 9	41.9	0.496 6	43.6	0.520 6

continued

Refractive Brix (Bx)	g sucrose/mL (In the air)	Refractive Brix (Bx)	g sucrose/mL (In the air)	Refractive Brix (Bx)	g sucrose/mL (In the air)
40.3	0.474 3	42.0	0.498 0	43.7	0.522 1
40.4	0.475 7	42.1	0.499 4	43.8	0.523 5
40.5	0.477 1	42.2	0.500 8	43.9	0.524 9
40.6	0.478 5	42.3	0.502 2	44.0	0.526 3
40.7	0.479 9	42.4	0.503 6	44.1	0.527 8
40.8	0.481 2	42.5	0.505 1	44.2	0.529 2
40.9	0.482 6	42.6	0.506 5	44.3	0.530 6
41.0	0.484 0	42.7	0.507 9	44.4	0.532 1
41.1	0.485 4	42.8	0.509 3	44.5	0.533 5
41.2	0.486 6	42.9	0.510 7	44.6	0.534 9
41.3	0.488 2	43.0	0.512 1	44.7	0.536 4
41.4	0.489 6	43.1	0.513 5	44.8	0.537 8
41.5	0.491 0	43.2	0.515 0	44.9	0.539 2
41.6	0.492 4	43.3	0.516 4		

[Example 3-3] 100.0 g of white granulated sugar was weighed and measured according to the specified method of colour, and the measured data are as follows:

(1) Brix of sample solution before filtration B_1: 43.48 °Bx

(2) Brix of filtered sample solution B_2: 43.29 °Bx

(3) Cuvette thickness: 3 cm.

(4) Absorbance of sample solution before filtration A_1: 0.225.

(5) Absorbance of filtered sample solution A_2: 0.184.

Find the color and turbidity of the sample.

[Solution]

The concentration correction sample solution before filtration: $0.986\ 2 \times 43.48 = 42.88$ (g/mL)

Refer to Table 3-3 for the concentration of sample solution before filtration: 0.5098 (g/mL)

Filtered sample solution concentration correction: $0.986\ 2 \times 43.29 = 42.69$ (g/mL)

Refer to Table 3-3 for the concentration of the filtered sample solution: 0.507 8 (g/mL).

The colour of sample solution before filtration:

$$IU_1 = \frac{0.225}{3 \times 0.509\ 8} \times 1\ 000 = 147$$

The colour of the filtered sample solution is:

$$IU_2 = \frac{0.184}{3 \times 0.5078} \times 1000 = 121$$

Turbidity 147 − 121 = 26 (MAU)

The colour of the white granulated sugar sample is 121 IU, and the turbidity is 26 MAU.

(4) Precautions.

1) The microporous membrane shall not be folded and shall be stored flatly. When using it, the front (smooth and bright) shall be upward and laid flat on the filter plate. Before using, apply neutral distilled water and soak at room temperature for about two hours to fully wet the film body. Since the membrane body contains water, the initial part of the filtrate should be discarded to reduce the influence on the concentration.

2) Because the pigment in the sugar solution is very sensitive to pH, the colour will increase by 25% when the pH increases by 0.1 unit, so the pH should be strictly controlled within the specified range.

3) When filtering the sugar solution, the vacuum degree should be controlled at 50−55 kPa. If the vacuum degree is too high, it is easy to damage the filter plate.

4) When measuring the absorbance of sugar solution, the cuvette should be placed close to one end of the photosensitive element to reduce the influence of light scattering on the measurement.

5) The colour of the same sample will decrease with the increase of wavelength, and the wavelength is affected by the accuracy of the instrument, so the wavelength of the instrument should be corrected frequently. In addition, the spectral width (color purity) of the instrument also affects the accuracy of the results. Usually, the spectral width of low-grade instruments is wider, and the measurement results will be smaller. Attention should be paid to this problem when upgrading the instrument.

Task 3.6 Determination of sulfur dioxide content in white granulated sugar

Enterprise case

Peer enterprises in the same area have seen a rapid decline in product quality recently. The main reason is that the sulfur dioxide content in white granulated sugar is too high. If you are a laboratory analyst, what method should you use to detect the sulfur dioxide content?

Mission objectives

Through the study of this task, students can achieve the following goals:

(1) Understand the determination principle of sulfur dioxide in white granulated sugar.

(2) Master the determination method and calculation of sulfur dioxide in white granulated sugar.

Quality objectives

Raise awareness of food safety.

Task description

Determination of sulfur dioxide content in a batch of samples.

Procedures and methods

(1) Ingredients: White granulated sugar.

(2) Equipment and instruments: Spectrophotometer, 6~12 pieces of 25 mL colorimetric tubes with stopper

(3) Reagents: Sodium tetrachloromercurate absorption solution, chromogenic agent (pararosaniline hydrochloride solution), 2 g/L ammonium sulfamate solution, 0.2% formaldehyde solution, 10 g/L starch indicator, 0.05 mol/L iodine solution, 0.1 mol/L sodium thiosulfate standard solution, 0.25 mol/L sulfuric acid solution, 0.5 mol/L sodium hydroxide solution, 0.5 sulfur dioxide standard solution (each milliliter contains about 1.5 mg of sulfur dioxide, and the accurate figure must be obtained by calibration), sulfur dioxide standard application solution (close to 5 µg per milliliter, not necessarily an integer, but know the exact number, preferably accuracy to 5 mg).

(4) Experimental steps:

1) Drawing of standard curve. Accurately suck 2.0 mL, 4.0 mL, 6.0 mL, 8.0 mL, 10.0 mL and 12.0 mL of sulfur dioxide standard application solution with a micropipette, and put them into a 25 mL colorimetric tube with a stopper respectively, and then add 18.0 mL, 16.0 mL, 14.0 mL, 12.0 mL, 10.0 mL and 8.0 mL of sodium tetrachloromercurate absorption solution in turn (the total volume is 20.0 mL, corresponding to 5 mg, 10 mg, 15 mg, 20 mg, 25 mg, 30 mg of sulfur dioxide per 10 mL), add 2 mL of 12 g/L ammonium sulfamate solution, 2 mL of 0.2%

formaldehyde solution and 2mL of chromogenic agent into each tube, cover tightly and shake well, place at room temperature of 15~25 °C (or control the temperature with aqueous solution) for 20 min, that is, measure the absorbance with a spectrophotometer at 550 nm wavelength and a 1cm cuvette filled with the solution, and adjust to zero with distilled water. The measured result is the same as that of standard application solution containing micrograms of sulfur dioxide per 10 mL. Draw the standard curve or obtain the regression equation by regression method.

2) Sample preparation and determination. Accurately weigh 10.0 g of white granulated sugar sample (increase or decrease depending on the content of sulfur dioxide in the sample), dissolve it in about 25 mL of distilled water, transfer it into a 50 mL volumetric flask, accurately add 2.0 mL of 0.5 mol/L sodium hydroxide solution, cover the volumetric flask tightly, shake it up, and accurately add 2.0 mL of 0.25 mol/L sulfuric acid solution 5 min later. Cover tightly and shake well, immediately add 10 mL of sodium tetrachloromercurate absorption solution, and add water to the marked line. Shake well and set aside.

Accurately suck 10 mL of the prepared sample solution, put it into a colorimetric tube with a stopper, add 10 mL of sodium tetrachloromercurate absorption solution, add 2mL of 12 g/L ammonium sulfamate solution, add 2 mL of 0.2% formaldehyde solution and 2 mL of chromogenic agent, cover the stopper and shake it up, place it for 20 minutes at the temperature of 15~25 °C, and immediately use 550 nm wavelength on the spectrophotometer. The sample solution was filled in a 1 cm cuvette, and the absorbance was determined with the sugar solution as the blank. Check the corresponding sulfur dioxide content from the standard curve (or substitute it into the regression equation to calculate the sulfur dioxide content) and then calculate it.

3) Data processing, calculation and curve drawing.

Project		Number					
		1	2	3	4	5	6
Plotting of standard curves	Sulfur dioxide standard application solution milliliters	2.0	4.0	6.0	8.0	10.0	12.0
	Volume of sodium tetrachloromercurate absorption solution	18.0	16.0	14.0	12.0	10.0	8.0
	Absorbance						
Determination	Absorbance						
	The equivalent micrograms of SO_2 are obtained from the curve (or calculated)						
	The number of grams of sample contained in the sample solution drawn during the determination						
	Sulfur dioxide (mg/kg)						

Reflection

What is the effect of high sulfur dioxide content on human body? In order to protect people's health, we should pay attention to food quality and safety, and produce white granulated sugar that meets health standards. What should we do at this time?

3.6.1 Task-related knowledge — determination of sulfur dioxide content in white granulated sugar

The content of SO_2 is an important quality index of white granulated sugar. The main methods to determine SO_2 content in white granulated sugar in sugar mills are pararosaniline hydrochloride colorimetry and iodometry.

1. Pararosaniline hydrochloride colorimetric method

(1) Basic principles. Sulfur dioxide is absorbed by sodium tetrachloromercurate to generate a stable dichlorosulfite, and then reacts with formaldehyde and pararosaniline hydrochloride to generate purplish red, which can be quantified by color comparison with the standard. The reaction process is as follows:

$$(HgCl_4)^{2-} + SO_2 + H_2O \longrightarrow (HgCl_2SO_3)^{2-} + 2Cl^- + 2H^+$$

$$(HgCl_2SO_3)^{2-} + HCHO + 2H^+ \longrightarrow HOCH_2SO_3H + HgCl_2$$

$$(H_2NC_6H_4)_2C \cdot C_6H_4 \cdot NH \cdot HCl + 3HCl \longrightarrow$$

(Colored)

$$(ClH_3N \cdot C_6H_4)_2C \cdot Cl \cdot C_6H_4 \cdot NH_2 \cdot HCl + 2HOCH_2SO_3H \longrightarrow$$

(Colorless)

$$(HSO_3 \cdot CH_2 \cdot HN \cdot C_6H4)_2C \cdot C_6H_4 \cdot NH \cdot HCl + 3HCl + 2H_2O$$

(Purplish red)

After the addition of basic fuchsin and hydrochloric acid, it becomes a non-quinone type of colorless compound. After the action with dichlorosulfite and formaldehyde, it is regenerated into a quinone type of coloured compound, which is purple-red.

Because white granulated sugar contains some inorganic or organic salts, the solubility of calcium sulfite is lower. The addition products of sulfur dioxide and glucose must be reacted with sodium hydroxide to release sulfur dioxide, so that the total content of sulfur dioxide in white granulated sugar can be measured.

(2) Calculation method.

$$\text{Sulfur dioxide (mg/kg)} = A/W$$

Where A—equivalent micrograms of SO_2 obtained from the curve (or calculated);

W—grams of the sample contained in the sample solution drawn during the determination.

2. Iodometry

(1) Measurement method and procedure.

1) Add 150 mL of distilled water, 5 mL of 5 g/L starch solution and 10 mL of 3 mol/L hydrochloric acid solution respectively into a 250 mL conical flask, and shake well.

2) Titrate with 0.002 5 mol/L iodine standard solution until light blue appears to compensate for the iodine solution consumed by distilled water and reagents.

3) Weigh 50.0 g of white granulated sugar sample into the above conical flask, and shake until it is dissolved.

4) Titrate with 0.002 5 mol/L iodine standard solution until light blue appears, and record the volume V.

(2) Calculation method.

$$\text{Sulfur dioxide content (mg/kg)} = V \times N \times 64.06 \times 1\,000/m$$

In the formula V—iodine solution consumption (mL);

N—concentration of standard iodine solution (0.002 5 mol/L);

m—weigh the sample (g).

Extended learning—Determination of finished sugar by near infrared spectroscopy

The method for near infrared spectroscopy determination of the sucrose (%) of finished sugar is referred to 2.11 for near infrared spectroscopy determination of the sucrose (%) of semi-product.

1. Main equipment

This paper takes the SupNIR-4000 long-wave near-infrared analyzer of Focus Technology as an example, including near-infrared host, diffuse reflection probe, thickness detector, optical fiber accessories, and CM-2000 chemometrics software. The performance parameters adopted in the near-infrared host are as follows: the wave band range is 1 000 ~ 1 800 nm, the sampling interval is 1 m, the spectrometer resolution is less than 7 nm, the wavelength accuracy is less than ± 0.2 nm, the stray light is lower than 0.1% (1 692 nm), and the full spectrum scanning time is lower than 0.2 s. The CM-2000 chemometrics software was used for modeling, which followed the ASTM specifications for quantitative and qualitative analysis methods.

2. Main Reagents

Reagents required for the analysis of each intermediate product. See the main reagents required for the analysis method of each sample in this section for details.

3. Measurement steps

(1) Modeling.

1) Spectral preprocessing.

Firstly, the whole band spectrum was selected, and the model was established by partial least squares (PLS), and the advantages and disadvantages of 415 pretreatment methods were investigated. According to the results and experience, the combined processing of "standardization + Savitzky-Golay first derivation + orthogonal signal correction (OSC)" was selected as the preprocessing method to eliminate the influence of temperature fluctuation, wavelength shift and baseline shift, and to better extract particle information. The near infrared spectrum after treatment is shown in Fig. 3-2.

2) Band selection. Select full band modeling.

3) Model. PLS algorithm is used as a data decomposition method to extract the principal components from the complex spectral data as the input variables of the neural network, so as to reduce the input variables and improve the prediction accuracy of the network. Based on the principle of minimum RMSEC, the optimal value of each parameter was selected, and 58 models were established with different parameters for optimization and comparison. The optimal parameters are as follows: the number of nodes in the input layer is 4, the number of nodes in the hidden layer is 5, the conversion function tansig in the hidden layer, the conversion function

logsig in the output layer, the initial learning rate is 0.9, the momentum term is 0.9, and the maximum number of iterations is 200. The RMSEC of the best ANN model is 2.971 8.

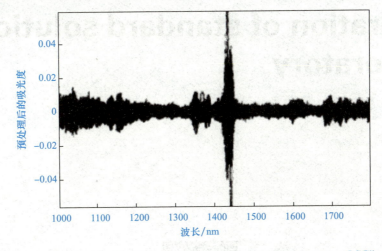

Fig. 3-2　Atlas after "Standardization + Savitzky-Golay First Derivation + OSC" processing

4) On-line measurement by ANN model.

The particle size model of white granulated sugar established by ANN algorithm was embedded into the near-infrared host, and the sample was analyzed and determined, which ran continuously for 168 h. The samples were sampled and analyzed every 1 h, and the deviation between the measured value by the inter national standard method and the predicted value by the near infrared spectroscopy was compared, and the predicted error of the particle size was required to be controlled within ± 5.

Determination of water-insoluble impurities

Determination of reducing sugar in brown granulated sugar

Module 4
Preparation of standard solution in laboratory

Module 4 Preparation of standard solution in laboratory

附录一　甘蔗制糖工业分析常用数据表

附表 1　观测锤度温度改正表（0～40 ℃）

标准温度 20 ℃

温度/℃	观测锤度													
	0	1	2	3	4	5	6	7	8	9	10	11	12	13
	---- 温度低于 20 ℃时读数应减之数 ----													
0	0.30	0.34	0.36	0.41	0.45	0.49	0.52	0.55	0.59	0.62	0.65	0.67	0.70	0.72
5	.36	.38	.40	.43	.45	.47	.49	.51	.52	.54	.56	.58	.60	.61
10	.32	.33	.34	.36	.37	.38	.39	.40	.41	.42	.43	.44	.45	.46
1/2	.31	.32	.33	.34	.35	.36	.37	.38	.39	.40	.41	.42	.43	.44
11	.31	.32	.33	.33	.34	.35	.36	.37	.38	.39	.40	.41	.42	.42
1/2	.30	.31	.31	.32	.32	.33	.34	.35	.36	.37	.38	.39	.40	.40
12	.29	.30	.30	.31	.31	.32	.33	.34	.34	.35	.36	.37	.38	.38
1/2	.27	.28	.28	.29	.29	.30	.31	.32	.32	.33	.34	.35	.35	.36
13	.26	.27	.27	.28	.28	.29	.30	.30	.31	.31	.32	.33	.33	.34
1/2	.25	.25	.25	.25	.26	.27	.28	.28	.29	.29	.30	.31	.31	.32
14	.24	.24	.24	.24	.25	.26	.27	.27	.28	.28	.29	.29	.30	.30
1/2	.22	.22	.22	.22	.23	.24	.24	.25	.25	.26	.26	.26	.27	.27
15	.20	.20	.20	.20	.22	.22	.22	.23	.23	.24	.24	.24	.25	.25
1/2	.18	.18	.18	.18	.19	.20	.20	.21	.21	.22	.22	.22	.23	.23
16	.17	.17	.17	.18	.18	.18	.18	.19	.19	.20	.20	.20	.21	.21
1/2	.15	.15	.15	.16	.16	.16	.16	.16	.17	.17	.17	.17	.18	.18
17	.13	.13	.13	.14	.14	.14	.14	.14	.15	.15	.15	.15	.16	.16
1/2	.11	.11	.11	.12	.12	.12	.12	.12	.12	.12	.12	.12	.12	.13
18	.09	.09	.09	.10	.10	.10	.10	.10	.10	.10	.10	.10	.10	.11
1/2	.07	.07	.07	.07	.07	.07	.07	.07	.07	.07	.07	.07	.07	.08
19	.05	.05	.05	.05	.05	.05	.05	.05	.05	.05	.05	.05	.05	.06
1/2	.03	.03	.03	.03	.03	.03	.03	.03	.03	.03	.03	.03	.03	.03
20	0	0	0	0	0	0	0	0	0	0	0	0	0	0

续表

温度/℃	观测锤度												
	14	15	16	17	18	19	20	21	22	23	24	25	30
	---- 温度低于 20 ℃时读数应减之数 ----												
0	0.75	0.77	0.79	0.82	0.84	0.87	0.89	0.91	0.93	0.95	0.97	0.99	1.08
5	.63	.65	.67	.68	.70	.71	.73	.74	.75	.76	.77	.80	.86
10	.47	.48	.49	.50	.50	.51	.52	.53	.54	.55	.56	.57	.60
1/2	.45	.46	.47	.48	.48	.49	.50	.51	.52	.52	.53	.54	.57
11	.43	.44	.45	.46	.46	.47	.48	.49	.49	.50	.50	.51	.55
1/2	.41	.42	.43	.43	.44	.44	.45	.46	.46	.47	.47	.48	.52
12	.39	.40	.41	.41	.42	.42	.43	.44	.44	.45	.45	.46	.50
1/2	.36	.37	.38	.38	.39	.39	.40	.41	.41	.42	.42	.43	.47
13	.34	.35	.36	.36	.37	.37	.38	.39	.39	.40	.40	.41	.44
1/2	.32	.33	.34	.34	.35	.35	.36	.36	.37	.37	.38	.38	.41
14	.31	.31	.32	.32	.33	.33	.34	.34	.35	.35	.36	.36	.38
1/2	.28	.28	.29	.29	.30	.30	.31	.31	.32	.32	.33	.33	.35
15	.26	.26	.26	.27	.27	.28	.28	.28	.29	.29	.30	.30	.32
1/2	.24	.24	.24	.24	.25	.25	.25	.25	.26	.26	.27	.27	.29
16	.22	.22	.22	.22	.23	.23	.23	.23	.24	.24	.25	.25	.26
1/2	.19	.19	.19	.19	.20	.20	.20	.20	.21	.21	.22	.22	.23
17	.16	.16	.16	.16	.17	.17	.18	.18	.19	.19	.19	.19	.20
1/2	.13	.13	.13	.13	.14	.14	.15	.15	.15	.16	.16	.16	.16
18	.11	.11	.11	.11	.12	.12	.12	.12	.12	.13	.13	.13	.13
1/2	.08	.08	.08	.08	.09	.09	.09	.09	.09	.09	.09	.09	.10
19	.06	.06	.06	.06	.06	.06	.06	.06	.06	.06	.06	.06	.07
1/2	.03	.03	.03	.03	.03	.03	.03	.03	.03	.03	.03	.03	.04
20	0	0	0	0	0	0	0	0	0	0	0	0	0

续表

温度/℃	观测锤度													
	0	1	2	3	4	5	6	7	8	9	10	11	12	13
	++++ 温度高于 20 ℃时读数应加之数 ++++													
20	0	0	0	0	0	0	0	0	0	0	0	0	0	0
1/2	0.02	0.02	0.02	0.03	0.03	0.03	0.03	0.03	0.03	0.03	0.03	0.03	0.03	0.03
21	.04	.04	.04	.05	.05	.05	.05	.05	.06	.06	.06	.06	.06	.06
1/2	.07	.07	.07	.08	.08	.08	.08	.08	.09	.09	.09	.09	.09	.09
22	.10	.10	.10	.10	.10	.10	.10	.10	.11	.11	.11	.11	.11	.12
1/2	.13	.13	.13	.13	.13	.13	.13	.13	.14	.14	.14	.14	.14	.15
23	.16	.16	.16	.16	.16	.16	.16	.16	.17	.17	.17	.17	.17	.17
1/2	.19	.19	.19	.19	.19	.19	.19	.19	.20	.20	.20	.20	.20	.21
24	.21	.21	.21	.22	.22	.22	.22	.22	.23	.23	.23	.23	.23	.24
1/2	.24	.24	.24	.25	.25	.25	.26	.26	.26	.27	.27	.27	.27	.28
25	.27	.27	.27	.28	.28	.28	.28	.29	.29	.30	.30	.30	.30	.31
1/2	.30	.30	.30	.31	.31	.31	.31	.32	.32	.33	.33	.33	.33	.34
26	.33	.33	.33	.34	.34	.34	.34	.35	.35	.36	.36	.36	.36	.37
1/2	.37	.37	.37	.38	.38	.38	.38	.38	.39	.39	.39	.39	.40	.40
27	.40	.40	.40	.41	.41	.41	.41	.41	.42	.42	.42	.42	.43	.43
1/2	.43	.43	.43	.44	.44	.44	.44	.45	.45	.46	.46	.46	.47	.47
28	.46	.46	.46	.47	.47	.47	.47	.48	.48	.49	.49	.49	.50	.50
1/2	.50	.50	.50	.51	.51	.51	.51	.52	.52	.53	.53	.53	.54	.54
29	.54	.54	.54	.55	.55	.55	.55	.55	.56	.56	.56	.56	.57	.58
1/2	.58	.58	.58	.59	.59	.59	.59	.59	.60	.60	.60	.60	.61	.62
30	.61	.61	.61	.62	.62	.62	.62	.62	.63	.63	.63	.63	.64	.65
1/2	.65	.65	.65	.66	.66	.66	.66	.66	.67	.67	.67	.67	.66	.69
31	.69	.69	.69	.70	.70	.70	.70	.70	.71	.71	.71	.71	.72	.73
1/2	.73	.73	.73	.74	.74	.74	.74	.74	.75	.75	.75	.75	.76	.77
32	.76	.76	.77	.77	.77	.77	.77	.77	.79	.79	.79	.79	.80	.81
1/2	.80	.80	.81	.81	.82	.82	.82	.83	.83	.83	.83	.83	.84	.85
33	.84	.84	.85	.85	.85	.85	.85	.86	.86	.86	.86	.86	.88	.88
1/2	.88	.88	.88	.89	.89	.89	.89	.89	.90	.90	.90	.90	.92	.92
34	.91	.91	.92	.92	.93	.93	.93	.93	.94	.94	.94	.94	.96	.96
1/2	.95	.95	.96	.96	.97	.97	.97	.97	.98	.98	.98	.98	.99	1.00
35	.99	.99	1.00	1.00	1.01	1.01	1.01	1.01	1.02	1.02	1.02	1.02	1.04	1.05
40	1.42	1.43	1.43	1.44	1.44	1.45	1.45	1.46	1.47	1.47	1.47	1.47	1.49	1.50

续表

温度/℃	观测锤度											
	14	15	16	17	18	19	20	21	22	23	24	25	30
	++++ 温度高于 20 ℃时读数应加之数 ++++											
20	0	0	0	0	0	0	0	0	0	0	0	0	0
1/2	0.03	0.03	0.03	0.03	0.03	0.03	0.03	0.03	0.03	0.03	0.04	0.04	0.04
21	.06	.06	.06	.06	.06	.06	.06	.06	.06	.07	.07	.07	.07
1/2	.09	.09	.09	.09	.09	.09	.09	.09	.09	.10	.10	.10	.11
22	.12	.12	.12	.12	.12	.12	.12	.12	.12	.13	.13	.13	.14
1/2	.15	.15	.15	.15	.16	.16	.16	.16	.16	.17	.17	.17	.18
23	.17	.17	.17	.18	.18	.20	.19	.19	.19	.20	.20	.20	.21
1/2	.21	.21	.21	.22	.22	.23	.23	.23	.23	.24	.24	.24	.25
24	.24	.24	.24	.25	.25	.26	.26	.26	.26	.27	.27	.27	.28
1/2	.28	.28	.28	.28	.29	.29	.29	.29	.30	.30	.31	.31	.32
25	.31	.31	.31	.31	.32	.32	.32	.32	.33	.33	.34	.34	.35
1/2	.34	.34	.34	.35	.35	.36	.36	.36	.36	.37	.37	.37	.39
26	.37	.37	.38	.38	.39	.39	.40	.40	.40	.40	.40	.40	.42
1/2	.41	.41	.41	.42	.42	.43	.43	.43	.43	.44	.44	.44	.46
27	.44	.44	.44	.45	.45	.46	.46	.46	.47	.47	.48	.48	.50
1/2	.48	.48	.48	.49	.49	.50	.50	.50	.51	.51	.52	.52	.54
28	.51	.51	.52	.52	.53	.53	.54	.54	.55	.55	.56	.56	.58
1/2	.55	.55	.56	.56	.57	.57	.58	.58	.59	.59	.60	.60	.62
29	.58	.59	.59	.60	.60	.61	.61	.61	.62	.62	.63	.63	.66
1/2	.62	.63	.63	.64	.64	.65	.65	.65	.66	.66	.67	.67	.70
30	.65	.66	.66	.67	.67	.68	.68	.68	.69	.69	.70	.70	.73
1/2	.69	.70	.70	.71	.71	.72	.72	.73	.73	.74	.74	.75	.78
31	.73	.74	.74	.75	.75	.76	.76	.77	.77	.78	.78	.79	.82
1/2	.77	.78	.79	.79	.80	.80	.81	.81	.82	.82	.83	.83	.86
32	.81	.82	.83	.83	.84	.84	.85	.85	.86	.86	.87	.87	.90
1/2	.85	.86	.87	.87	.88	.88	.89	.90	.90	.91	.91	.92	.95
33	.89	.90	.91	.91	.92	.92	.93	.94	.94	.95	.95	.96	.99
1/2	.93	.94	.95	.95	.96	.97	.98	.98	.99	.99	1.00	1.00	1.03
34	.97	.98	.99	1.00	1.00	1.01	1.02	1.02	1.03	1.03	1.04	1.04	1.07
1/2	1.01	1.02	1.03	1.04	1.04	1.05	1.06	1.07	1.07	1.08	1.08	1.09	1.12
35	1.05	1.06	1.07	1.08	1.08	1.09	1.10	1.11	1.11	1.12	1.12	1.13	1.16
40	1.50	1.51	1.52	1.53	1.53	1.54	1.54	1.55	1.55	1.56	1.56	1.57	1.62

附表 2　观测锤度温度改正表（0～80 ℃）
标准温度 20 ℃

温度/℃	观测锤度													
	0	5	10	15	20	25	30	35	40	45	50	55	60	70
---- 温度低于 20 ℃时读数应减之数 ----														
0	0.30	0.49	0.65	0.77	0.89	0.99	1.08	1.16	1.24	1.31	1.37	1.41	1.44	1.49
5	0.36	0.47	0.56	0.65	0.73	0.80	0.86	0.91	0.97	1.01	1.05	1.08	1.10	1.14
10	.32	.38	.48	.48	.52	.57	.60	.64	.67	.70	.72	.74	.75	.77
11	.31	.35	.44	.44	.48	.51	.55	.58	.60	.63	.65	.66	.68	.70
12	.29	.32	.40	.40	.43	.46	.50	.52	.54	.56	.58	.59	.60	.62
13	.26	.29	.35	.35	.38	.41	.44	.46	.48	.49	.51	.52	.53	.55
14	.24	.26	.31	.31	.34	.36	.38	.40	.41	.42	.44	.45	.46	.47
15	.20	.22	.26	.26	.28	.30	.32	.33	.34	.36	.36	.37	.38	.39
16	.17	.18	.22	.22	.23	.25	.26	.27	.28	.28	.29	.30	.31	.32
17	.13	.14	.16	.16	.18	.19	.20	.20	.21	.21	.22	.23	.23	.24
18	.09	.10	.11	.11	.12	.13	.13	.14	.14	.14	.15	.15	.15	.16
19	.05	.05	.06	.06	.06	.06	.07	.07	.07	.07	.08	.08	.08	.08
++++ 温度高于 20 ℃时读数应加之数 ++++														
21	0.04	0.05	0.06	0.06	0.06	0.07	0.07	0.07	0.07	0.08	0.08	0.08	0.08	0.09
22	.10	.10	.11	.12	.12	.13	.14	.14	.15	.15	.16	.16	.16	.16
23	.16	.16	.17	.17	.19	.20	.21	.21	.22	.23	.24	.24	.24	.24
24	.21	.22	.23	.24	.26	.27	.28	.29	.30	.31	.32	.32	.32	.32
25	.27	.28	.30	.31	.32	.34	.35	.36	.38	.38	.39	.39	.40	.39
26	.33	.34	.36	.37	.40	.40	.42	.44	.46	.47	.47	.48	.48	.48
27	.40	.41	.42	.44	.46	.48	.50	.52	.54	.54	.55	.56	.56	.56
28	.46	.47	.49	.51	.54	.56	.58	.60	.61	.62	.63	.64	.64	.64
29	.54	.55	.56	.59	.61	.63	.66	.68	.70	.70	.71	.72	.72	.72
30	.61	.62	.63	.66	.68	.71	.73	.76	.78	.78	.79	.80	.80	.81
35	.99	1.01	1.06	1.10	1.13	1.16	1.18	1.20	1.21	1.22	1.22	1.23	1.23	1.22
40	1.42	1.45	1.47	1.51	1.54	1.57	1.62	1.62	1.64	1.65	1.65	1.65	1.66	1.65
45	1.91	1.94	1.96	2.00	2.03	2.05	2.07	2.09	2.10	2.10	2.10	2.10	2.10	2.08
50	2.46	2.48	2.50	2.53	2.56	2.57	2.58	2.59	2.59	2.58	2.58	2.57	2.56	2.52
55	3.05	3.07	3.09	3.12	3.12	3.12	3.12	3.11	3.10	3.08	3.07	3.05	3.03	2.97
60	3.69	3.72	3.73	3.73	3.72	3.70	3.67	3.65	3.62	3.60	3.57	3.54	3.50	3.43
65	4.4	4.4	4.4	4.4	4.4	4.4	4.3	4.2	4.2	4.1	4.1	4.0	4.0	3.9
70	5.1	5.1	5.1	5.0	5.0	5.0	4.9	4.8	4.8	4.7	4.7	4.6	4.6	4.4
75	6.1	6.0	6.0	5.9	5.8	5.8	5.7	5.6	5.5	5.4	5.4	5.3	5.2	5.0
80	7.1	7.0	7.0	6.9	6.8	6.7	6.6	6.4	6.3	6.2	6.1	6.0	5.9	5.6

附表3 糖液折光锤度温度改正表（10～30 ℃）
标准温度 20 ℃

温度/℃	锤度														
	0	5	10	15	20	25	30	35	40	45	50	55	60	65	70
————温度低于 20 ℃时读数应减之数————															
10	0.50	0.54	0.58	0.61	0.64	0.66	0.68	0.70	0.72	0.73	0.74	0.75	0.76	0.78	0.79
11	.46	.49	.53	.55	.58	.60	.62	.64	.65	.66	.67	.68	.69	.70	.71
12	.42	.45	.48	.50	.52	.54	.56	.57	.58	.59	.60	.61	.61	.63	.63
13	.37	.40	.42	.44	.46	.48	.49	.50	.51	.52	.53	.54	.54	.55	.55
14	.33	.35	.37	.39	.40	.41	.42	.43	.44	.45	.45	.46	.46	.47	.48
15	.27	.29	.31	.33	.34	.34	.35	.36	.37	.37	.38	.39	.39	.40	.40
16	.22	.24	.25	.26	.27	.28	.28	.29	.30	.30	.30	.31	.31	.32	.32
17	.17	.18	.19	.20	.21	.21	.21	.22	.22	.23	.23	.23	.23	.24	.24
18	.12	.13	.13	.14	.14	.14	.14	.15	.15	.15	.15	.16	.16	.16	.16
19	.06	.06	.06	.07	.07	.07	.07	.08	.08	.08	.08	.08	.08	.08	.08
++++ 温度高于 20 ℃时读数应加之数 ++++															
21	0.06	0.07	0.07	0.07	0.07	0.08	0.08	0.08	0.08	0.08	0.08	0.08	0.08	0.08	0.08
22	.13	.13	.14	.14	.15	.15	.15	.15	.15	.16	.16	.16	.16	.16	.16
23	.19	.20	.21	.22	.22	.23	.23	.23	.23	.24	.24	.24	.24	.24	.24
24	.26	.27	.28	.29	.30	.30	.31	.31	.31	.31	.31	.32	.32	.32	.32
25	.33	.35	.36	.37	.38	.38	.39	.40	.40	.40	.40	.40	.40	.40	.40
26	.40	.42	.43	.44	.45	.46	.47	.48	.48	.48	.48	.48	.48	.48	.48
27	.48	.50	.52	.53	.54	.55	.55	.56	.56	.56	.56	.56	.56	.56	.56
28	.56	.57	.60	.61	.62	.63	.63	.64	.64	.64	.64	.64	.64	.64	.64
29	.64	.66	.68	.69	.71	.72	.72	.73	.73	.73	.73	.73	.73	.73	.73
30	.72	.74	.77	.78	.79	.80	.80	.81	.81	.81	.81	.81	.81	.81	.81

附表4 糖度（或蔗糖分）因数检索表

糖度（或蔗糖分）因数＝26.000/［100×糖液的视密度（20℃）］
例：旋光读数＝52.7，观测锤度＝16.5
糖度＝52.7×0.244 23＝12.87

锤度	糖度（或蔗糖分）因数									
	0.00	0.01	0.02	0.03	0.04	0.05	0.06	0.07	0.08	0.09
0.1	.260 64	.260 83	.260 62	.260 61	.260 60	.260 59	.260 57	.260 56	.260 55	.260 54
.2	.260 53	.260 52	.260 51	.260 50	.260 49	.260 48	.260 47	.260 46	.260 45	.260 44
.3	.260 43	.260 42	.260 41	.260 40	.260 39	.260 38	.260 37	.260 36	.260 35	.260 34
.4	.260 33	.260 32	.260 31	.260 30	.260 29	.260 28	.260 27	.260 26	.260 25	.260 24
.5	.260 23	.260 22	.260 21	.260 20	.260 19	.260 18	.260 17	.260 16	.260 15	.260 14
.6	.260 13	.260 12	.260 11	.260 10	.260 09	.260 08	.260 07	.260 06	.260 05	.260 04
.7	.260 30	.260 02	.260 01	.260 00	.259 99	.259 97	.259 96	.259 95	.259 94	.259 93
.8	.259 92	.259 91	.259 90	.259 89	.259 88	.259 87	.259 86	.259 85	.259 84	.259 83
.9	.259 82	.259 81	.259 80	.259 79	.259 78	.259 77	.259 76	.259 75	.259 74	.259 73
1.0	.259 72	.259 71	.259 70	.259 69	.259 68	.259 67	.259 66	.259 65	.259 64	.259 63
.1	.259 62	.259 61	.259 60	.259 59	.259 58	.259 57	.259 56	.259 55	.259 54	.259 53
.2	.259 52	.259 51	.259 50	.259 49	.259 48	.259 47	.259 46	.259 45	.259 44	.259 43
.3	.259 42	.259 41	.259 40	.259 39	.259 38	.259 37	.259 36	.259 35	.259 34	.259 33
.4	.259 32	.259 31	.259 30	.259 29	.259 28	.259 27	.259 26	.259 25	.259 24	.259 23
.5	.259 22	.259 21	.259 20	.259 19	.259 18	.259 17	.259 16	.259 15	.259 14	.259 13
.6	.259 12	.259 11	.259 10	.259 09	.259 08	.259 07	.259 06	.259 05	.259 04	.259 03
.7	.259 02	.259 01	.259 00	.258 99	.258 98	.258 97	.258 96	.258 95	.258 94	.258 93
.8	.258 92	.258 91	.258 90	.258 89	.258 88	.258 87	.258 86	.258 85	.258 84	.258 83
.9	.258 82	.258 81	.258 80	.258 79	.258 78	.258 77	.258 76	.258 75	.258 74	.258 73
2.0	.258 72	.258 71	.258 70	.258 69	.258 68	.258 67	.258 66	.258 65	.258 64	.258 63
.1	.258 62	.258 61	.258 60	.258 59	.258 58	.258 57	.258 56	.258 55	.258 54	.258 53
.2	.258 52	.258 51	.258 50	.258 49	.258 48	.258 46	.258 45	.258 44	.258 43	.258 42
.3	.258 41	.258 40	.258 39	.258 38	.258 37	.258 36	.258 35	.258 34	.258 33	.258 32
.4	.258 31	.258 30	.258 29	.258 28	.258 27	.258 26	.258 25	.258 24	.258 23	.258 22
.5	.258 21	.258 20	.258 19	.258 18	.258 17	.258 16	.258 15	.258 14	.258 13	.258 12
.6	.258 11	.258 10	.258 09	.258 08	.258 07	.258 06	.258 05	.258 04	.258 03	.258 02
.7	.258 01	.258 00	.257 99	.257 98	.257 97	.257 96	.257 95	.257 94	.257 93	.257 92
.8	.257 91	.257 90	.257 89	.257 88	.257 87	.257 86	.257 85	.257 84	.257 83	.257 82
.9	.257 81	.257 80	.257 79	.257 78	.257 77	.257 76	.257 75	.257 74	.257 73	.257 72

续表

锤度	糖度（或蔗糖分）因数									
	0.00	0.01	0.02	0.03	0.04	0.05	0.06	0.07	0.08	0.09
3.0	.257 71	.257 70	.257 69	.257 68	.257 67	.257 66	.257 65	.257 64	.257 63	.257 62
.1	.257 61	.257 60	.257 59	.257 58	.257 57	.257 56	.257 55	.257 54	.257 53	.257 52
.2	.257 51	.257 50	.257 49	.257 48	.257 47	.257 46	.257 45	.257 44	.257 43	.257 42
.3	.257 41	.257 40	.257 39	.257 38	.257 37	.257 36	.257 35	.257 34	.257 33	.257 32
.4	.257 31	.257 30	.257 29	.257 28	.257 27	.257 26	.257 25	.257 24	.257 23	.257 22
.5	.257 21	.257 20	.257 19	.257 18	.257 17	.257 16	.257 15	.257 14	.257 13	.257 12
.6	.257 11	.257 10	.257 09	.257 08	.257 07	.257 06	.257 05	.257 04	.257 03	.257 02
.7	.257 01	.257 00	.256 99	.256 98	.256 97	.256 96	.256 95	.256 94	.256 93	.256 92
.8	.256 91	.256 90	.256 89	.256 88	.256 87	.256 86	.256 85	.256 84	.256 83	.256 82
.9	.256 81	.256 80	.256 79	.256 78	.256 77	.256 76	.256 75	.256 74	.256 73	.256 72
4.0	.256 71	.256 70	.256 69	.256 68	.256 67	.256 66	.256 65	.256 64	.256 63	.256 62
.1	.256 61	.256 60	.256 59	.256 58	.256 57	.256 56	.256 55	.256 54	.256 53	.256 52
.2	.256 51	.256 50	.256 49	.256 48	.256 47	.256 46	.256 45	.256 44	.256 43	.256 42
.3	.256 41	.256 40	.256 39	.256 38	.256 37	.256 36	.256 35	.256 34	.256 33	.256 32
.4	.256 31	.256 30	.256 29	.256 28	.256 27	.256 26	.256 25	.256 24	.256 23	.256 22
.5	.256 21	.256 20	.256 19	.256 18	.256 17	.256 16	.256 15	.256 14	.256 13	.256 12
.6	.256 10	.256 09	.256 08	.256 07	.256 06	.256 05	.256 04	.256 03	.256 02	.256 01
.7	.256 00	.255 99	.255 98	.255 97	.255 96	.255 98	.255 94	.255 93	.255 92	.255 91
.8	.255 90	.255 89	.255 88	.255 87	.255 86	.255 85	.255 84	.255 83	.255 82	.255 81
.9	.255 80	.255 79	.255 78	.255 77	.255 76	.255 75	.255 74	.255 73	.255 72	.255 71
5.0	.255 70	.255 69	.255 68	.255 67	.255 66	.255 65	.255 64	.255 63	.255 62	.255 61
.1	.255 60	.255 59	.255 58	.255 57	.255 56	.255 55	.255 54	.255 53	.255 52	.255 51
.2	.255 50	.255 49	.255 48	.255 47	.255 46	.255 45	.255 44	.255 43	.255 42	.255 41
.3	.255 40	.255 39	.255 38	.255 37	.255 36	.255 35	.255 34	.255 33	.255 32	.255 31
.4	.255 30	.255 29	.255 28	.255 27	.255 26	.255 25	.255 24	.255 23	.255 22	.255 21
.5	.255 20	.255 19	.255 18	.255 17	.255 16	.255 15	.255 14	.255 13	.255 12	.255 11
.6	.255 10	.255 09	.255 08	.255 07	.255 06	.255 05	.255 04	.255 03	.255 02	.255 01
.7	.255 00	.254 99	.254 98	.254 97	.254 96	.254 95	.254 94	.254 93	.254 92	.254 91
.8	.254 90	.254 89	.254 88	.254 87	.254 86	.254 85	.254 84	.254 83	.254 82	.254 81
.9	.254 80	.254 79	.254 78	.254 77	.254 76	.254 75	.254 74	.254 73	.254 72	.254 71
6.0	.254 70	.254 69	.254 68	.254 67	.254 66	.254 65	.254 64	.254 63	.254 62	.254 61
.1	.254 60	.254 59	.254 58	.254 57	.254 56	.254 55	.254 54	.254 53	.254 52	.254 51
.2	.254 50	.254 49	.254 48	.254 47	.254 46	.254 45	.254 44	.254 43	.254 42	.254 41
.3	.254 40	.254 39	.254 38	.254 37	.254 36	.254 35	.254 34	.254 33	.254 32	.254 31
.4	.254 30	.254 29	.254 28	.254 27	.254 26	.254 25	.254 24	.254 23	.254 22	.254 21
.5	.254 20	.254 19	.254 18	.254 17	.254 16	.254 15	.254 14	.254 13	.254 12	.254 11
.6	.254 10	.254 09	.254 08	.254 07	.254 06	.254 00	.254 04	.254 03	.254 02	.254 01
.7	.254 00	.253 99	.253 98	.253 97	.253 96	.253 95	.253 94	.253 93	.253 92	.253 91
.8	.253 90	.253 89	.253 88	.253 87	.253 86	.253 85	.253 84	.253 83	.253 82	.253 81
.9	.253 80	.253 79	.253 78	.253 77	.253 76	.253 75	.253 74	.253 73	.253 72	.253 71

续表

锤度	糖度（或蔗糖分）因数									
	0.00	0.01	0.02	0.03	0.04	0.05	0.06	0.07	0.08	0.09
7.0	.253 70	.253 69	.253 68	.253 67	.253 66	.253 65	.253 64	.253 63	.253 62	.253 61
.1	.253 60	.253 59	.253 58	.253 57	.253 56	.253 55	.253 54	.253 53	.253 52	.253 51
.2	.253 50	.253 49	.253 48	.253 47	.253 46	.253 45	.253 44	.253 43	.253 42	.253 41
.3	.253 40	.253 39	.253 38	.253 37	.253 36	.253 35	.253 34	.253 33	.253 32	.253 31
.4	.253 30	.253 29	.253 28	.253 27	.253 26	.253 25	.253 24	.253 23	.253 22	.253 21
.5	.253 20	.253 19	.253 18	.253 17	.253 16	.253 15	.253 14	.253 13	.253 12	.253 11
.6	.253 10	.253 09	.253 08	.253 07	.253 06	.253 05	.253 04	.253 03	.253 02	.253 01
.7	.253 00	.252 99	.252 98	.252 97	.252 96	.252 95	.252 94	.252 93	.252 92	.252 91
.8	.252 90	.252 89	.252 88	.252 87	.252 86	.252 85	.252 84	.252 83	.252 82	.252 81
.9	.252 80	.252 79	.252 78	.252 77	.252 76	.252 75	.252 74	.252 73	.252 72	.252 71
8.0	.252 70	.252 69	.252 68	.252 67	.252 66	.252 65	.252 64	.252 63	.252 62	.252 61
.1	.252 60	.252 59	.252 58	.252 57	.252 56	.252 55	.252 54	.252 53	.252 52	.252 51
.2	.252 50	.252 49	.252 48	.252 47	.252 46	.252 45	.252 44	.252 43	.252 42	.252 41
.3	.252 40	.252 39	.252 38	.252 37	.252 36	.252 35	.252 34	.252 33	.252 32	.252 31
.4	.252 30	.252 29	.252 28	.252 27	.252 26	.252 25	.252 24	.252 23	.252 22	.252 21
.5	.252 20	.252 19	.252 18	.252 17	.252 16	.252 15	.252 14	.252 13	.252 12	.252 11
.6	.252 10	.252 09	.252 08	.252 07	.252 06	.252 05	.252 04	.252 03	.252 02	.252 01
.7	.252 00	.251 99	.251 98	.251 97	.251 96	.251 95	.251 94	.251 93	.251 92	.251 91
.8	.251 90	.251 89	.251 88	.251 87	.251 86	.251 85	.251 84	.251 83	.251 82	.251 81
.9	.251 80	.251 79	.251 78	.251 77	.251 76	.251 75	.251 74	.251 73	.251 72	.251 71
9.0	.251 70	.251 69	.251 68	.251 67	.251 66	.251 65	.251 64	.251 63	.251 62	.251 61
.1	.251 60	.251 59	.251 58	.251 57	.251 56	.251 55	.251 54	.251 53	.251 52	.251 51
.2	.251 50	.251 49	.251 48	.251 47	.251 46	.251 45	.251 44	.251 43	.251 42	.251 41
.3	.251 40	.251 39	.251 38	.251 37	.251 36	.251 35	.251 34	.251 33	.251 32	.251 31
.4	.251 30	.251 29	.251 28	.251 27	.251 26	.251 25	.251 24	.251 23	.251 22	.251 21
.5	.251 20	.251 19	.251 18	.251 17	.251 16	.251 15	.251 14	.251 13	.251 12	.251 11
.6	.251 10	.251 09	.251 08	.251 07	.251 06	.251 05	.251 04	.251 03	.251 02	.251 01
.7	.251 00	.250 99	.250 98	.250 97	.250 96	.250 95	.250 94	.250 93	.250 92	.250 91
.8	.250 90	.250 89	.250 88	.250 87	.250 86	.250 85	.250 84	.250 83	.250 82	.250 81
.9	.250 80	.250 79	.250 78	.250 77	.250 76	.250 75	.250 74	.250 73	.250 72	.250 71
10.0	.250 70	.250 69	.250 68	.250 67	.250 66	.250 65	.250 64	.250 63	.250 62	.250 61
.1	.250 60	.250 59	.250 58	.250 57	.250 56	.250 55	.250 54	.250 53	.250 52	.250 51
.2	.250 50	.250 49	.250 48	.250 47	.250 46	.250 45	.250 44	.250 43	.250 42	.250 41
.3	.250 40	.250 39	.250 38	.250 37	.250 36	.250 35	.250 34	.250 33	.250 32	.250 31
.4	.250 30	.250 29	.250 28	.250 27	.250 26	.250 25	.250 24	.250 23	.250 22	.250 21
.5	.250 20	.250 19	.250 18	.250 17	.250 16	.250 15	.250 14	.250 13	.250 12	.250 11
.6	.250 10	.250 09	.250 08	.250 07	.250 06	.250 05	.250 04	.250 03	.250 02	.250 01
.7	.250 00	.249 99	.249 98	.249 97	.249 96	.249 95	.249 94	.249 93	.249 92	.249 91
.8	.249 90	.249 89	.249 88	.249 87	.249 86	.249 85	.249 84	.249 83	.249 82	.249 81
.9	.249 80	.249 79	.249 78	.249 77	.249 76	.249 75	.249 74	.249 73	.249 72	.249 71

续表

锤度	糖度（或蔗糖分）因数									
	0.00	0.01	0.02	0.03	0.04	0.05	0.06	0.07	0.08	0.09
11.0	.249 70	.249 69	.249 68	.249 67	.249 66	.249 65	.249 64	.249 63	.249 62	.249 61
.1	.249 60	.249 59	.249 58	.249 57	.249 56	.249 55	.249 54	.249 53	.249 52	.249 51
.2	.249 50	.249 49	.249 48	.249 47	.249 46	.249 45	.249 44	.249 43	.249 42	.249 41
.3	.249 40	.249 39	.249 38	.249 37	.249 36	.249 35	.249 34	.249 33	.249 32	.249 31
.4	.249 30	.249 29	.249 28	.249 27	.249 26	.249 25	.249 24	.249 23	.249 22	.249 21
.5	.249 20	.249 19	.249 18	.249 17	.249 16	.249 15	.249 14	.249 13	.249 12	.249 11
.6	.249 10	.249 09	.249 08	.249 07	.249 06	.249 05	.249 04	.249 03	.249 02	.249 01
.7	.249 00	.248 99	.248 98	.248 97	.248 96	.248 95	.248 94	.248 93	.248 92	.248 91
.8	.248 90	.248 89	.248 88	.248 87	.248 86	.248 85	.248 85	.248 84	.248 83	.248 82
.9	.248 81	.248 80	.248 79	.248 78	.248 77	.248 76	.248 75	.248 74	.248 73	.248 72
12.0	.248 71	.248 70	.248 69	.248 68	.248 67	.248 66	.248 65	.248 64	.248 63	.248 62
.1	.248 61	.248 60	.248 59	.248 58	.248 57	.248 56	.248 55	.248 54	.248 53	.248 52
.2	.248 51	.248 50	.248 49	.248 48	.248 47	.248 46	.248 45	.248 44	.248 43	.248 42
.3	.248 41	.248 40	.248 39	.248 38	.248 37	.248 36	.248 35	.248 34	.248 33	.248 32
.4	.248 31	.248 30	.248 29	.248 28	.248 27	.248 26	.248 25	.248 24	.248 23	.248 22
.5	.248 21	.248 20	.248 19	.248 18	.248 17	.248 16	.248 15	.248 14	.248 13	.248 12
.6	.248 11	.248 10	.248 09	.248 08	.248 07	.248 06	.248 05	.248 04	.248 03	.248 02
.7	.248 01	.248 00	.247 99	.247 98	.247 97	.247 96	.247 95	.247 94	.247 93	.247 92
.8	.247 91	.247 90	.247 89	.247 88	.247 87	.247 86	.247 85	.247 84	.247 83	.247 82
.9	.247 81	.247 80	.247 79	.247 78	.247 77	.247 76	.247 75	.247 74	.247 73	.247 72
13.0	.247 71	.247 70	.247 69	.247 68	.247 67	.247 66	.247 65	.247 64	.247 63	.247 62
.1	.247 61	.247 60	.247 59	.247 58	.247 57	.247 56	.247 55	.247 54	.247 53	.247 52
.2	.247 51	.247 50	.247 49	.247 48	.247 47	.247 46	.247 45	.247 44	.247 43	.247 42
.3	.247 41	.247 40	.247 39	.247 38	.247 37	.247 36	.247 35	.247 34	.247 33	.247 32
.4	.247 31	.247 30	.247 29	.247 28	.247 27	.247 26	.247 25	.247 24	.247 23	.247 22
.5	.247 21	.247 20	.247 19	.247 18	.247 17	.247 16	.247 15	.247 14	.247 13	.247 12
.6	.247 11	.247 10	.247 09	.247 08	.247 07	.247 06	.247 05	.247 04	.247 03	.247 02
.7	.247 01	.247 00	.246 99	.246 98	.246 97	.246 96	.246 95	.246 94	.246 93	.246 92
.8	.246 91	.246 90	.246 89	.246 88	.246 87	.246 86	.246 85	.246 84	.246 83	.246 82
.9	.246 81	.246 80	.246 79	.246 78	.246 77	.246 76	.246 75	.246 74	.246 73	.246 72
14.0	.246 71	.246 70	.246 69	.246 68	.246 67	.246 66	.246 65	.246 64	.246 63	.246 62
.1	.246 61	.246 60	.246 59	.246 58	.246 57	.246 56	.246 55	.246 54	.246 53	.246 52
.2	.246 51	.246 50	.246 49	.246 48	.246 47	.246 46	.246 46	.246 45	.246 44	.246 43
.3	.246 42	.246 41	.246 40	.246 39	.246 38	.246 37	.246 36	.246 35	.246 34	.246 33
.4	.246 32	.246 31	.246 30	.246 29	.246 28	.246 27	.246 26	.246 25	.246 24	.246 23
.5	.246 22	.246 21	.246 20	.246 19	.246 18	.246 17	.246 16	.246 15	.246 14	.246 13
.6	.246 12	.246 11	.246 10	.246 09	.246 08	.246 07	.246 06	.246 05	.246 04	.246 03
.7	.246 02	.246 01	.246 00	.245 99	.245 98	.245 97	.245 96	.245 95	.245 94	.245 93
.8	.245 92	.245 91	.245 90	.245 89	.245 88	.245 87	.245 86	.245 85	.245 84	.245 83
.9	.245 82	.245 81	.245 80	.245 79	.245 78	.245 77	.245 76	.245 75	.245 74	.245 73

续表

锤度	糖度（或蔗糖分）因数									
	0.00	0.01	0.02	0.03	0.04	0.05	0.06	0.07	0.08	0.09
15.0	.245 72	.245 71	.245 70	.245 69	.245 68	.245 67	.245 66	.245 65	.245 64	.245 63
.1	.245 62	.245 61	.245 60	.245 59	.245 58	.245 57	.245 56	.245 55	.245 54	.245 53
.2	.245 52	.245 51	.245 50	.245 49	.245 48	.245 47	.245 46	.245 45	.245 44	.245 43
.3	.245 42	.245 41	.245 40	.245 39	.245 38	.245 37	.245 36	.245 35	.245 34	.245 33
.4	.245 32	.245 31	.245 30	.245 29	.245 28	.245 27	.245 26	.245 25	.245 24	.245 23
.5	.245 22	.245 21	.245 20	.245 19	.245 18	.245 17	.245 16	.245 15	.245 14	.245 13
.6	.245 12	.245 11	.245 10	.245 09	.245 08	.245 07	.245 06	.245 05	.245 04	.245 03
.7	.245 02	.245 01	.245 00	.244 99	.244 98	.244 97	.244 96	.244 96	.244 95	.244 94
.8	.244 93	.244 92	.244 91	.244 90	.244 88	.244 87	.244 86	.244 85	.244 84	.244 83
.9	.244 82	.244 81	.244 80	.244 79	.244 78	.244 77	.244 77	.244 76	.244 75	.244 74
16.0	.244 73	.244 72	.244 71	.244 70	.244 69	.244 68	.244 67	.244 66	.244 65	.244 64
.1	.244 63	.244 62	.244 61	.244 60	.244 59	.244 58	.244 57	.244 56	.244 55	.244 54
.2	.244 53	.244 52	.244 51	.244 50	.244 49	.244 48	.244 47	.244 46	.244 45	.244 44
.3	.244 43	.244 42	.244 41	.244 40	.244 39	.244 38	.244 37	.244 36	.244 35	.244 34
.4	.244 33	.244 32	.244 31	.244 30	.244 29	.244 28	.244 27	.244 26	.244 25	.244 24
.5	.244 23	.244 22	.244 21	.244 20	.244 19	.244 18	.244 17	.244 16	.244 15	.244 14
.6	.244 13	.244 12	.244 11	.244 10	.244 09	.244 08	.244 07	.244 06	.244 05	.244 04
.7	.244 03	.244 02	.244 01	.244 00	.243 99	.243 98	.243 97	.243 96	.243 95	.243 94
.8	.243 93	.243 92	.243 91	.243 90	.243 89	.243 88	.243 87	.243 86	.243 85	.243 84
.9	.243 83	.243 82	.243 81	.243 80	.243 79	.243 78	.243 77	.243 76	.243 75	.243 74
17.0	.243 73	.243 72	.243 71	.243 70	.243 69	.243 68	.243 67	.243 66	.243 65	.243 64
.1	.243 63	.243 62	.243 61	.243 60	.243 59	.243 58	.243 57	.243 56	.243 55	.243 54
.2	.243 53	.243 52	.243 51	.243 50	.243 49	.243 48	.243 47	.243 46	.243 45	.243 44
.3	.243 43	.243 42	.243 41	.243 40	.243 39	.243 38	.243 37	.243 36	.243 35	.243 34
.4	.243 33	.243 32	.243 31	.243 30	.243 29	.243 28	.243 28	.243 27	.243 26	.243 25
.5	.243 24	.243 23	.243 22	.243 21	.243 20	.243 19	.243 18	.243 17	.243 16	.243 15
.6	.243 14	.243 13	.243 12	.243 11	.243 10	.243 09	.243 08	.243 07	.243 06	.243 05
.7	.243 04	.243 03	.243 02	.243 01	.243 00	.242 99	.242 98	.242 97	.242 96	.242 95
.8	.242 94	.242 93	.242 92	.242 91	.242 90	.242 89	.242 88	.242 87	.242 86	.242 85
.9	.242 84	.242 83	.242 82	.242 81	.242 89	.242 79	.242 78	.242 77	.242 76	.242 75
18.0	.242 74	.242 73	.242 72	.242 71	.242 70	.242 69	.242 68	.242 67	.242 66	.242 65
.1	.242 64	.242 63	.242 62	.242 61	.242 60	.242 59	.242 58	.242 57	.242 56	.242 55
.2	.242 54	.242 53	.242 52	.242 51	.242 50	.242 49	.242 48	.242 47	.242 46	.242 45
.3	.242 44	.242 43	.242 42	.242 41	.242 40	.242 39	.242 38	.242 37	.242 36	.242 35
.4	.242 34	.242 33	.242 32	.242 31	.242 30	.242 29	.242 28	.242 27	.242 26	.242 25
.5	.242 24	.242 23	.242 22	.242 21	.242 20	.242 19	.242 19	.242 18	.242 17	.242 16
.6	.242 15	.242 14	.242 13	.242 12	.242 11	.242 10	.242 09	.242 08	.242 07	.242 06
.7	.242 05	.242 04	.242 03	.242 02	.242 01	.242 00	.241 99	.241 98	.241 97	.241 96
.8	.241 95	.241 94	.241 93	.241 92	.241 91	.241 90	.241 89	.241 88	.241 87	.241 86
.9	.241 85	.241 84	.241 83	.241 82	.241 81	.241 80	.241 79	.241 78	.241 77	.241 76

续表

锤度	糖度（或蔗糖分）因数									
	0.00	0.01	0.02	0.03	0.04	0.05	0.06	0.07	0.08	0.09
19.0	.241 75	.241 74	.241 73	.241 72	.241 71	.241 70	.241 69	.241 68	.241 67	.241 66
.1	.241 65	.241 64	.241 63	.241 62	.241 61	.241 60	.241 59	.241 58	.241 57	.241 56
.2	.241 55	.241 54	.241 53	.241 52	.241 51	.241 50	.241 49	.241 48	.241 47	.241 46
.3	.241 45	.241 44	.241 43	.241 42	.241 41	.241 40	.241 39	.241 38	.241 37	.241 36
.4	.241 35	.241 34	.241 33	.241 32	.241 31	.241 30	.241 29	.241 28	.241 27	.241 26
.5	.241 25	.241 24	.241 23	.241 22	.241 21	.241 20	.241 20	.241 19	.241 18	.241 17
.6	.241 16	.241 15	.241 14	.241 13	.241 12	.241 11	.241 10	.241 09	.241 08	.241 07
.7	.241 06	.241 05	.241 04	.241 03	.241 02	.241 01	.241 00	.240 99	.240 98	.240 97
.8	.240 96	.240 95	.240 94	.240 93	.240 92	.240 91	.240 90	.240 89	.240 88	.240 87
.9	.240 86	.240 85	.240 84	.240 83	.240 82	.240 81	.240 80	.240 79	.240 78	.240 77
20.0	.240 76	.240 75	.240 74	.240 73	.240 72	.240 71	.240 70	.240 69	.240 68	.240 67
.1	.240 66	.240 65	.240 64	.240 63	.240 62	.240 61	.240 60	.240 59	.240 58	.240 57
.2	.240 56	.240 55	.240 54	.240 53	.240 52	.240 51	.240 50	.240 49	.240 48	.240 47
.3	.240 46	.240 45	.240 44	.240 43	.240 42	.240 41	.240 40	.240 39	.240 38	.240 37
.4	.240 36	.240 35	.240 34	.240 33	.240 32	.240 31	.240 30	.240 29	.240 28	.240 27
.5	.240 26	.240 25	.240 24	.240 23	.240 22	.240 21	.240 21	.240 20	.240 19	.240 18
.6	.240 17	.240 16	.240 15	.240 14	.240 13	.240 12	.240 11	.240 10	.240 09	.240 08
.7	.240 07	.240 06	.240 05	.240 04	.240 03	.240 02	.240 01	.240 00	.239 99	.239 98
.8	.239 97	.239 96	.239 95	.239 94	.239 93	.239 92	.239 91	.239 90	.239 89	.239 88
.9	.239 87	.239 86	.239 85	.239 84	.239 83	.239 82	.239 81	.239 80	.239 79	.239 78
21.0	.239 77	.239 76	.239 75	.239 74	.239 73	.239 72	.239 71	.239 70	.239 69	.239 68
.1	.239 67	.239 66	.229 65	.229 64	.239 63	.239 62	.239 61	.239 60	.239 59	.239 58
.2	.239 57	.239 56	.239 55	.239 54	.239 53	.239 52	.239 51	.239 50	.239 49	.239 48
.3	.239 17	.239 46	.239 45	.239 44	.239 43	.239 42	.239 42	.239 41	.239 40	.239 39
.4	.239 38	.239 37	.239 36	.239 35	.239 34	.239 33	.239 32	.239 31	.239 30	.239 29
.5	.239 28	.239 27	.239 26	.239 25	.239 24	.239 23	.239 22	.239 21	.239 20	.239 19
.6	.239 18	.239 17	.239 16	.239 15	.239 14	.239 12	.239 12	.239 11	.239 10	.239 09
.7	.239 08	.239 07	.239 06	.239 05	.239 04	.239 03	.239 02	.239 01	.239 00	.238 99
.8	.238 98	.238 97	.233 96	.238 95	.238 94	.238 93	.238 92	.238 91	.238 90	.238 89
.9	.238 88	.238 87	.238 86	.238 85	.238 84	.238 83	.238 82	.238 81	.238 80	.238 79
22.0	.238 78	.238 77	.238 76	.238 75	.238 74	.238 73	.238 72	.238 71	.238 70	.238 69
.1	.238 68	.238 67	.238 66	.238 65	.238 64	.238 63	.238 63	.23S 62	.238 61	.238 60
.2	.238 59	.238 58	.238 57	.238 56	.238 55	.238 54	.238 53	.22S 52	.238 51	.238 50
.3	.238 49	.238 48	.238 47	.238 46	.238 45	.238 44	.238 43	.238 42	.238 41	.238 40
.4	.238 39	.238 38	.238 37	.238 36	.238 35	.238 34	.238 33	.238 32	.238 31	.238 30
.5	.238 29	.238 28	.238 27	.238 26	.238 25	.238 24	.238 23	.238 22	.238 21	.238 20
.6	.238 19	.238 18	.238 17	.238 16	.238 15	.238 14	.238 13	.238 12	.238 11	.238 10
.7	.238 09	.238 08	.238 07	.238 06	.238 05	.238 04	.238 03	.238 02	.238 01	.238 00
.8	.237 99	.237 98	.237 97	.237 96	.237 95	.237 94	.237 93	.237 92	.237 91	.237 90
.9	.237 89	.237 88	.237 87	.237 86	.237 85	.237 84	.237 84	.237 83	.237 82	.237 81

续表

锤度	糖度（或蔗糖分）因数									
	0.00	0.01	0.02	0.03	0.04	0.05	0.06	0.07	0.08	0.09
23.0	.237 80	.237 79	.237 78	.237 77	.237 76	.237 75	.237 74	.237 73	.237 72	.237 71
.1	.237 70	.237 69	.237 68	.237 67	.237 66	.237 65	.237 64	.237 63	.237 62	.237 61
.2	.237 60	.237 59	.237 58	.237 57	.237 56	.237 55	.237 54	.237 53	.237 52	.237 51
.3	.237 50	.237 49	.237 48	.237 47	.237 46	.237 45	.234 44	.237 43	.237 42	.237 41
.4	.237 40	.237 39	.237 38	.237 37	.237 36	.237 35	.237 34	.237 23	.237 32	.237 31
.5	.237 30	.237 29	.227 28	.237 27	.237 26	.237 25	.237 24	.237 23	.237 22	.237 21
.6	.237 20	.237 19	.237 18	.237 17	.237 16	.237 15	.237 15	.237 14	.237 13	.237 12
.7	.237 11	.237 10	.237 09	.237 08	.237 07	.237 06	.237 05	.237 04	.237 03	.237 02
.8	.237 01	.237 00	.236 99	.236 98	.236 97	.236 96	.236 95	.236 94	.236 93	.236 92
.9	.236 91	.236 90	.236 89	.236 88	.236 87	.236 86	.236 85	.236 84	.236 83	.236 82
24.0	.236 81	.236 80	.236 79	.236 78	.236 77	.236 76	.236 75	.236 74	.236 73	.236 72
.1	.236 71	.236 70	.236 69	.236 68	.236 67	.236 66	.236 65	.236 64	.236 63	.236 62
.2	.236 61	.236 60	.236 59	.236 58	.236 57	.236 56	.236 55	.236 54	.236 53	.236 52
.3	.236 51	.236 50	.236 49	.236 48	.236 47	.236 46	.236 46	.236 45	.236 44	.236 43
.4	.236 42	.236 41	.236 40	.236 39	.236 38	.236 37	.236 36	.236 35	.236 34	.236 33
.5	.236 22	.236 31	.236 30	.236 29	.236 28	.236 27	.236 26	.236 25	.236 24	.236 23
.6	.236 22	.236 21	.236 20	.226 19	.236 18	.236 17	.236 16	.236 15	.236 14	.236 13
.7	.236 12	.236 11	.236 10	.236 09	.236 08	.236 07	.236 06	.236 05	.236 04	.236 03
.8	.236 02	.236 01	.236 00	.235 99	.235 98	.235 97	.235 96	.235 95	.235 94	.235 93
.9	.235 92	.235 91	.235 90	.235 89	.235 88	.235 87	.235 86	.235 85	.235 84	.235 83
25.0	.235 82	.235 81	.235 80	.235 79	.235 78	.235 77	.235 76	.235 75	.235 74	.235 73
.1	.235 72	.235 71	.235 70	.235 69	.235 68	.235 67	.235 67	.235 66	.235 65	.235 64
.2	.235 63	.235 62	.235 61	.235 60	.235 59	.235 58	.235 57	.235 56	.235 55	.235 54
.3	.235 53	.235 52	.235 51	.233 50	.235 49	.235 48	.235 47	.235 46	.235 45	.235 44
.4	.235 43	.235 42	.235 41	.235 40	.235 39	.235 38	.235 37	.235 36	.235 35	.235 34
.5	.235 33	.235 32	.235 31	.235 30	.235 29	.235 28	.235 27	.235 26	.235 25	.235 24
.6	.235 23	.235 22	.235 21	.235 20	.235 19	.235 18	.235 18	.235 17	.235 16	.235 15
.7	.235 14	.235 13	.235 12	.235 11	.235 10	.235 09	.235 08	.235 07	.235 06	.235 05
.8	.235 04	.235 03	.235 02	.235 01	.235 00	.234 99	.234 98	.234 97	.234 96	.234 95
.9	.234 94	.234 93	.234 92	.234 91	.234 90	.234 89	.234 88	.234 87	.234 86	.234 85

附表5 蔗汁克来杰除数检索表（20℃）

锤度	0.0	0.1	0.2	0.3	0.4	0.5	0.6	0.7	0.8	0.9
12	132.026	132.030	132.035	132.039	132.043	132.048	132.052	132.056	132.061	132.065
13	132.070	132.074	132.078	132.083	132.087	132.092	132.096	132.100	132.105	132.109
14	132.114	132.118	132.122	132.127	132.131	132.136	132.140	132.145	132.149	132.153
15	132.158	132.162	122.167	132.171	132.176	132.180	132.185	132.189	132.194	132.198
16	132.203	132.207	132.212	132.216	132.221	132.225	132.230	132.234	132.239	132.243
17	132.248	132.252	132.257	132.261	132.266	132.270	132.275	132.280	132.284	132.289
18	132.293	132.298	132.302	132.307	132.311	132.316	132.321	132.325	132.330	132.334
19	132.339	132.344	132.348	132.353	132.357	132.362	132.367	132.371	132.376	132.381
20	132.385	132.390	132.395	132.399	132.404	132.408	132.413	132.418	132.422	132.427
21	132.432	132.437	132.441	132.446	132.451	132.155	132.460	132.465	132.469	132.474
22	132.479	132.484	132.488	132.49	132.498	132.502	132.507	132.512	132.517	132.521

计算公式：132.56−0.0794（13−g）

式中：$g = 1/2 \times$ 锤度 $\times$ 相应的视密度（20℃）

附表6 1规定量、1/2规定量、1/3规定量糖液克来杰除数检索表（20℃）

锤度	克来杰除数			锤度	克来杰除数		
	1规定量	1/2规定量	1/3规定量		1规定量	1/2规定量	1/3规定量
12.00	132.271	131.899		13.25	132.348	131.937	131.801
.05	.274	.900		.30	.351	.939	.802
.10	.277	.902		.35	.354	.940	.803
.15	.280	.903		.40	.357	.942	.804
.20	.283	.905		.45	.360	.943	.805
.25	.286	.906		.50	.364	.945	.806
.30	.289	.908		.55	.367	.947	.807
.35	.292	.909		.60	.370	.948	.808
.40	.295	.911		.65	.373	.950	.810
.45	.298	.912		.70	.376	.951	.811
.50	.302	.914		.75	.379	.953	.812
.55	.305	.915		.80	.382	.954	.813
.60	.308	.917		.85	.385	.956	.814
.65	.311	.918		.90	.388	.957	.815
.70	.314	.920		.95	.392	.959	.816
.75	.317	.922		14.00	.395	.961	.817
.80	.320	.923		.05	.398	.962	.818
.85	.323	.925		.10	.401	.964	.819
.90	.326	.926		.15	.404	.965	.820
.95	.329	.928		.20	.407	.967	.821

续表

锤度	克来杰除数			锤度	克来杰除数		
	1 规定量	1/2 规定量	1/3 规定量		1 规定量	1/2 规定量	1/3 规定量
13.00	.333	.930	131.796	.25	.410	.968	
.05	.336	.931	.797	.30	.413	.970	
.10	.339	.933	.798	.35	.416	.971	
.15	.342	.934	.799	.40	.419	.973	
.20	.345	.936	.800	.45	.422	.975	
14.50	132.426	131.976	131.827	15.80	132.506	132.016	
.55	.429	.977	.828	.85	.509	.018	
.60	.432	.979	.829	.90	.512	.019	
65	.435	.981	.830	.95	.516	.021	
.70	.438	.982	.831	16.00	.519	.023	.822
.75	.441	.984	.832	.05	.522	.024	.823
.80	.444	.985	.833	.10	.525	.026	.824
.85	.447	.987	.834	.15	.528	.027	.825
.90	.450	.988	.835	.20	.531	.029	.826
.95	.453	.990	.836	.25	.534	.030	131.854
15.00	.457	.992	.837	.30	.537	.032	.855
.05	.460	.993	.839	.35	.540	.034	.856
.10	.463	.995	.840	.40	.543	.035	.857
.15	.466	.996	.841	.45	.547	.037	.858
.20	.469	.998	.842	.50	.550	.038	
.25	.472	.999	.843	.55	.553	.040	
.30	.475	132.001	.844	.60	.556	.041	
.35	.478	.002	.845	.65	.559	.043	
.40	.481	.004	.846	.70	.562	.044	
.45	.485	.005	.847	.75	.565	.046	
.50	.488	.007	.848	.80	.568	.047	
.55	.491	.009	.849	.85	.571	.049	
.60	.494	.010	.850	.90	.575	.050	
.65	.497	.012	.851	.95	.578	.052	
.70	.500	.014	.852	17.00	132.581	132.054	
.75	.503	.015	.853				

附注：1. 表中锤度为样品 6 倍稀释后的锤度。

2. 计算公式：$132.56-0.0794(13-g)$

式中：$g=$ 每 100 mL 糖液含样品质量 × 锤度 × 稀释倍数／2

附表 7　克来杰除数温度改正表（4～35 ℃）

温度/℃	———温度高于 20 ℃时应减之数———									
	0.0	0.1	0.2	0.3	0.4	0.5	0.6	0.7	0.8	0.9
20	0.00	0.05	0.11	0.16	0.21	0.27	0.32	0.37	0.42	0.48
21	0.53	0.58	0.64	0.69	0.74	0.80	0.85	0.90	0.95	1.01
22	1.06	1.11	1.17	1.22	1.27	1.33	1.38	1.43	1.48	1.54
23	1.59	1.64	1.70	1.75	1.80	1.86	1.91	1.96	2.01	2.07
24	2.12	2.17	2.23	2.28	2.33	2.39	2.44	2.49	2.54	2.60
25	2.65	2.70	2.76	2.81	2.86	2.92	2.97	3.02	3.07	3.13
26	3.18	3.23	3.29	3.34	3.39	3.44	3.50	3.55	3.60	3.66
27	3.71	3.76	3.82	3.87	3.92	3.98	4.03	4.08	4.13	4.19
28	4.24	4.29	4.35	4.40	4.45	4.51	4.56	4.61	4.66	4.72
29	4.77	4.82	4.88	4.93	4.98	5.04	5.09	5.14	5.19	5.25
30	5.30	5.35	5.41	5.46	5.51	5.57	5.62	5.67	5.72	5.78
31	5.83	5.88	5.94	5.99	6.04	6.10	6.15	6.20	6.25	6.31
32	6.36	6.41	6.47	6.52	6.57	6.63	6.68	6.73	6.78	6.84
33	6.89	6.94	7.00	7.05	7.10	7.16	7.21	7.26	7.31	7.37
34	7.42	7.47	7.53	7.58	7.63	7.69	7.74	7.79	7.84	7.90
35	7.95	8.00	8.06	8.11	8.16	8.22	8.27	8.32	8.37	8.43
	++++ 温度低于 20 ℃时应加之数 ++++									
19	0.53	0.48	0.42	0.37	0.32	0.27	0.21	0.16	0.11	0.05
18	1.06	1.01	0.95	0.90	0.85	0.80	0.74	0.69	0.64	0.58
17	1.59	1.54	1.48	1.43	1.38	1.33	1.27	1.22	1.17	1.11
16	2.12	2.07	2.01	1.96	1.91	1.86	1.80	1.75	1.70	1.64
15	2.65	2.60	2.54	2.49	2.44	2.39	2.33	2.28	2.23	2.17
14	3.18	3.13	3.07	3.02	2.97	2.92	2.86	2.81	2.76	2.70
13	3.71	3.66	3.60	3.55	3.50	3.44	3.39	3.54	3.29	3.23
12	4.24	4.19	4.13	4.08	4.03	3.98	3.92	3.87	3.82	3.76
11	4.77	4.72	4.66	4.61	4.56	4.51	4.45	4.40	4.35	4.29
10	5.30	5.25	5.19	5.14	5.09	5.04	4.98	4.93	4.88	4.82
9	5.83	5.78	5.72	5.67	5.62	5.57	5.51	5.46	5.41	5.35
8	6.36	6.31	6.25	6.20	6.15	6.10	6.04	5.99	5.94	5.88
7	6.89	6.84	6.78	6.73	6.68	6.63	6.57	6.52	6.47	6.41
6	7.42	7.37	7.31	7.26	7.21	7.16	7.10	7.05	7.00	6.94
5	7.95	7.90	7.84	7.79	7.74	7.69	7.63	7.58	7.53	7.47
4	8.48	8.43	8.37	8.32	8.27	8.22	8.16	8.11	8.06	8.00

附表 8　还原糖因数表

（适用于兰－艾农法）

滴定配制糖液体积/mL	每 100 mL 配制糖液含蔗糖克数											
	0 g	0.5 g	1 g	2 g	3 g	4 g	5 g		10 g		25 g	
	因　数						因数	100 mL 配制糖液含还原糖量（mg）	因数	100 mL 配制糖液含还原糖量（mg）	因数	100 mL 配制糖液含还原糖量（mg）
15	50.5	50.2	49.9	49.4	48.8	48.3	47.6	317	46.1	307	43.4	289
16	50.6	50.3	50.0	49.4	48.8	48.3	47.6	297	46.1	288	43.4	271
17	50.7	50.4	50.1	49.4	48.8	48.3	47.6	280	46.1	271	43.4	255
18	50.8	50.4	50.1	49.4	48.8	48.3	47.6	264	46.1	256	43.3	240
19	50.8	50.4	50.2	49.5	48.9	48.3	47.6	250	46.1	243	43.3	227
20	50.9	50.5	50.2	49.5	48.9	48.3	47.6	238.0	46.1	230.5	43.2	216
21	51.0	50.6	50.2	49.5	48.9	48.3	47.6	226.7	46.1	219.5	43.2	206
22	51.0	50.6	50.3	49.5	48.9	48.4	47.6	216.4	46.1	209.5	43.1	196
23	51.1	50.7	50.3	49.6	49.0	48.4	47.6	207.0	46.1	200.4	43.0	187
24	51.2	50.7	50.3	49.6	49.0	48.4	47.6	198.3	46.1	192.1	42.9	179
25	51.2	50.8	50.4	49.6	49.0	48.4	47.6	190.4	46.0	184.0	42.8	171
26	51.3	50.8	50.4	49.6	49.0	48.4	47.6	183.1	46.0	176.9	42.8	164
27	51.4	50.9	50.4	49.6	49.0	48.4	47.6	176.4	46.0	170.4	42.7	158
28	51.4	50.9	50.5	49.7	49.1	48.4	47.7	170.3	46.0	164.3	42.7	152
29	51.5	51.0	50.5	49.7	49.1	48.4	47.7	164.5	46.0	158.6	42.6	147

续表

滴定配制糖液体积/mL	每100 mL配制糖液含蔗糖克数												
	0 g	0.5 g	1 g	2 g	3 g	4 g	5 g		10 g		25 g	
	因 数						因数	100 mL配制糖液含还原糖量（mg）	因数	100 mL配制糖液含还原糖量（mg）	因数	100 mL配制糖液含还原糖量（mg）
30	51.5	51.0	50.5	49.7	49.1	48.4	47.7	159.0	46.0	153.3	42.5	142
31	51.6	51.1	50.6	49.8	49.2	48.5	47.7	153.9	45.9	148.1	42.5	137
32	51.6	51.1	50.6	49.8	49.2	48.5	47.7	149.1	45.9	143.4	42.4	132
33	51.7	51.2	50.6	49.8	49.2	48.5	47.7	144.5	45.9	139.1	42.3	128
34	51.7	51.2	50.6	49.8	49.2	48.5	47.7	140.3	45.8	134.9	42.2	124
35	51.8	51.3	50.7	49.9	49.2	48.5	47.7	136.3	45.8	130.9	42.2	121
36	51.8	51.3	50.7	49.9	49.2	48.5	47.7	132.5	45.8	127.1	42.1	117
37	51.9	51.3	50.7	49.9	49.2	48.5	47.7	128.9	45.7	123.5	42.0	114
38	51.9	51.3	50.7	49.9	49.2	48.5	47.7	125.5	45.7	120.3	42.0	111
39	52.0	51.4	50.8	50.0	49.2	48.5	47.7	122.3	45.7	117.1	41.9	107
40	52.0	51.4	50.8	50.0	49.2	48.5	47.7	119.2	45.6	114.1	41.8	104
41	52.1	51.4	50.8	50.0	49.2	48.5	47.7	116.3	45.6	111.2	41.8	102
42	52.1	51.5	50.8	50.0	49.2	48.5	47.7	113.5	45.6	108.5	41.7	99
43	52.2	51.5	50.8	50.0	49.3	48.5	47.7	110.9	45.5	105.8	41.6	97
44	52.2	51.5	50.9	50.0	49.3	48.5	47.7	108.4	45.5	103.4	41.5	94
45	52.3	51.5	50.9	50.1	49.3	48.6	47.7	106.0	45.4	101.0	41.4	92
46	52.3	51.5	50.9	50.1	49.3	48.6	47.7	103.7	45.4	98.7	41.4	90
47	52.4	51.6	50.9	50.1	49.3	48.6	47.7	101.5	45.3	96.4	41.3	88
48	52.4	51.6	50.9	50.1	49.3	48.6	47.7	99.4	45.3	94.3	41.2	86
49	52.5	51.7	51.0	50.2	49.3	48.6	47.7	97.4	45.2	92.3	41.1	84
50	52.5	51.7	51.0	50.2	49.4	48.6	47.7	95.4	45.2	90.4	41.0	82

附表 9　兰-艾农恒容法测定还原糖校正系数表

沸腾混合液中含蔗糖质量 /g	校正系数 (f)	
	1986 年（新）	1978 年（旧）
0	1.000	1.000
0.5		0.985
1.0		0.972
1.5		0.964
2.0	0.946	0.956
2.5		0.949
3.0		0.942
3.5		0.936
4.0	0.912	0.930
4.5		0.924
5.0		0.918
5.5		0.913
6.0	0.887	0.908
6.5		0.903
7.0		0.898
7.5		0.895
8.0	0.865	0.891
8.5		0.887
9.0		0.883
9.5		0.880
10.0	0.849	0.876
12.0	0.828	—
14.0	0.811	—
16.0	0.802	—
18.0	0.791	—
20.0	0.780	—

附表10 奥夫纳尔法还原糖改正表

消耗 0.016 15mol/L 碘液体积/mL	蔗糖质量/g									
	1	2	3	4	5	6	7	8	9	10
1	0.11	0.22	0.34	0.45	0.55	0.66	0.77	0.89	1.00	1.11
2	0.17	0.28	0.40	0.51	0.61	0.72	0.84	0.95	1.06	1.16
3	0.22	0.34	0.45	0.57	0.67	0.78	0.90	1.01	1.12	1.22
4	0.28	0.39	0.51	0.62	0.73	0.84	0.95	1.07	1.18	1.28
5	0.33	0.45	0.56	0.68	0.78	0.90	1.01	1.12	1.24	1.33
6	0.39	0.50	0.61	0.73	0.83	0.95	1.06	1.18	1.29	1.39
7	0.44	0.55	0.67	0.78	0.88	1.00	1.11	1.23	1.34	1.44
8	0.49	0.60	0.72	0.83	0.94	1.05	1.16	1.28	1.39	1.50
9	0.54	0.65	0.76	0.88	0.99	1.10	1.21	1.33	1.44	1.55
10	0.59	0.70	0.82	0.93	1.03	1.15	1.26	1.37	1.49	1.60
11	0.63	0.75	0.86	0.98	1.08	1.20	1.31	1.42	1.54	1.65
12	0.67	0.78	0.90	1.02	1.12	1.24	1.35	1.47	1.58	1.69
13	0.70	0.82	0.93	1.05	1.16	1.27	1.39	1.51	1.62	1.72
14	0.74	0.85	0.97	1.09	1.19	1.31	1.42	1.54	1.65	1.76
15	0.77	0.88	1.00	1.12	1.22	1.34	1.45	1.57	1.69	1.79
16	0.80	0.91	1.03	1.15	1.25	1.37	1.48	1.60	1.72	1.82
17	0.82	0.94	1.05	1.18	1.28	1.40	1.51	1.63	1.74	1.85
18	0.84	0.96	1.08	1.20	1.30	1.42	1.54	1.66	1.77	1.88
19	0.86	0.98	1.10	1.22	1.32	1.45	1.56	1.68	1.79	1.90
20	0.88	1.00	1.11	1.24	1.34	1.46	1.58	1.70	1.81	1.92
21	0.89	1.01	1.13	1.25	1.35	1.48	1.59	1.71	1.83	1.94
22	0.86	0.98	1.11	1.23	1.34	1.47	1.59	1.71	1.84	1.95

附表 11　糖液锤度、视密度、视比重、每 100 mL 含蔗糖质量及波美度对照表

蔗糖质量百分率（Bx）	视密度（20℃）	视比重（20°/20℃）	蔗糖（g）100 mL（在真空中）	波美度（°Bé）	蔗糖质量百分率（Bx）	视密度（20℃）	视比重（20°/20℃）	蔗糖（g）100 mL（在真空中）	波美度（°Bé）
0.0	0.997 17	1.000 00	0.000	0.00	3.0	1.008 87	1.011 73	3.030	1.68
.1	.997 56	.000 39	.100	0.06	.1	.009 27	.012 13	.132	1.74
.2	.997 95	.000 78	.200	0.11	.2	.009 66	.012 52	.234	1.79
.3	.998 34	.001 17	.300	0.17	.3	.010 06	.012 92	.337	1.85
.4	.998 72	.001 56	.400	0.22	.4	.010 45	.013 31	.439	1.90
.5	.999 11	.001 94	.500	0.28	.5	.010 84	.013 71	.542	1.96
.6	.999 50	.002 33	.600	0.34	.6	.011 24	.014 10	.644	2.02
.7	.999 89	.002 73	.701	0.39	.7	.011 63	.014 50	.747	2.07
.8	1.000 28	.003 12	.801	0.45	.8	.012 03	.014 90	.850	2.13
.9	.000 67	.003 51	.902	0.51	.9	.012 43	.015 29	.953	2.18
1.0	1.001 06	1.003 90	1.002	0.56	4.0	1.012 82	1.015 69	4.056	2.24
.1	.001 45	.004 29	.103	0.62	.1	.013 22	.016 09	.159	2.29
.2	.001 84	.004 68	.203	0.67	.2	.013 61	.016 49	.262	2.35
.3	.002 23	.005 07	.304	0.73	.3	.014 01	.016 88	.365	2.40
.4	.002 61	.005 46	.405	0.79	.4	.014 41	.017 28	.468	2.46
.5	.003 00	.005 85	.506	0.84	.5	.014 80	.017 68	.571	2.52
.6	.003 39	.006 24	.607	0.90	.6	.015 20	.018 08	.675	2.57
.7	.003 78	.006 63	.708	0.95	.7	.015 60	.018 48	.778	2.63
.8	.004 17	.007 02	.809	1.01	.8	.016 00	.018 88	.882	2.68
.9	.004 56	.007 41	.911	1.07	.9	.016 40	.019 28	.986	2.74
2.0	1.004 95	1.007 80	2.012	1.12	5.0	1.016 80	1.019 68	5.089	2.79
.1	.005 34	.008 19	.113	1.18	.1	.017 19	.020 08	.193	2.85
.2	.005 74	.008 59	.215	1.23	.2	.017 59	.020 48	.297	2.91
.3	.006 13	.008 98	.317	1.29	.3	.017 99	.020 88	.401	2.96
.4	.006 52	.009 37	.418	1.34	.4	.018 39	.021 28	.506	3.02
.5	.006 91	.009 77	.520	1.40	.5	.018 79	.021 68	.609	3.07
.6	.007 30	.010 16	.622	1.46	.6	.019 19	.022 08	.713	2.13
.7	.007 69	.010 55	.724	1.51	.7	.019 55	.022 48	.818	3.18
.8	.008 09	.010 94	.826	1.57	.8	.019 99	.022 89	.922	3.24
.9	.008 40	.011 34	.928	1.62	.9	.020 40	.023 29	6.027	3.30

注：1. 第 2 列视密度，是 1 mL 糖液在 20 ℃大气中，用黄铜砝码称得的质量（克）。

2. 第 3 列视比重，是第 2 列视密度除以水在 20 ℃时的视密度 0.997 174。

3. 第 4 列为 100 mL 糖液含蔗糖质量（g）——在真空中称重，等于第 1 列 Bx 乘蔗糖溶液的真密度。

续表

蔗糖质量百分率(Bx)	视密度(20℃)	视比重(20°/20℃)	蔗糖(g)100 mL(在真空中)	波美度(°Bé)	蔗糖质量百分率(Bx)	视密度(20℃)	视比重(20°/20℃)	蔗糖(g)100 mL(在真空中)	波美度(°Bé)
6.0	1.020 80	1.023 69	6.131	3.35	10.0	1.037 09	1.040 03	10.381	5.57
.1	.021 20	.024 09	.236	3.41	.1	.037 50	.040 44	.489	5.63
.2	.021 60	.024 50	.340	3.46	.2	.037 91	.040 86	.597	5.68
.3	.022 00	.024 90	.445	3.52	.3	.038 33	.041 27	.706	5.74
.4	.022 41	.025 30	.550	3.57	.4	.038 74	.041 69	.814	5.80
.5	.022 81	.025 71	.655	3.63	.5	.039 16	.042 10	.922	5.85
.6	.023 21	.026 11	.760	3.69	.6	.039 57	.042 52	11.031	5.91
.7	.023 62	.026 52	.865	3.74	.7	.039 99	.042 93	.139	5.96
.8	.024 02	.026 92	.971	3.80	.8	.040 40	.043 35	.248	6.02
.9	.024 42	.027 33	7.076	3.85	.9	.040 82	.043 77	356	6.07
7.0	1.024 83	1.027 73	7.181	3.91	11.0	1.041 23	1.044 18	11.465	6.13
.1	.025 23	.028 14	.287	3.96	.1	.041 65	.044 60	.574	6.18
.2	.025 64	.028 54	.392	4.02	.2	.042 07	.045 02	.683	6.24
.3	.026 04	.028 95	.498	4.08	.3	.042 48	.045 44	.792	6.30
.4	.026 45	.029 36	.604	4.13	.4	.042 90	.045 85	.901	6.35
.5	.026 85	.029 76	.709	4.19	.5	.043 32	.046 27	12.010	6.41
.6	.027 26	.030 17	.815	4.24	.6	.043 73	.046 69	.120	6.46
.7	.027 66	.030 58	.921	4.30	.7	.044 15	.047 11	.229	6.52
.8	.028 07	.030 98	8.027	4.35	.8	.044 57	.047 53	.338	6.57
.9	.028 48	.031 39	.133	4.41	.9	.044 99	.047 95	.448	6.63
8.0	1.028 88	1.031 80	8.240	4.46	12.0	1.045 41	1.048 37	12.558	6.68
.1	.029 29	.032 21	.346	4.52	.1	.045 83	.048 79	.667	6.74
.2	.029 70	.032 62	.452	4.58	.2	.046 25	.049 21	.777	6.79
.3	.030 11	.033 03	.559	4.63	.3	.046 67	.049 63	.887	6.85
.4	.030 52	.033 44	.665	4.69	.4	.047 09	.050 05	.997	6.90
.5	.030 93	.033 85	.772	4.74	.5	.047 50	.050 47	13.107	6.96
.6	.031 33	.034 26	.879	4.80	.6	.047 93	.050 90	.217	7.02
.7	.031 74	.034 67	.985	4.85	.7	.048 35	.051 32	.327	7.07
.8	.032 15	.035 08	9.092	4.91	.8	.048 77	.051 74	438	7.13
.9	.032 56	.035 49	.199	4.96	.9	.049 19	.052 16	.548	7.18
9.0	1.032 97	1.035 90	9.306	5.02	13.0	1.049 61	1.052 59	13.659	7.24
.1	.033 38	.036 31	.413	5.07	.1	.050 03	.053 01	.769	7.29
.2	.033 79	.036 72	.521	5.13	.2	.050 46	.053 43	.880	7.35
.3	.034 20	.037 13	.628	5.19	.3	.050 88	.053 86	.991	7.40
.4	.034 61	.037 55	.735	5.24	.4	.051 30	.054 28	14.102	7.46
.5	.035 03	.037 96	.843	5.30	.5	.051 72	.054 70	.213	7.51
.6	.035 44	.038 37	.950	5.35	.6	.052 15	.055 13	.324	7.57
.7	.035 85	.038 79	10.058	5.41	.7	.052 57	.055 56	.435	7.62
.8	.036 26	.039 20	.166	5.46	.8	.053 00	.055 98	.546	7.68
.9	.036 67	.039 61	.274	5.52	.9	.053 42	.056 41	.657	7.73

续表

蔗糖质量百分率（Bx）	视密度（20℃）	视比重（20°/20℃）	蔗糖（g）100 mL（在真空中）	波美度（°Bé）	蔗糖质量百分率（Bx）	视密度（20℃）	视比重（20°/20℃）	蔗糖（g）100 mL（在真空中）	波美度（°Bé）
14.0	1.053 85	1.056 83	14.769	7.79	18.0	1.07 110	1.074 13	19.299	10.00
.1	.054 27	.057 26	.880	7.84	.1	.071 53	.074 57	.414	10.05
.2	.054 70	.057 69	.992	7.90	.2	.071 97	.075 01	.529	10.11
.3	.055 12	.058 11	15.103	7.96	.3	.072 41	.075 45	.644	10.16
.4	.055 55	.058 54	.215	8.01	.4	.072 85	.075 89	.760	10.22
.5	.055 98	.058 97	.327	8.07	.5	.073 29	.076 33	.875	10.27
.6	.056 40	.059 40	.439	8.12	.6	.073 73	.076 77	.991	10.33
.7	.056 83	.059 82	.551	8.18	.7	.074 17	.077 21	20.107	10.38
.8	.057 26	.060 25	.663	8.23	.8	.074 61	.077 65	.222	10.44
.9	.057 68	.060 68	.775	8.29	.9	.075 05	.078 09	.338	10.49
15.0	1.058 11	1.061 11	15.887	8.34	19.0	1.075 49	1.078 53	20.454	10.55
.1	.058 54	.061 54	16.000	8.40	.1	.075 93	.078 98	.570	10.60
.2	.058 97	.061 97	.112	8.45	.2	.076 37	.079 42	.686	10.66
.3	.059 40	.062 40	.225	8.51	.3	.076 81	.079 86	.803	10.71
.4	.059 83	.062 83	.338	8.56	.4	.077 25	.080 30	.919	10.77
.5	.060 26	.063 26	.450	8.62	.5	.077 69	.080 75	21.036	10.82
.6	.0C0 69	.063 69	.563	8.67	.6	.078 14	.081 19	.152	10.88
.7	.061 12	.064 12	.676	8.73	.7	.078 58	.081 64	.269	10.93
.8	.061 55	.064 55	.789	8.78	.8	.079 02	.082 08	.385	10.99
.9	.061 98	.064 99	.902	8.84	.9	.079 47	.082 52	.502	11.04
16.0	1.062 41	1.065 42	17.015	8.89	20.0	1.079 91	1.082 97	21.619	11.10
.1	.062 84	.065 85	.129	8.95	.1	.080 35	.083 42	.736	11.15
.2	.063 27	.066 29	.242	9.00	.2	.080 80	.083 86	.853	11.21
.3	.063 70	.066 72	.356	9.06	.3	.081 24	.084 31	.971	11.26
.4	.064 14	.067 15	.469	9.11	.4	.081 69	.084 75	22.088	11.32
.5	.064 57	.067 59	.583	9.17	.5	.082 13	.085 20	.205	11.37
.6	.065 00	.668 02	.697	9.22	.6	.082 58	.085 65	.323	11.43
.7	.065 44	.068 45	.810	9.28	.7	.083 02	.086 09	.440	11.48
.8	.065 87	.068 89	.924	9.33	.8	.083 47	.086 54	.558	11.54
.9	.066 30	.069 33	18.038	9.39	.9	.083 92	.086 99	.676	1159
17.0	1.066 74	1.069 76	18.152	9.45	21.0	1.084 36	1.087 44	22.794	11.65
.1	.067 17	.070 20	.267	9.50	.1	.084 81	.087 89	.912	11.70
.2	.067 61	.070 63	.381	9.56	.2	.085 26	.088 34	23.030	11.76
.3	.068 04	.071 07	.495	9.61	.3	.085 71	.088 79	.148	11.81
.4	.068 48	.071 51	.610	9.67	.4	.086 16	.089 23	.266	1187
.5	.068 91	.071 94	.724	9.72	.5	.086 60	.089 68	.385	11.92
.6	.069 35	.072 38	.839	9.78	.6	.087 05	.090 13	.503	11.98
.7	.069 78	.072 82	.954	9.83	.7	.087 50	.090 58	.622	12.03
.8	.070 22	.073 25	19.069	9.89	.8	.087 95	.091 03	.740	12.09
.9	.070 66	.073 69	.184	9.94	.9	.088 40	.091 49	.859	12.14

续表

蔗糖质量百分率（Bx）	视密度（20℃）	视比重（20°/20℃）	蔗糖（g）100 mL（在真空中）	波美度（°Bé）	蔗糖质量百分率（Bx）	视密度（20℃）	视比重（20°/20℃）	蔗糖（g）100 mL（在真空中）	波美度（°Bé）
22.0	1.088 85	1.091 94	23.978	12.20	26.0	1.107 13	1.110 27	28.813	14.39
.1	.089 30	.092 39	24.097	12.25	.1	.107 59	.110 73	.935	14.44
.2	.089 75	.092 84	.216	12.31	.2	.108 06	.111 20	29.059	14.49
.3	.090 20	.093 29	.335	12.36	.3	.108 52	.111 66	.182	14.55
.4	.090 66	.093 75	.454	12.42	.4	.108 99	.112 13	.305	14.60
.5	.091 11	.094 20	.573	12.47	.5	.109 45	.112 60	.428	14.66
.6	.091 56	.094 65	.693	12.52	.6	.109 92	.113 06	.522	14.71
.7	.092 01	.095 11	.812	12.58	.7	.110 38	.113 53	.675	14.77
.8	.092 47	.095 56	.932	12.63	.8	.110 85	.114 00	.799	14.82
.9	.092 92	.096 02	25.052	12.69	.9	.111 31	.114 47	.923	14.88
23.0	1.093 37	1.096 47	25.172	12.74	27.0	1.111 78	1.114 93	30.046	14.93
.1	.093 83	.096 93	.292	12.80	.1	.112 25	.115 40	.170	14.99
.2	.094 28	.097 38	.412	12.85	.2	.112 72	.115 87	.294	15.04
.3	.094 73	.097 84	.532	12.91	.3	.113 18	.116 34	.418	15.09
.4	.095 19	.098 29	.652	12.96	.4	.113 65	.116 81	.543	15.15
.5	.095 64	.098 75	.772	13.02	.5	.114 12	.117 28	.667	15.20
.6	.096 10	.099 21	.893	13.07	.6	.114 59	.117 75	.792	15.26
.7	.096 56	.099 66	26.013	13.13	.7	.115 06	.118 22	.916	15.31
.8	.097 01	.100 12	.134	13.18	.8	.115 53	.118 69	31.041	15.37
.9	.097 47	.100 58	.255	13.24	.9	.116 00	.119 16	.165	15.42
24.0	1.097 92	1.101 04	26.375	13.29	28.0	1.116 47	1.119 63	31.290	15.48
.1	.098 38	.101 49	.496	13.35	.1	.116 94	.120 10	.415	15.53
.2	.098 84	.101 95	.617	13.40	.2	.117 41	.120 58	.540	15.59
.3	.099 30	.102 41	.738	13.46	.3	.117 88	.121 05	.666	15.64
.4	.099 76	.102 87	.860	13.51	.4	.118 35	.121 52	.791	15.69
.5	.100 21	.103 33	.981	13.57	.5	.118 82	.121 99	.916	15.75
.6	.100 67	.103 79	27.102	13.62	.6	.119 29	.122 47	32.042	15.80
.7	.101 13	.104 25	.224	13.67	.7	.119 77	.122 94	.167	15.86
.8	.101 59	.104 71	.345	13.73	.8	.120 24	.123 41	.293	15.91
.9	.102 05	.105 17	.467	13.78	.9	.120 71	.123 89	.419	15.97
25.0	1.102 51	1.105 64	27.589	13.84	29.0	1.121 19	1.124 36	32.545	16.02
.1	.102 97	.106 10	.710	13.89	.1	.121 66	.124 84	.671	16.08
.2	.103 43	.106 56	.833	13.95	.2	.122 14	.125 32	.797	16.13
.3	.103 89	.107 02	.955	14.00	.3	.122 61	.125 79	.923	16.18
.4	.104 35	.107 48	28.077	14.06	.4	.123 08	.126 27	33.049	16.24
.5	.104 82	.107 95	.199	14.11	.5	.123 56	.126 74	.176	16.29
.6	.105 28	.108 41	.322	14.17	.6	.124 04	.127 22	.302	16.35
.7	.105 74	.108 87	.444	14.22	.7	.124 51	.127 70	.429	16.40
.8	.106 20	.109 34	.567	14.28	.8	.124 99	.128 17	.556	16.46
.9	.106 67	.109 80	.690	14.33	.9	.125 46	.128 65	.683	16.51

续表

蔗糖质量百分率（Bx）	视密度（20℃）	视比重（20℃/20℃）	蔗糖（g）100 mL（在真空中）	波美度（°Bé）	蔗糖质量百分率（Bx）	视密度（20℃）	视比重（20℃/20℃）	蔗糖（g）100 mL（在真空中）	波美度（°Bé）
30.0	1.125 94	1.129 13	33.810	16.57	34.0	1.145 30	1.148 55	38.976	18.73
.1	.126 42	.129 61	.937	16.62	.1	.145 89	.149 04	39.107	18.79
.2	.126 90	.130 09	34.064	16.67	.2	.146 29	.149 54	.239	18.84
.3	.127 37	.130 57	.191	16.73	.3	.146 78	.150 03	.370	18.90
.4	.127 85	.131 05	.318	16.78	.4	.147 27	.150 52	.502	18.95
.5	.128 33	.131 53	.446	16.84	.5	.147 76	.151 02	.634	19.00
.6	.128 81	.132 01	.574	16.89	.6	.148 26	.151 51	.767	19.06
.7	.129 29	.132 49	.701	16.95	.7	.148 75	.152 01	.898	19.11
.8	.129 77	.132 97	.829	17.00	.8	.149 25	.152 50	40.030	19.17
.9	.130 25	.133 45	.957	17.05	.9	.149 74	.153 00	.162	19.22
31.0	1.130 73	1.133 94	35.085	17.11	35.0	1.150 24	1.153 50	40.295	19.28
.1	.131 21	.134 42	.213	17.16	.1	.150 73	.153 99	.427	19.33
.2	.131 69	.134 90	.341	17.22	.2	.151 23	.154 49	.560	19.38
.3	.132 17	.135 38	.470	17.27	.3	.151 72	.154 98	.692	19.44
.4	.132 66	.135 87	.598	17.33	.4	.152 22	.155 48	.825	19.49
.5	.133 14	.136 35	.727	17.38	.5	.152 71	.155 98	.958	19.55
.6	.133 62	.136 83	.855	17.43	.6	.153 21	.156 48	41.091	19.60
.7	.134 10	.137 32	.984	17.49	.7	.153 71	.156 98	.224	19.65
.8	.134 59	.137 80	36.113	17.54	.8	.154 20	.157 47	.358	19.71
.9	.135 07	.138 29	.242	17.60	.9	.154 70	.157 97	.491	19.76
32.0	1.135 55	1.138 77	36.371	17.65	36.0	1.155 20	1.158 47	41.625	19.81
.1	.136 04	.139 26	.500	17.70	.1	.155 70	.158 97	.758	19.87
.2	.136 52	.139 74	.630	17.76	.2	.156 20	.159 47	.892	19.92
.3	.137 01	.140 23	.759	17.81	.3	.156 69	.159 97	42.026	19.98
.4	.137 49	.140 72	.889	17.87	.4	.157 19	.160 47	.160	20.03
.5	.137 98	.141 20	37.018	17.92	.5	.157 69	.160 98	.294	20.08
.6	.138 46	.141 69	.148	17.98	.6	.158 19	.161 48	.428	20.14
.7	.138 95	.142 18	.278	18.03	.7	.158 69	.161 98	.562	20.19
.8	.139 44	.142 67	.408	18.08	.8	.159 19	.162 48	.697	20.25
.9	.139 92	.143 16	.538	18.14	.9	.159 70	.162 98	.831	20.30
33.0	1.140 41	1.143 64	37.668	18.19	37.0	1.160 20	1.163 49	42.966	20.35
.1	.140 90	.144 13	.798	18.25	.1	.160 70	.163 99	43.100	20.41
.2	.141 39	.144 62	.929	18.30	.2	.161 20	.164 49	.235	20.46
.3	.141 88	.145 11	38.059	18.36	.3	.161 70	.165 00	.370	20.52
.4	.142 36	.145 60	.190	18.41	.4	.162 21	.165 50	.505	20.57
.5	.142 85	.146 09	.320	18.46	.5	.162 70	.166 01	.641	20.62
.6	.143 34	.146 58	.451	18.52	.6	.163 21	.166 51	.776	20.68
.7	.143 83	.147 08	.582	18.57	.7	.163 72	.167 02	.911	20.73
.8	.144 32	.147 57	.713	18.63	.8	.164 22	.167 52	44.047	20.78
.9	.144 81	.148 06	.844	18.68	.9	.164 73	.168 03	.182	20.84

续表

蔗糖质量百分率（Bx）	视密度（20℃）	视比重（20°/20℃）	蔗糖（g）100 mL（在真空中）	波美度（°Bé）	蔗糖质量百分率（Bx）	视密度（20℃）	视比重（20°/20℃）	蔗糖（g）100 mL（在真空中）	波美度（°Bé）
38.0	1.165 23	1.168 53	44.318	20.89	42.0	1.185 74	1.189 10	49.845	23.04
.1	.165 74	.169 04	.454	20.94	.1	.186 26	.189 62	.985	23.09
.2	.166 24	.169 55	.590	21.00	.2	.186 78	.190 14	50.126	23.14
.3	.166 75	.170 06	.726	21.05	.3	.187 30	.190 62	.267	23.20
.4	.167 26	.170 56	.862	21.11	.4	.187 82	.191 19	.408	23.25
.5	.167 76	.171 07	.999	21.16	.5	.188 35	.191 71	.549	23.30
.6	.168 27	.171 58	45.135	21.21	.6	.188 87	.192 24	.690	23.36
.7	.168 78	.172 09	.272	21.27	.7	.189 39	.192 76	.831	23.41
.8	.169 29	.172 60	.408	21.32	.8	.189 91	.193 29	.973	23.46
.9	.169 79	.173 11	.545	21.38	.9	.190 44	.193 81	51.114	23.52
39.0	1.170 30	1.173 62	45.682	21.43	43.0	1.190 96	1.194 34	51.256	23.57
.1	.170 81	.174 13	.819	21.48	.1	.191 48	.194 86	.398	23.62
.2	.171 32	.174 64	.956	21.54	.2	.192 01	.195 39	.539	23.68
.3	.171 83	.175 15	46.094	21.59	.3	.192 53	.195 91	.681	23.73
.4	.172 34	.175 66	.213	21.64	.4	.193 06	.196 44	.824	23.78
.5	.172 85	.176 18	.369	21.70	.5	.193 58	.196 97	.966	23.84
.6	.173 36	.176 69	.506	21.75	.6	.194 11	.197 49	52.108	23.89
.7	.173 87	.177 20	.644	21.80	.7	.194 83	.198 02	.251	23.94
.8	.174 39	.177 72	.782	21.86	.8	.195 16	.198 55	.393	24.00
.9	.174 90	.178 23	.920	21.91	.9	.195 69	.199 08	.536	24.05
40.0	1.175 41	1.178 74	47.058	21.97	44.0	1.196 22	1.199 61	52.679	24.10
.1	.175 93	.179 26	.196	22.02	.1	.196 74	.200 13	.822	24.16
.2	.176 44	.179 77	.334	22.07	.2	.197 27	.200 66	.965	24.21
.3	.176 95	.180 29	.473	22.13	.3	.197 80	.201 19	53.108	24.26
.4	.177 47	.180 80	.611	22.18	.4	.198 33	.201 72	.252	24.32
.5	.177 98	.181 32	.750	22.23	.5	.198 86	.202 26	.395	24.37
.6	.178 49	.181 83	.889	22.29	.6	.199 39	.202 79	.539	24.42
.7	.179 01	.182 35	48.028	22.34	.7	.199 92	.203 32	.683	24.48
.8	.179 53	.182 87	.167	22.39	.8	.200 45	.203 85	.826	24.53
.9	.180 04	.183 39	.306	22.45	.9	.200 98	.204 38	.970	24.58
41.0	1.180 56	1.183 90	48.445	22.50	45.0	1.201 51	1.204 91	54.114	24.63
.1	.181 07	.184 42	.585	22.55	.1	.202 04	.205 45	.259	24.69
.2	.181 59	.184 94	.723	22.61	.2	.202 57	.205 98	.403	24.74
.3	.182 11	.185 46	.864	22.66	.3	.203 11	.206 51	.547	24.79
.4	.182 63	.185 98	49.004	22.72	.4	.203 64	.207 05	.692	24.85
.5	.183 14	.186 50	.143	22.77	.5	.204 17	.207 58	.837	24.90
.6	.183 56	.187 02	.283	22.82	.6	.204 70	.208 12	.981	24.95
.7	.184 18	.187 54	.424	22.88	.7	.205 24	.208 65	55.126	25.01
.8	.184 70	.188 06	.564	22.93	.8	.205 77	.209 19	.272	25.06
.9	.185 22	.188 58	.704	22.98	.9	.206 30	.209 72	.417	25.11

续表

蔗糖质量百分率（Bx）	视密度（20℃）	视比重（20°/20℃）	蔗糖（g）100 mL（在真空中）	波美度（°Bé）	蔗糖质量百分率（Bx）	视密度（20℃）	视比重（20°/20℃）	蔗糖（g）100 mL（在真空中）	波美度（°Bé）
46.0	1.206 84	1.210 26	55.562	25.17	50.0	1.228 54	1.232 02	61.478	27.28
.1	.207 37	.210 80	.708	25.22	.1	.229 09	.232 57	.629	27.33
.2	.207 91	.211 33	.853	25.27	.2	.229 64	.233 13	.780	27.39
.3	.208 45	.211 87	.999	25.32	.3	.230 19	.233 68	.930	27.44
.4	.208 98	.212 41	56.145	25.38	.4	.230 74	.234 23	62.081	27.49
.5	.209 62	.212 95	.291	25.43	.5	.231 30	.234 78	.232	27.54
.6	.210 06	.213 40	.437	25.48	.6	.231 85	.235 34	.383	27.60
.7	.210 59	.214 02	.583	25.54	.7	.232 40	.235 89	.535	27.65
.8	.211 13	.214 56	.729	25.59	.8	.232 95	.236 45	.686	27.70
.9	.211 67	.215 10	.876	25.64	.9	.233 51	.237 00	.838	27.75
47.0	1.212 21	1.215 64	57.022	25.70	51.0	1.234 06	1.237 56	62.989	27.81
.1	.212 75	.216 18	.169	25.75	.1	.234 61	.238 11	63.141	27.86
.2	.213 29	.216 73	.316	25.80	.2	.235 17	.238 67	.293	27.91
.3	.213 83	.217 27	.463	25.86	.3	.235 72	.239 22	.445	27.96
.4	.214 37	.217 81	.610	25.91	.4	.236 28	.239 78	.579	28.02
.5	.214 91	.218 35	.757	25.96	.5	.236 83	.240 34	.750	28.07
.6	.215 45	.218 89	.904	26.01	.6	.237 39	.240 89	.902	28.12
.7	.215 99	.219 43	58.052	26.07	.7	.237 94	.241 45	64.005	28.17
.8	.216 53	.219 98	.199	26.12	.8	.238 50	.242 01	.208	28.23
.9	.217 07	.220 52	.347	26.17	.9	.239 06	.242 57	.360	28.28
48.0	1.217 61	1.221 06	58.495	26.23	52.0	1.239 62	1.243 13	64.513	28.33
.1	.218 16	.221 61	.643	26.28	.1	.240 17	.243 69	.666	28.38
.2	.218 70	.222 15	.791	26.33	.2	.240 73	.244 25	.820	28.44
.3	.219 24	.222 70	.939	26.38	.3	.241 29	.244 81	.973	28.49
.4	.219 79	.223 24	58.087	26.44	.4	.241 85	.245 37	65.127	28.54
.5	.220 33	.223 79	.236	26.49	.5	.242 41	.245 93	.280	28.59
.6	.220 88	.224 34	.385	26.54	.6	.242 97	.246 49	.433	28.65
.7	.221 42	.224 88	.533	26.59	.7	.243 53	.247 05	.588	28.70
.8	.221 97	.225 43	.682	26.65	.8	.244 09	.247 61	.742	28.75
.9	.222 51	.225 98	.831	26.70	.9	.244 65	.248 18	.896	28.80
49.0	1.223 06	1.226 52	59.980	26.75	53.0	1.245 21	1.248 74	66.050	28.86
.1	.223 60	.222 07	60.129	26.81	.1	.245 77	.249 30	.205	28.91
.2	.224 15	.227 62	.279	26.86	.2	.246 33	.249 87	.359	28.96
.3	.224 70	.228 17	.428	26.91	.3	.246 90	.250 43	.514	29.01
.4	.225 25	.228 72	.578	26.96	.4	.247 46	.250 99	.669	29.06
.5	.225 80	.229 27	.728	27.02	.5	.248 02	.251 56	.824	29.12
.6	.226 34	.229 82	.878	27.07	.6	.248 58	.252 12	.979	29.17
.7	.226 89	.230 37	61.028	27.12	.7	.249 15	.252 69	67.134	29.22
.8	.227 44	.230 92	.178	27.18	.8	.249 71	.253 25	.290	29.27
.9	.227 99	.231 47	.328	27.23	.9	.250 28	.253 82	.445	29.32

续表

蔗糖质量百分率（Bx）	视密度（20℃）	视比重（20°/20℃）	蔗糖（g）100 mL（在真空中）	波美度（°Bé）	蔗糖质量百分率（Bx）	视密度（20℃）	视比重（20°/20℃）	蔗糖（g）100 mL（在真空中）	波美度（°Bé）
54.0	1.250 84	1.254 39	67.601	29.38	58.0	1.273 75	1.277 36	73.937	31.46
.1	.251 41	.254 95	.757	29.43	.1	.274 33	.277 94	74.098	31.51
.2	.251 97	.255 52	.912	29.48	.2	.274 92	.278 53	.260	31.56
.3	.252 54	.256 09	68.069	29.53	.3	.275 50	.279 11	.421	31.61
.4	.253 11	.256 66	.225	29.59	.4	.276 08	.279 69	.583	31.66
.5	.253 67	.257 23	.381	29.64	.5	.276 64	.280 28	.744	31.71
.6	.254 24	.257 80	.537	29.69	.6	.277 24	.280 86	.906	31.76
.7	.254 81	.258 36	.694	29.74	.7	.277 82	.281 45	75.068	31.82
.8	.255 38	.258 93	.851	29.80	.8	.278 41	.282 03	.230	31.87
.9	.255 94	.259 50	69.008	29.85	.9	.278 92	.282 62	.393	31.92
55.0	1.256 51	1.260 07	69.164	29.90	59.0	1.279 58	1.283 20	75.555	31.97
.1	.257 08	.260 64	.322	29.95	.1	.280 17	.283 79	.718	32.02
.2	.257 65	.261 22	.479	30.00	.2	.280 75	.284 37	.880	32.07
.3	.258 22	.261 79	.636	30.06	.3	.281 34	.284 97	76.043	32.13
.4	.258 79	.262 36	.794	30.11	.4	.281 93	.285 56	.207	32.18
.5	.259 36	.262 93	.951	30.16	.5	.282 51	.286 14	.369	32.23
.6	.259 93	.263 50	70.109	30.21	.6	.283 09	.286 72	.533	32.28
.7	.260 50	.264 08	.267	30.26	.7	.283 67	.287 31	.696	32.33
.8	.261 08	.264 65	.425	30.32	.8	.284 26	.287 89	.860	32.38
.9	.261 65	.265 22	.583	30.37	.9	.284 85	.288 49	77.024	32.43
56.0	1.262 22	1.265 80	70.742	30.42	60.0	1.285 44	1.289 08	77.188	32.49
.1	.262 79	.266 37	.900	30.47	.1	.286 02	.289 66	.351	32.54
.2	.263 37	.266 95	71.059	30.52	.2	.286 61	.290 25	.515	32.59
.3	.263 94	.267 52	.217	30.57	.3	.287 20	.290 84	.680	32.64
.4	.264 52	.268 10	.376	30.63	.4	.287 79	.291 43	.844	32.69
.5	.265 09	.268 68	.535	30.68	.5	.288 38	.292 03	78.009	32.74
.6	.265 66	.269 25	.694	30.73	.6	.288 97	.292 62	.173	32.79
.7	.266 24	.269 83	.854	30.78	.7	.289 65	.293 21	.338	32.85
.8	.266 82	.270 41	72.013	30.83	.8	.290 15	.293 80	.503	32.90
.9	.267 39	.270 98	.173	30.89	.9	.290 74	.294 39	.628	32.95
57.0	1.267 97	1.271 56	72.332	30.94	61.0	1.291 33	1.294 98	78.833	33.00
.1	.268 54	.272 14	.492	30.99	.1	.291 93	.295 59	.999	33.05
.2	.269 12	.272 72	.652	31.04	.2	.292 52	.296 18	79.165	33.10
.3	.269 70	.273 30	.812	31.09	.3	.293 11	.296 77	.330	33.15
.4	.270 28	.273 88	.973	31.15	.4	.293 70	.297 36	.496	33.20
.5	.270 86	.274 46	73.133	31.20	.5	.294 30	.297 96	.662	33.26
.6	.271 43	.275 04	.293	31.25	.6	.294 89	.298 55	.828	33.31
.7	.272 01	.275 62	.454	31.30	.7	.295 48	.299 15	.995	33.36
.8	.272 59	.276 20	.615	31.35	.8	.296 08	.299 75	80.161	33.41
.9	.273 17	.276 78	.776	31.40	.9	.296 67	.300 34	.328	33.46

续表

蔗糖质量百分率（Bx）	视密度（20℃）	视比重（20°/20℃）	蔗糖（g）100 mL（在真空中）	波美度（°Bé）	蔗糖质量百分率（Bx）	视密度（20℃）	视比重（20°/20℃）	蔗糖（g）100 mL（在真空中）	波美度（°Bé）
62.0	1.297 26	1.300 93	80.494	33.51	66.0	1.321 42	1.325 16	87.280	35.55
.1	.297 86	.301 53	.661	33.56	.1	.322 03	.325 77	.453	35.60
.2	.298 45	.302 12	.828	33.61	.2	.322 64	.326 38	.626	35.65
.3	.299 05	.302 73	.995	33.67	.3	.323 25	.326 99	.798	35.70
.4	.299 66	.303 34	81.162	33.72	.4	.323 85	.327 59	.971	35.75
.5	.300 25	.303 93	.329	33.77	.5	.324 46	.328 20	88.142	35.80
.6	.300 85	.304 53	.497	33.82	.6	.325 09	.328 84	.318	35.85
.7	.301 45	.305 13	.665	33.87	.7	.325 70	.329 45	.492	35.90
.8	.302 05	.305 73	.833	33.92	.8	.326 32	.330 07	.666	35.95
.9	.302 65	.306 33	82.001	33.97	.9	.326 93	.330 68	.839	36.00
63.0	1.303 25	1.306 94	82.169	34.02	67.0	1.327 54	1.331 29	89.012	36.05
.1	.303 85	.307 54	.337	34.07	.1	.328 16	.331 92	.187	36.10
.2	.304 46	.308 15	.506	34.12	.2	.328 78	.322 54	.361	36.15
.3	.305 06	.308 75	.674	34.18	.3	.329 39	.323 15	.536	36.20
.4	.305 66	.309 36	.843	34.23	.4	.330 01	.333 77	.711	36.25
.5	.306 26	.309 94	83.012	34.28	.5	.330 62	.334 38	.885	36.30
.6	.306 86	.310 55	.180	34.33	.6	.331 24	.335 00	90.060	36.35
.7	.307 47	.311 17	.350	34.38	.7	.331 86	.335 62	.235	36.40
.8	.308 07	.311 77	.519	34.43	.8	.332 48	.336 25	.411	36.45
.9	.308 67	.312 37	.688	34.48	.9	.333 09	.336 86	.585	36.50
64.0	1.309 27	1.312 97	83.858	34.53	68.0	1.333 71	1.337 48	90.761	36.55
.1	.309 88	.313 59	84.028	34.58	.1	.334 33	.338 10	.937	36.61
.2	.310 48	.314 18	.198	34.63	.2	.324 95	.328 72	91.112	36.66
.3	.311 08	.314 79	.367	34.68	.3	.325 57	.329 35	.288	36.71
.4	.311 69	.315 40	.538	34.74	.4	.336 19	.339 97	.464	36.76
.5	.312 29	.316 00	84.708	34.79	.5	.336 81	.340 59	.641	36.81
.6	.312 90	.316 61	.879	34.84	.6	.337 43	.341 21	.817	36.86
.7	.313 50	.317 23	85.049	34.89	.7	.338 05	.341 83	.993	36.91
.8	.314 12	.317 84	.220	34.94	.8	.338 67	.342 45	92.169	36.96
.9	.314 73	.318 45	.391	34.99	.9	.339 30	.343 09	.347	37.01
65.0	1.315 33	1.319 05	85.561	35.04	69.0	1.339 92	1.343 71	92.524	37.06
.1	.315 94	.319 66	.733	35.09	.1	.340 54	.344 33	.701	37.11
.2	.316 55	.320 28	.904	35.14	.2	.341 16	.344 95	.878	37.16
.3	.317 16	.320 89	86.076	35.19	.3	.341 79	.345 58	93.056	37.21
.4	.317 77	.321 50	.248	35.24	.4	.342 41	.346 21	.233	37.26
.5	.318 37	.322 10	.419	35.29	.5	.343 04	.346 84	.411	37.31
.6	.318 98	.322 71	.591	35.34	.6	.343 66	.347 46	.589	37.36
.7	.319 59	.323 32	.763	35.39	.7	.344 29	.348 09	.767	37.41
.8	.320 19	.323 93	.935	35.45	.8	.344 91	.348 71	.945	37.46
.9	.320 81	.324 55	87.107	35.50	.9	.345 54	.349 34	94.123	37.51

续表

蔗糖质量百分率（Bx）	视密度（20℃）	视比重（20°/20℃）	蔗糖（g）100 mL（在真空中）	波美度（°Bé）	蔗糖质量百分率（Bx）	视密度（20℃）	视比重（20°/20℃）	蔗糖（g）100 mL（在真空中）	波美度（°Bé）
70.0	1.346 16	1.349 97	94.302	37.56	74.0	1.371 53	1.375 41	101.568	39.54
.1	.346 79	.350 60	.481	37.61	.1	.372 17	.376 05	.753	39.59
.2	.347 42	.351 23	.660	37.66	.2	.372 81	.376 69	.937	39.64
.3	.348 05	.351 86	.839	37.71	.3	.373 45	.377 33	102.122	39.69
.4	.348 67	.352 48	95.017	37.76	.4	.374 10	.377 98	.308	39.74
.5	.349 30	.353 11	.197	37.81	.5	.374 75	.378 64	.493	39.79
.6	.349 93	.353 75	.376	37.86	.6	.375 39	.379 28	.679	39.84
.7	.350 56	.354 38	.556	37.91	.7	.376 04	.379 93	.865	39.89
.8	.351 19	.355 01	.736	37.96	.8	.376 88	.380 57	103.050	39.94
.9	.351 82	.355 64	.916	38.01	.9	.377 33	.381 22	.237	39.99
71.0	1.352 45	1.356 27	96.096	38.06	75.0	1.377 97	1.381 87	103.423	40.03
.1	.353 08	.356 91	.276	38.11	.1	.378 62	.382 52	.609	40.08
.2	.353 71	.357 54	.456	38.16	.2	.379 26	.383 16	.796	40.13
.3	.354 34	.358 17	.636	38.21	.3	.379 91	.383 81	.983	40.18
.4	.354 98	.358 81	.817	38.26	.4	.380 55	.384 45	104.170	40.23
.5	.355 61	.359 44	.998	38.3	.5	.381 19	.385 10	.356	40.28
.6	.356 25	.360 08	97.179	38.35	.6	.381 84	.385 75	.543	40.33
.7	.356 88	.360 72	.360	38.40	.7	.382 49	.386 40	.731	40.38
.8	.357 51	.361 35	.541	38.45	.8	.383 14	.387 05	.919	40.43
.9	.358 14	.361 98	.722	38.50	.9	.383 79	.387 70	105.106	40.48
72.0	1.358 77	1.362 61	97.904	38.55	76.0	1.384 44	1.388 35	105.294	40.53
.1	.359 40	.363 24	98.085	38.60	.1	.385 10	.389 02	.482	40.57
.2	.360 04	.363 89	.268	38.65	.2	.385 75	.389 67	.670	40.62
.3	.360 67	.364 52	.449	38.70	.3	.386 40	.390 32	.859	40.67
.4	.361 31	.365 16	.632	38.75	.4	.387 05	.390 97	106.047	40.72
.5	.361 94	.365 79	.814	38.80	.5	.387 70	.391 62	.236	40.77
.6	.362 58	.366 43	.997	38.85	.6	.388 35	.392 28	.424	40.82
.7	.363 22	.367 07	99.179	38.90	.7	.389 00	.392 93	.613	40.87
.8	.363 85	.367 71	.362	38.95	.8	.389 65	.393 58	.802	40.92
.9	.364 50	.368 36	.545	39.00	.9	.390 30	.394 23	.991	40.97
73.0	1.365 14	1.369 00	99.728	39.05	77.0	1.390 96	1.394 89	107.181	41.01
.1	.365 78	.369 64	.912	39.10	.1	.391 61	.395 54	.370	41.06
.2	.366 42	.370 28	100.095	39.15	.2	.392 25	.396 19	.560	41.11
.3	.367 05	.370 92	.278	39.20	.3	.392 91	.396 85	.750	41.16
.4	.367 69	.371 56	.462	39.25	.4	.393 56	.397 50	.940	41.21
.5	.368 33	.372 20	.646	39.30	.5	.394 22	.398 16	108.130	41.26
.6	.368 96	.372 83	.827	39.35	.6	.394 88	.398 82	.320	41.31
.7	.369 60	.373 47	101.014	39.39	.7	.395 54	.399 49	.511	41.36
.8	.370 24	.374 11	.198	39.44	.8	.396 19	.400 14	.701	41.40
.9	.370 88	.374 76	.383	39.49	.9	.396 85	.400 80	.892	41.45

续表

蔗糖质量百分率（Bx）	视密度（20℃）	视比重（20°/20℃）	蔗糖（g）100 mL（在真空中）	波美度（°Bé）	蔗糖质量百分率（Bx）	视密度（20℃）	视比重（20°/20℃）	蔗糖（g）100 mL（在真空中）	波美度（°Bé）
78.0	1.397 51	1.401 46	109.084	41.50	82.0	1.424 07	1.428 10	116.856	43.43
.1	.398 16	.402 11	.274	41.55	.1	.424 75	.428 78	117.053	43.48
.2	.398 82	.402 73	.466	41.60	.2	.425 43	.429 46	.252	43.53
.3	.399 48	.403 44	.657	41.65	.3	.426 10	.430 13	.449	43.57
.4	.400 13	.404 09	.848	41.70	.4	.426 77	.430 80	.647	43.62
.5	.400 79	.404 75	110.041	41.74	.5	.427 44	.431 48	.845	43.67
.6	.401 45	.405 41	.232	41.79	.6	.428 11	.432 14	118.044	43.72
.7	.402 11	.406 07	.425	41.84	.7	.428 78	.432 82	.243	43.77
.8	.402 77	.406 74	.617	41.89	.8	.429 46	.433 50	.442	43.81
.9	.403 43	.407 40	.809	41.94	.9	.430 13	.434 17	.641	43.86
79.0	1.404 09	1.408 06	111.022	41.99	83.0	1.430 81	1.434 86	118.840	43.91
.1	.404 75	.408 72	.195	42.03	.1	.431 48	.435 53	119.039	43.96
.2	.405 41	.409 38	.388	42.08	.2	.432 16	.436 21	.239	44.00
.3	.406 07	.410 05	.581	42.13	.3	.432 83	.436 88	.438	44.05
.4	.406 74	.410 72	.775	42.18	.4	.433 51	.437 56	.638	44.10
.5	.407 40	.411 38	.968	42.23	.5	.434 19	.438 24	.838	44.15
.6	.408 06	.412 04	112.161	42.28	.6	.434 88	.438 94	120.039	44.19
.7	.408 72	.412 70	.354	42.32	.7	.435 55	.439 61	.238	44.24
.8	.409 39	.413 37	.549	42.37	.8	.436 23	.440 29	.439	44.29
.9	.410 05	.414 04	.743	42.42	.9	.436 91	.440 97	.640	44.34
80.0	1.410 72	1.414 71	112.938	42.47	84.0	1.437 58	1.441 65	120.841	44.38
.1	.411 38	.415 37	113.131	42.52	.1	.438 26	.442 34	121.042	44.43
.2	.412 04	.416 03	.326	42.57	.2	.438 94	.443 02	.243	44.48
.3	.412 71	.416 70	.521	42.61	.3	.439 62	.443 70	.444	44.53
.4	.413 37	.417 37	.715	42.66	.4	.440 30	.444 38	.646	44.57
.5	.414 04	.418 04	.911	42.71	.5	.440 98	.445 07	.847	44.62
.6	.414 72	.418 72	114.106	42.76	.6	.441 66	.445 75	122.049	44.67
.7	.415 37	.419 37	.301	42.81	.7	.442 34	.446 43	.251	44.72
.8	.416 04	.420 04	.497	42.85	.8	.443 03	.447 12	.453	44.76
.9	.416 71	.420 72	.692	42.90	.9	.443 71	.447 80	.655	44.81
81.0	1.417 37	1.421 38	114.888	42.95	85.0	1.444 39	1.448 48	122.858	44.86
.1	.418 04	.422 05	115.084	43.00	.1	.445 07	.449 17	123.061	44.91
.2	.418 71	.422 72	.280	43.05	.2	.445 76	.449 85	.263	44.95
.3	.419 38	.423 39	.477	43.10	.3	.446 44	.450 54	.466	45.00
.4	.420 05	.424 06	.673	43.14	.4	.447 12	.451 23	.670	45.05
.5	.420 72	.424 74	.870	43.19	.5	.447 81	.451 91	.873	45.09
.6	.421 39	.425 41	116.067	43.24	.6	.448 49	.452 60	124.076	45.14
.7	.422 06	.426 08	.264	43.29	.7	.449 18	.453 29	.280	45.19
.8	.422 73	.426 75	.461	43.33	.8	.449 86	.453 97	.484	45.24
.9	.423 40	.427 42	.658	43.38	.9	.450 55	.454 66	.688	45.28

续表

蔗糖质量百分率（Bx）	视密度（20℃）	视比重（20°/20℃）	蔗糖（g）100 mL（在真空中）	波美度（°Bé）	蔗糖质量百分率（Bx）	视密度（20℃）	视比重（20°/20℃）	蔗糖（g）100 mL（在真空中）	波美度（°Bé）
86.0	1.451 24	1.455 35	124.892	45.33	90.0	1.478 98	1.483 17	133.198	47.20
.1	.451 92	.456 04	125.096	45.38	.1	.479 68	.483 88	.409	47.24
.2	.452 61	.456 73	.301	45.42	.2	.480 39	.484 58	.420	47.29
.3	.453 30	.457 41	.505	45.47	.3	.481 09	.485 29	.832	47.34
.4	.453 98	.458 10	.710	45.52	.4	.481 79	.485 99	134.043	47.38
.5	.454 67	.458 79	.915	45.57	.5	.482 49	.486 69	.255	47.43
.6	.455 36	.459 49	126.121	45.61	.6	.483 20	.487 40	.467	47.48
.7	.456 05	.460 18	.326	45.66	.7	.483 90	.488 10	.680	47.52
.8	.456 74	.460 87	.531	45.71	.8	.484 60	.488 81	.892	47.57
.9	.457 43	.461 56	.737	45.75	.9	.485 31	.489 51	135.104	47.61
87.0	1.458 12	1.462 25	126.943	45.80	91.0	1.486 01	1.490 22	135.317	47.66
.1	.458 81	.462 94	127.149	45.85	.1	.486 72	.490 93	.530	47.71
.2	.459 50	.463 64	.355	45.89	.2	.487 42	.491 64	.743	47.75
.3	.460 19	.464 33	.562	45.94	.3	.488 13	.492 34	.956	47.80
.4	.460 88	.465 02	.768	45.99	.4	.488 83	.493 05	136.170	47.84
.5	.461 57	.465 72	.975	46.03	.5	.489 54	.493 76	.383	47.89
.6	.462 27	.466 41	128.182	46.08	.6	.490 24	.494 47	.597	47.94
.7	.462 96	.467 10	.389	46.13	.7	.490 95	.495 18	.811	47.98
.8	.463 65	.467 80	.596	46.17	.8	.491 66	.495 88	137.025	48.03
.9	.464 34	.468 49	.803	46.22	.9	.492 36	.496 59	.239	48.08
88.0	1.465 04	1.469 19	129.011	46.27	92.0	1.493 07	1.497 30	137.454	48.12
.1	.465 73	.469 89	.219	46.31	.1	.493 78	.498 01	.668	48.17
.2	.466 43	.470 58	.426	46.36	.2	.494 49	.498 72	.883	48.21
.3	.467 12	.471 28	.635	46.41	.3	.495 20	.499 44	138.098	48.26
.4	.467 82	.471 98	.843	46.45	.4	.495 91	.500 15	.313	48.30
.5	.468 51	.472 67	130.051	46.50	.5	.496 62	.500 86	.529	48.35
.6	.469 21	.473 37	.260	46.55	.6	.497 33	.501 57	.744	48.40
.7	.469 90	.474 07	.468	46.59	.7	.498 04	.502 28	.960	48.44
.8	.470 60	.474 77	.677	46.64	.8	.498 75	.502 99	139.176	48.49
.9	.471 30	.475 47	.886	46.69	.9	.499 46	.503 71	.392	48.53
89.0	1.471 99	1.476 16	131.096	46.73	93.0	1.500 17	1.504 42	139.608	48.58
.1	.472 69	.476 86	.305	46.78	.1	.500 88	.505 13	.824	48.62
.2	.473 39	.477 56	.515	46.83	.2	.501 59	.505 85	140.041	48.67
.3	.474 09	.478 26	.725	46.87	.3	.502 30	.506 56	.257	48.72
.4	.474 79	.478 97	.935	46.92	.4	.503 02	.507 28	.474	48.76
.5	.475 48	.479 67	132.145	46.97	.5	.503 73	.507 99	.691	48.81
.6	.476 18	.480 37	.355	47.01	.6	.504 44	.508 71	.908	48.85
.7	.476 88	.481 07	.365	47.06	.7	.505 16	.509 42	141.126	48.90
.8	.477 58	.481 77	.776	47.11	.8	.505 87	.510 14	.343	48.94
.9	.478 28	.482 47	.987	47.15	.9	.506 59	.510 86	.561	48.99

续表

蔗糖质量百分率（Bx）	视密度（20℃）	视比重（20°/20℃）	蔗糖（g）100 mL（在真空中）	波美度（°Bé）	蔗糖质量百分率（Bx）	视密度（20℃）	视比重（20°/20℃）	蔗糖（g）100 mL（在真空中）	波美度（°Bé）
94.0	1.507 30	1.511 57	141.779	49.03	95.0	1.514 47	1.518 76	143.968	49.49
.1	.508 02	.512 29	.997	49.08					
.2	.508 73	.513 01	142.216	49.12					
.3	.509 45	.513 72	.434	49.17					
.4	.510 16	.514 44	.653	49.22					
.5	.510 88	.515 16	.872	49.26					
.6	.511 60	.515 88	143.091	49.31					
.7	.512 31	.516 60	.310	49.35					
.8	.513 03	.517 32	.529	49.40					
.9	.513 75	.518 04	.749	49.44					

附表 12　在制品测定色值样液配制表

（配成样液：5 °Bx，200 mL）

锤度	需吸取样液体积 /mL									
	.0	.1	.2	.3	.4	.5	.6	.7	.8	.9
11	88.8	87.9	87.1	86.3	85.5	84.7	84.0	83.2	82.5	81.8
12	81.1	80.4	79.7	79.0	78.3	77.7	77.0	76.4	75.7	75.1
13	74.5	73.9	73.3	72.7	72.2	71.6	71.1	70.5	70.0	69.4
14	68.9	68.4	67.9	67.4	66.9	66.4	65.9	65.5	65.0	64.5
15	64.1	63.6	63.2	62.7	62.3	61.9	61.5	61.0	60.6	60.2
16	59.8	59.4	59.0	58.6	58.3	57.9	57.5	57.1	56.8	56.4
17	56.1	55.7	55.4	55.0	54.7	54.4	54.0	53.7	53.4	53.1
18	52.7	52.4	52.1	51.8	51.5	51.2	50.9	50.6	50.3	50.0
19	49.8	49.5	49.2	48.9	48.7	48.4	48.1	47.9	47.6	47.3
20	47.1	46.8	46.6	46.3	46.1	45.8	45.6	45.4	45.1	44.9

附表 13　纯蔗糖在水中的溶解度表

按公式：$S = 64.397+0.072\,51t+0.002\,056\,9t^2-9.035\times 10^{-6}t^3$

温度/℃	蔗糖质量百分数/%	每100份水溶解蔗糖分数	温度/℃	蔗糖质量百分数/%	每100份水溶解蔗糖分数	温度/℃	蔗糖质量百分数/%	每100份水溶解蔗糖分数
0	64.40	180.9	35	69.07	223.3	70	76.45	324.7
1	64.47	181.5	36	69.25	225.2	71	76.68	328.8
2	64.55	182.1	37	69.44	227.2	72	76.91	333.1
3	64.63	182.7	38	69.63	229.2	73	77.14	337.4
4	64.72	183.4	39	69.82	231.3	74	77.36	341.8
5	64.81	184.2	40	70.01	233.4	75	77.59	346.3
6	64.90	184.9	41	70.20	235.6	76	77.82	350.9
7	65.00	185.7	42	70.40	237.8	77	78.05	355.6
8	65.10	186.6	43	70.60	240.1	78	78.28	360.4
9	65.21	187.5	44	70.80	242.5	79	78.51	365.3
10	65.32	188.4	45	71.00	244.8	80	78.74	370.3
11	65.43	189.3	46	71.20	247.3	81	78.96	375.4
12	65.55	190.3	47	71.41	249.8	82	79.19	380.6
13	65.67	191.3	48	71.62	252.3	83	79.42	385.9
14	65.79	192.3	49	71.83	254.9	84	79.65	391.3
15	65.92	193.4	50	72.04	257.6	85	79.87	396.8
16	66.05	194.5	51	72.25	260.3	86	80.10	402.5
17	66.18	195.7	52	72.46	263.1	87	80.32	408.3
18	66.32	196.9	53	72.67	265.9	88	80.55	414.1
19	66.45	198.1	54	72.89	268.8	89	80.77	420.1
20	66.60	199.4	55	73.10	271.8	90	81.00	426.2
21	66.74	200.7	56	73.32	274.8			
22	66.89	202.0	57	73.54	277.9			
23	67.04	203.4	58	73.76	281.1			
24	67.20	204.8	59	73.98	284.3			
25	67.35	206.3	60	74.20	287.6			
26	67.51	207.8	61	74.42	291.0			
27	67.68	209.4	62	74.65	294.4			
28	67.84	211.0	63	74.87	297.9			
29	68.01	212.6	64	75.09	301.5			
30	68.18	214.3	65	75.62	305.2			
31	68.35	216.0	66	75.54	308.9			
32	68.53	217.7	67	75.77	312.7			
33	68.70	219.5	68	76.00	316.6			
34	68.88	221.4	69	76.22	320.6			

附录二 糖生产许可证审查细则(2006年版)

附录二 糖生产许可证审查细则(2006年版)

参考文献

[1] 李埔,郑长庚.甘蔗制糖化学管理分析方法 [M].广州:中国轻工总会甘蔗糖业质量监督检测中心,1995.

[2] 张笃思.甘蔗制粮工业分析 [M].北京:中国轻工业出版社,1991.

[3] 华南工学院.制糖工业分析 [M].北京:中国轻工业出版社,1981.

[4] 李埔.甘蔗制糖日常分析方法 [M].广州:全国甘蔗糖业标准化与质量检测中心,1985.